D1274938

INTERNATIONAL SERIES OF MONOGRAPHS IN

NATURAL PHILOSOPHY

GENERAL EDITOR: D. TER HAAR

VOLUME 24

FUNDAMENTAL PRINCIPLES OF MODERN THEORETICAL PHYSICS

OTHER TITLES IN THE SERIES IN NATURAL PHILOSOPHY

Vol. 1. Davydov – Quantum Mechanics
Vol. 2. Fokker – Time and Space, Weight and Inertia
Vol. 3. Kaplan – Interstellar Gas Dynamics
Vol. 4. Abrikosov, Gor'kov and Dzyaloshinskii – Quantum Field Theoretical Methods in Statistical Physics
Vol. 5. Okun' – Weak Interaction of Elementary Particles
Vol. 6. Shklovskii – Physics of the Solar Corona
Vol. 7. Akhiezer *et al.* – Collective Oscillations in a Plasma
Vol. 8. Kirzhnits – Field Theoretical Methods in Many-body Systems
Vol. 9. Klimontovich – Statistical Theory of Non-equilibrium Processes in a Plasma
Vol. 10. Kurth – Introduction to Stellar Statistics
Vol. 11. Chalmers – Atmospheric Electricity (2nd Edition)
Vol. 12. Renner – Current Algebras and their Applications
Vol. 13. Fain and Khanin – Quantum Electronics, Volume 1 – Basic Theory
Vol. 14. Fain and Khanin – Quantum Electronics, Volume 2 – Maser Amplifiers and Oscillators
Vol. 15. March – Liquid Metals
Vol. 16. Hori – Spectral Properties of Disordered Chains and Lattices
Vol. 17. Saint James, Thomas and Sarma – Type II Superconductivity
Vol. 18. Margenau and Kestner – Theory of Intermolecular Forces
Vol. 19. Jancel – Foundations of Classical and Quantum Statistical Mechanics
Vol. 20. Takahashi – An Introduction to Field Quantization
Vol. 21. Yvon – Correlations and Entropy in Classical Statistical Mechanics
Vol. 22. Penrose – Foundations of Statistical Mechanics
Vol. 23. Visconti – Quantum Field Theory, Volume 1

FUNDAMENTAL PRINCIPLES OF MODERN THEORETICAL PHYSICS

BY

R. H. FURTH

PERGAMON PRESS

OXFORD · LONDON · EDINBURGH · NEW YORK

TORONTO · SYDNEY · PARIS · BRAUNSCHWEIG

Pergamon Press Ltd., Headington Hill Hall, Oxford
4 & 5 Fitzroy Square, London W.1

Pergamon Press (Scotland) Ltd., 2 & 3 Teviot Place, Edinburgh 1

Pergamon Press Inc., Maxwell House, Fairview Park, Elmsford, New York 10523

Pergamon of Canada Ltd., 207 Queen's Quay West, Toronto 1

Pergamon Press (Aust.) Pty. Ltd., 19a Boundary Street, Rushcutters Bay, N.S.W. 2011, Australia

Pergamon Press S.A.R.L., 24 rue des Écoles, Paris 5e

Vieweg & Sohn GmbH, Burgplatz 1, Braunschweig

Copyright © 1970 Pergamon Press Ltd.

All Rights Reserved. No part of this publication may be reproduced, stored in a retrieval system, or transmitted, in any form or by any means, electronic, mechanical, photocopying, recording or otherwise, without the prior permission of Pergamon Press Ltd.

First edition 1970

Library of Congress Catalog Card No. 72-79112

Library
I.U.P.
Indiana, Pa.

530.1 F984f
c. 1

Printed in Germany

08 013375 4

Contents

Preface

THIS book is mainly based on lecture courses which the author used to give to special (honours) students of physics at Birkbeck College, University of London, in their last year of studies for the special B.Sc. degree of that university. But the range of material covered has been widened considerably so that the book could also be of value to postgraduate students of physics and related subjects.

It is not a textbook of theoretical physics but, as the title indicates, is meant to give a fairly elementary presentation of the fundamental principles which underlie modern theoretical physics (as distinct from classical physics). For this reason the main emphasis was put on the explanation, development, and consequences of these principles, in a logical manner, and on the basis of experimental facts rather than axioms. Applications of the general principles to special problems are therefore confined to illustrations of these principles for their better understanding.

At the end of each chapter the reader will find a number of exercises and problems relating to the subjects of these chapters. The former are meant to give him some practice in the manipulation of the mathematical methods involved. The latter will give him the opportunity to use his own initiative in applying the general theory to special cases apart from those dealt with in the text; they are all fairly simple and devised in such a way that any reader, having followed the text with understanding, should be able to solve them without great difficulty. The solutions of the problems are given in an appendix at the end of the book.

The reader is supposed to be acquainted with classical theoretical physics up to first-degree standards, that is Newtonian dynamics, Maxwell's electromagnetic theory, and classical thermodynamics. But special sections on classical principles are included in which these principles are briefly re-stated and the relevant equations are set out. In order to be able to follow the mathematical arguments the reader has to have a fairly good knowledge of infinitesimal calculus, partial differential equations, vector algebra and analysis, and complex functions, as usually demanded from students of physics in their last undergraduate year at British universities.

The author is well aware of the fact that his treatment of the fundamental principles is incomplete in many respects and that important aspects, especially regarding recent developments, have hardly been touched upon. However, packing too much material into a textbook of this level is only likely to discourage and confuse the reader. Those who, after having studied the book, wish to go deeper into the matter will have no difficulty in finding suitable reading material from the wide selection of excellent special text-books available.

Introduction

CLASSICAL THEORETICAL PHYSICS is usually understood to be concerned with the mathematical formulation of the laws of physics based on three fundamental principles. These are:

(1) *The principle of causation.* It is assumed that the state of a closed physical system at a certain time t is completely determined by the numerical values of a (finite or infinite) set of variables, and that the state at some particular time, say $t = 0$, completely determines the state at all subsequent times. The laws of physics must therefore have the form of *equations* between the state variables and time.

(2) *The principle of continuity.* It is assumed that all changes in the states of physical systems occur *continuously* or, in other words, that the state variables f_n are continuous functions $f_n(t)$ of time. It is therefore possible to consider infinitely small time intervals dt and the corresponding changes df_n of the state variables. The laws of physics can thus be expressed in the form of equations between the quantities df_n and dt; that is to say they have the form of *differential* equations.

(3) *The existence of a universal frame of reference.* For the full description of the state of a physical system the positions in space of certain recognizable physical entities have to be known. A particular class of state variables are therefore the position variables or *co-ordinates* which can only be expressed numerically if they refer to a properly defined "*system of co-ordinates*" or "*frame of reference*". The mathematical form of the equations of physics will, of course, in general depend on the choice of the reference system. It is believed that there exists a universal preferential frame of reference in nature with respect to which the laws of physics assume a particularly simple and universal form. This preferential frame of reference may either be identified with Newton's metaphysical "absolute space" or with a physical substance, the "ether", which is supposed to fill the whole of the universe.

An overwhelming amount of experimental evidence, especially from investigations in the realm of very small dimensions in space and of very large velocities, has forced us to abandon the above formulated fundamental principles and to replace them by different ones. This does not mean that the whole building of classical theoretical physics had to be broken up. Fortunately, a very substantial part of the original structure, although apparently built on a faulty foundation, is still perfectly sound and adequate for very many problems of pure and applied physics. But to this old building completely new ones have been added, founded on the new principles; for lack of a more appropriate name we shall call this new structure "*modern theoretical physics*".

The first indication that the principle of causation may not be valid in physics came from the difficulties which were encountered when one tried to deduce the *macroscopic* laws—governing physical systems consisting of a very large number of identical particles—from the *microscopic* laws pertaining to the behaviour of individual particles. In order to solve this problem it was found necessary to make use of *statistical methods*, that is to say to introduce an element of *chance* into the microscopic laws. This procedure turned out to be highly successful and led to the development of statistical mechanics in the first place and statistical physics in general. However, the necessity of having to combine causal and chance aspects into one and the same theory was rather unsatisfactory from the conceptual point of view. These difficulties were overcome only after the advent of quantum theory, when it at last became evident that the principle of causation in its classical form could not be valid for *individual* physical systems. It emerged, however, that a modified law of causation apparently holds for so-called "*ensembles*" of identical systems.

A further consequence of the establishment of quantum theory was that the classical principle of continuity had to be given up. Whereas classical theory had been unable to explain most of the observations on atomic phenomena, quantum theory made it possible to account, at least in principle, for all the known facts. But this theory is not compatible with the notion of a continuous movement of the constituent fundamental particles of matter in space, and one has to assume that the interchange of energy between them takes place in discontinuous "jumps", the occurrence of which is governed by *probabilities*. Nevertheless, the laws of quantum physics are still expressible in the form of differential equations from which these probabilities may be derived; these equations therefore hold for the physical behaviour of *ensembles* only and not for that of individual systems, in accordance with the statement made in the preceding paragraph.

Finally it became necessary to discard the notion of an absolute universal frame of reference, for all attempts to identify such a reference system, or to detect and measure the movement of material bodies against it, have failed. It therefore seems highly improbable that something like an ether exists at all, and correspondingly the laws of physics ought to have such a form that they are valid with respect to any arbitrary frame of reference. This postulate is called the "*principle of relativity*", and as evidently the laws of classical physics are not compatible with this principle they had to be suitably modified. The *theory of relativity* which was developed along these lines has been able to account for the phenomena connected with very high speeds which were found to be in contradiction to the predictions from classical laws. From relativity theory also arose a new theory of gravitation which differs characteristically from Newton's theory and is borne out by observation. Moreover, relativity theory and quantum theory were linked up with each other right from their beginnings, and "relativistic quantum mechanics" is now a well-established branch of modern theoretical physics.

In this book it is attempted to present the fundamental principles of modern theoretical physics, as briefly outlined above, in three main parts devoted to quantum theory, relativity theory, and statistical mechanics. It is meant as a guide for beginners in the study of these subjects, and although it contains a certain number of applications of the general theory to special problems, these are mainly intended as illustrations for the better understanding of the general principles. Those readers who are mostly interested in such applications, and those who wish to acquire a knowledge of the subject-matter on a higher level, are referred to the many excellent specialized textbooks available.

PART I

Quantum Theory

CHAPTER 1

The Fundamental Principles of Classical Physics

ALTHOUGH, as explained in the introduction, the fundamental principles of quantum theory are in many respects different from those of classical physics and the quantum laws are formulated differently from the classical laws, quantum theory has, nevertheless, grown organically from classical theory. It was therefore considered to be necessary to give in this chapter first a brief review of the main methods employed in classical physics to an extent to which they will subsequently be needed.

1.1. Discontinuum or particle theory

In the "*discontinuum*" or "*particle*" *theories* it is assumed that material physical systems fill space discontinuously, being built up of particles which, on any observable scale, are so small that their position in space is completely determined by a *position vector* $\mathbf{r}$ with respect to a suitable system of coordinates.† The state variables of a closed system are therefore identical with the position vectors $\mathbf{r}_i$ and the *velocity vectors* $\mathbf{v}_i = \dot{\mathbf{r}}_i$ of all its constituent particles. The particles are acted upon by *forces* $\mathbf{F}_i$ which are partly external, having their origin outside the system, and partly internal, acting between the particles. The task of the theory is to set up equations between the coordinates, the forces, and the time which make it possible to determine the motion of the particles at any time when the state of the system at time $t = 0$ is given.

In Newtonian dynamics the equations of motion have the form

$$m_i \ddot{\mathbf{r}}_i = m_i \dot{\mathbf{v}}_i = \mathbf{F}_i \quad (i = 1, 2, \ldots, N) \tag{1.1.1}$$

where m_i is the (constant) mass of the ith particle, $\mathbf{F}_i$ the force on this particle, which in general is a function of the position and velocity of that particle and may also depend on the positions and velocities of the other

† Throughout this book vectors will always be denoted by symbols in heavy type.

particles and on time explicitly. The equations are valid only with respect to a preferential system of coordinates which is called an "*inertial frame of reference*" (more about this will be found in Part II). By twice integrating the N second-order differential equations (1.1.1), provided the functions $\mathbf{F}_i$ are given, it is possible to determine the quantities $\mathbf{r}_i$ and $\mathbf{v}_i$ completely for any time t if their values at $t = 0$ are known.

The *momenta* $\mathbf{p}_i$ and the *kinetic energies* K_i of the particles are defined by

$$\left.\begin{aligned} \mathbf{p}_i &= m_i \mathbf{v}_i = m_i \dot{\mathbf{r}}_i, \\ K_i &= \frac{m_i v_i^2}{2} = \frac{p_i^2}{2m_i}. \end{aligned}\right\} \tag{1.1.2}$$

The total momentum $\mathbf{p}$ and the total kinetic energy K of the system are the sums of the expressions (1.1.2) over all N particles of the system. It follows from (1.1.2) that

$$\mathbf{p}_i = \nabla_i^v K, \tag{1.1.3}$$

where ∇_i^v is the "gradient operator" in the velocity components of the ith particle, and that the equations of motion (1.1.1) can be written in the form

$$\dot{\mathbf{p}}_i = \mathbf{F}_i. \tag{1.1.4}$$

Of particular importance is the case where the forces $\mathbf{F}_i$ can be obtained from scalar functions U_i of the *positions* of the particles (and possibly the time explicitly) but not of their velocities, by the operation

$$\mathbf{F}_i = -\nabla_i^r U_i, \tag{1.1.5}$$

where ∇_i^r is the gradient operator in the position coordinates of the ith particle. The quantities U_i are the "potential energies" of the particles, and their total energies E_i are the sums of their kinetic and potential energies

$$E_i = K_i + U_i. \tag{1.1.6}$$

The total energy E of the system is then the sum of the total kinetic energy K and the total potential energy U.

By means of (1.1.2) E can be expressed as a function of the $2N$ independent variables $\mathbf{r}_i$ and $\mathbf{p}_i$ and t and is then called the "*Hamiltonian function*" of the system, denoted by $H(\mathbf{r}_i, \ldots \mathbf{r}_N, \mathbf{p}_i, \ldots \mathbf{p}_N; t)$ or briefly by $H(\mathbf{r}, \mathbf{p}; t)$. It follows from the equations (1.1.2) to (1.1.6) that

$$\dot{\mathbf{r}}_i = \nabla_i^p H, \quad \dot{\mathbf{p}}_i = -\nabla_i^r H \quad (i_1 = 1, 2, \ldots, N); \tag{1.1.7}$$

here ∇_i^p is the gradient operator in the momentum components of the ith particle. The integration of the $2N$ first-order partial differential equations (1.1.7) yields again the values of the parameters $\mathbf{r}_i$ and $\mathbf{p}_i$ as functions of time, provided the Hamiltonian function of the system is known and the initial values of the $\mathbf{r}_i$ and $\mathbf{p}_i$ are given.

The total derivative of H with respect to time is, from (1.1.7),

$$\frac{dH}{dt} = \sum_{i=1}^{N} (\nabla_i^r H \cdot \dot{\mathbf{r}}_i + \nabla_i^p H \cdot \dot{\mathbf{p}}_i) + \frac{\partial H}{\partial t} = \frac{\partial H}{\partial t}. \tag{1.1.8}$$

Thus, if H does not explicitly depend on time, that is if $\partial H/\partial t = 0$, the energy E of the system remains constant in the course of time. Such a system is called "*conservative*" and the equation

$$H(\mathbf{r}_1, \ldots, \mathbf{r}_N, p_1, \ldots, p_N) = E \tag{1.1.9}$$

with a constant E is then valid.

The above outlined alternative scheme for solving the basic problem of classical particle dynamics has been shown by Hamilton to be capable of considerable generalization which also holds if the system is under some sort of "constraint". It is then always possible to give a full description of the "configuration" of the system at any moment in terms of the values of a finite number n of "*generalized coordinates*" $q_1, q_2, \ldots, q_n$ which can be chosen arbitrarily as long as they determine the configuration uniquely with respect to an inertial frame of reference. The number n, which is independent of the particular choice of the generalized coordinates, is called the "*number of degrees of freedom*" of the system. The quantities $\dot{q}_1, \dot{q}_2, \ldots, \dot{q}_n$ are called the "*generalized velocities*".

The kinetic energy K can be expressed in terms of the quantities q_i and $\dot{q}_i$, and we shall again assume the existence of a potential energy U, depending on the q_i and on t only. We can then define "*generalized momenta*" p_1, $p_2, \ldots, p_n$ by means of the equations

$$p_i = \frac{\partial K}{\partial \dot{q}_i} \quad (i = 1, \ldots, n), \tag{1.1.10}$$

which are generalizations of (1.1.3), and further "*generalized forces*" F_1, $F_2, \ldots, F_n$ by the equations

$$F_i = -\frac{\partial U}{\partial q_i} \tag{1.1.11}$$

which are generalizations of (1.1.5)

It is now possible to express the total energy of the system as a function of the $2n$ independent variables $q_1, \ldots, q_n, p_1, \ldots, p_n$ and t, and this will again be called the Hamiltonian function H (or in short the Hamiltonian) of the system. It can then be shown that a set of partial differential equations, similar to the set (1.1.7), namely

$$\dot{q}_i = \frac{\partial H}{\partial p_i}, \quad \dot{p}_i = -\frac{\partial H}{\partial q_i} \quad (i = 1, \ldots, n) \tag{1.1.12}$$

are valid, which are known under the name of *Hamilton's "canonical" equations of motion.* Again, the basic problem of classical dynamics can be solved for any system whose Hamiltonian function $H(q, p; t)$ is known, by solving the equations (1.1.12) for given initial conditions.

It follows from (1.1.12) that

$$\frac{dH}{dt} = \sum_{i=1}^{n} \left(\frac{\partial H}{\partial q_i} \dot{q}_i + \frac{\partial H}{\partial p_i} \dot{p}_i \right) + \frac{\partial H}{\partial t} = \frac{\partial H}{\partial t}, \tag{1.1.13}$$

in analogy to (1.1.8), and that therefore E remains constant in time if the Hamiltonian does not explicitly depend on time. The conservation of energy is then contained in the generalization of equation (1.1.9), namely

$$H(q_1, \ldots, q_n, p_1, \ldots, p_n) = E. \tag{1.1.14}$$

1.2. Continuum or field theory

The so-called *"continuum theories"* are based on the concept of a space continuously filled with matter. In order to give a mathematical description of a physical system of this kind it is necessary to subdivide space into small volume elements dV and to define the state within each of these volume elements at a certain time by the numerical values of some observable parameters. This means that we now have an *infinite* set of state variables which are functions of time. This is equivalent to saying that at any given time the state of the system is described by a finite number of continuous functions of space. The change of the state of a system in time is therefore described when the observable parameters are known functions $\varphi(x, y, z, t)$ of the space coordinates and the time, and the laws of physics are accordingly expressed by partial differential equations for these functions.

This concept can be extended to non-material systems and non-material parameters. The corresponding functions φ are then said to define *"fields"* which are called *"stationary"* if the functions are independent of time and *"non-stationary"* if they depend on time.

A typical example of a differential equation for a stationary field is the *Poisson equation* for a *"scalar potential"* φ or a *"vector potential"* $\mathbf{A}$:

$$\left.\begin{aligned} \nabla^2 \varphi &= -4\pi\varrho, \\ \nabla^2 \mathbf{A} &= -4\pi\mathbf{i}, \end{aligned}\right\} \tag{1.2.1}$$

in which ϱ and $\mathbf{i}$ are *"mass"* or *"charge densities"* and *"current densities"* respectively. It can be shown that the equations (1.2.1) have unique solutions $\varphi(x, y, z)$ or $\mathbf{A}(x, y, z)$ if ϱ or $\mathbf{i}$ are given functions of space within a certain domain at the surface of which the values of φ or $\mathbf{A}$ are determined by *"boundary conditions"*. The problem of solving the equations (1.2.1) for a given boundary condition is called the "*boundary problem*" of the theory of potentials. If the domain is infinitely large and the boundary at infinity the solution of (1.2.1) can be found if the functions ϱ or $\mathbf{i}$ are given everywhere in space and the limiting behaviour of the functions φ or $\mathbf{A}$ at infinity are defined.

One of the most important examples of a non-stationary field is the *wave field* whose differential equations, the *wave equations*, have the form

$$\left.\begin{aligned}\frac{\partial^2\varphi}{\partial t^2} &= u^2\,\nabla^2\varphi,\\ \frac{\partial^2\mathbf{A}}{\partial t^2} &= u^2\,\nabla^2\mathbf{A}.\end{aligned}\right\}\tag{1.2.2}$$

Here u is a constant, and φ and $\mathbf{A}$ are called "*wave functions*", which may have the meaning of scalar or vector potentials but can have also other meanings, for example pressure in a fluid or electric or magnetic field strength in an electromagnetic field.

The boundary problem for the wave equation can be formulated as follows: If the values of the wave function at the boundary of a finite space domain are given for all time *(boundary condition)* and if its values are given in the interior of the domain at some specific time, say $t = 0$ *(initial condition)* then the solution of the wave equation is unique inside the domain for all time. Again, if there is no finite boundary, the solution is unique if the wave function is known in the whole space at $t = 0$ and its behaviour at infinity is defined for all time.

A particular solution of the wave equation is a "*harmonic wave*" whose wave function is defined by one of the equations

$$\left.\begin{aligned}\varphi\,(x,y,z;t) &= \psi\,(x,y,z)\,e^{2\pi i\nu t},\\ \mathbf{A}\,(x,y,z;t) &= \mathbf{B}\,(x,y,z)\,e^{2\pi i\nu t},\end{aligned}\right\}\tag{1.2.3}$$

where ν is a constant, called the "*frequency*", and the space functions ψ and $\mathbf{B}$ are called "*amplitude functions*". Substituting from (1.2.3) into (1.2.2) one finds that the amplitude functions have to satisfy differential equations of the form

$$\left.\begin{aligned}\nabla^2\psi + 4\pi^2k^2\psi &= 0,\\ \nabla^2\mathbf{B} + 4\pi^2k^2\mathbf{B} &= 0,\end{aligned}\right\}\tag{1.2.4}$$

called "*amplitude equations*"; the constant k, called "*wave number*", is connected to u and ν by the relation

$$k = \nu/u.\tag{1.2.5}$$

The differential equations (1.2.4) provide another boundary problem which consists in obtaining solutions ψ or $\mathbf{B}$ of these equations, satisfying certain boundary conditions. It turns out that this (and similar problems) can in general be solved only if the parameter k has one of an infinite series of discrete values k_n which are called the "*eigenvalues*" or "*proper values*" of the differential equation; but there are also boundary conditions for which the eigenvalues form a continuous and not a discrete "spectrum". The corresponding values ν_n of the frequency are called the "*eigenfrequencies*" of the particular boundary problem in question. To each eigenvalue k_n belongs in general one solution ψ_n or $\mathbf{B}_n$ of the differential equation which is called an "*eigenfunction*" or "*proper function*". There are, however, cases where more than one eigenfunction belongs to a certain eigenvalue, which is then called "*degenerate*". In the particular case of the amplitude equations (1.2.4) the eigenfunctions are usually referred to as "*normal modes of vibration*".

More general solutions of the wave equation (1.2.2) for given boundary conditions can be constructed by superposing solutions of the type (1.2.3) in the form of a finite or infinite series with arbitrary complex coefficients c_n, constituting a *Fourier series*, for discrete eigenvalue spectra,

$$\left.\begin{aligned} \varphi(x,y,z;t) &= \sum_n c_n\psi_n(x,y,z)\,e^{2\pi i\nu_n t} \\ \mathbf{A}(x,y,z;t) &= \sum_n c_n\mathbf{B}_n(x,y,z)\,e^{2\pi i\nu_n t}, \end{aligned}\right\} \quad (1.2.6)$$

or a *Fourier integral*, for continuous eigenvalue spectra,

$$\left.\begin{aligned} \varphi(x,y,z;t) &= \int f(\nu)\,\psi(x,y,z;\nu)\,e^{2\pi i\nu t}\,d\nu, \\ \mathbf{A}(x,y,z;t) &= \int f(\nu)\,\mathbf{B}(x,y,z;\nu)\,e^{2\pi i\nu t}\,d\nu, \end{aligned}\right\} \quad (1.2.7)$$

where $f(\nu)$ is an arbitrary (real or complex) function of ν. By a suitable choice of the coefficients c_n in (1.2.6) or the functions $f(\nu)$ in (1.2.7) these solutions can be made to satisfy a given initial condition as well.

By direct substitution it can be proved at once that amplitude functions of the form

$$\left.\begin{aligned} \psi &= \gamma e^{-2\pi i(\mathbf{k}\cdot\mathbf{r})}, \\ \mathbf{B} &= \mathbf{C}e^{-2\pi i(\mathbf{k}\cdot\mathbf{r})}, \end{aligned}\right\} \quad (1.2.8)$$

in which γ and $\mathbf{C}$ are constants and $\mathbf{k}$ is a vector with magnitude k, called "*wave vector*", are always solutions of the equations (1.2.4) when no special boundary conditions are imposed. As $(\mathbf{k}\cdot\mathbf{r})$ remains constant for all points on any plane normal to $\mathbf{k}$ these solutions represent *plane waves.* $(\mathbf{k}\cdot\mathbf{r})$ is equal to k times the perpendicular distance of a wave surface from the origin of coordinates and therefore changes by unity if that distance is increased by $\lambda = 1/k$, leaving the exponential function unchanged. Thus λ is seen to be the *wavelength* of the harmonic wave, satisfying the relation

$$\lambda = u/\nu; \quad (1.2.9)$$

consequently u is the *wave velocity.*

Combining (1.2.8) with (1.2.3) we obtain the complete expressions for plane harmonic waves of frequency ν progressing in the direction of the wave vector **k**:

$$\left.\begin{aligned} \varphi(\mathbf{r};t) &= \gamma\, e^{2\pi i(\nu t - \mathbf{k}\cdot\mathbf{r})}, \\ \mathbf{A}(\mathbf{r};t) &= \mathbf{C}\, e^{2\pi i(\nu t - \mathbf{k}\cdot\mathbf{r})}. \end{aligned}\right\} \quad 1.2.10)$$

More general solutions of the wave equation with no special boundary condition imposed can again be constructed by the superposition of a finite or infinite number of such plane waves.

1.3. The Hamiltonian analogy between dynamics and geometrical optics

Although it would appear from the contents of sections 1.1 and 1.2 that the mathematical techniques for the solution of problems of particle dynamics are quite different from those applied in continuum physics, it is nevertheless possible, as was first shown by Hamilton, to treat particle dynamics by a method strictly analogous to the one used in *geometrical optics*, which is a branch of continuum physics.

We shall first consider the movement of a single particle of mass m in a potential field of force with a potential energy function $U(\mathbf{r})$. Assume that this particle starts its movement from a point P on a closed surface S in the direction of the unit vector **n**, pointing outwards normal to the surface, and that its energy is E (which will, of course, remain constant during the subsequent movement of the particle). We assume E to be sufficiently large so that all the trajectories outside S extend to infinity. At each of its points the momentum **p** of the particle will therefore be determined uniquely.

If we now construct all the trajectories starting from all the points P of S under the above specified conditions we see that to all points in space within a certain domain outside S definite values of **p** will be assigned. We can thus define a scalar function $W(\mathbf{r})$ in this domain such that

$$\mathbf{p} = \nabla W. \quad (1.3.1)$$

For reasons which will become plain later we shall call $W(\mathbf{r})$ the "*action function*".

The energy equation (1.1.9) can now be written in the form

$$p^2 = |\nabla W|^2 = 2m(E - U) \quad (1.3.2)$$

or, in Cartesian coordinates,

$$\left(\frac{\partial W}{\partial x}\right)^2 + \left(\frac{\partial W}{\partial y}\right)^2 + \left(\frac{\partial W}{\partial z}\right)^2 = 2m\,[E - U(x,y,z)]. \quad (1.3.3)$$

This partial differential equation of the first order for W is a special form of the so-called "*Hamilton–Jacobi equation*" of classical dynamics, and one notices that it has the characteristic form of the differential equations of continuum physics, discussed in section 1.2. Thus it can be solved uniquely if the function $U(\mathbf{r})$ is known and the energy E and the shape of the surface S are given. For, according to (1.3.1), the trajectories must everywhere be normal to the surfaces $W(\mathbf{r}) = \text{const}$, and hence S must also be one of these surfaces; this constitutes the boundary condition for which (1.3.3) is to be solved.

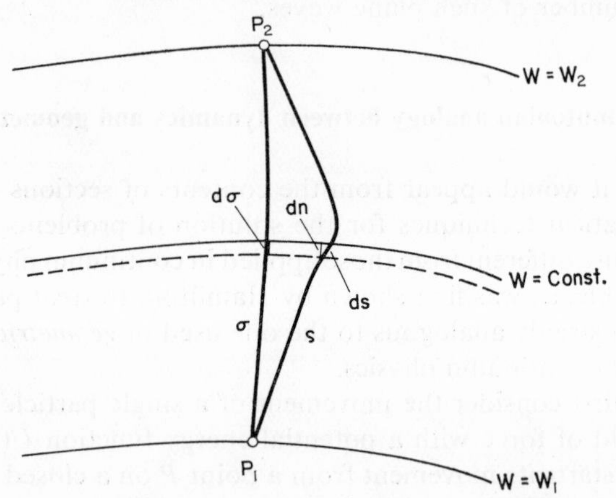

FIG. 1. Diagram showing surfaces of constant value of the action function W, and illustrating the derivation of Maupertuis' principle of least action.

Once the action function $W(x, y, z)$ has been found the complete solution of the problem, namely the determination of $\mathbf{r}(t)$ for any of the starting points P can be obtained by integration of the ordinary differential equations

$$\dot{x} = \frac{1}{m}\frac{\partial W}{\partial x}, \quad \dot{y} = \frac{1}{m}\frac{\partial W}{\partial y}, \quad \dot{z} = \frac{1}{m}\frac{\partial W}{\partial z}. \tag{1.3.4}$$

The Hamilton–Jacobi equation (1.3.3) may also be solved by a *variational principle*. Consider the part of a trajectory σ (Fig. 1) between the points P_1, P_2 outside S, lying on the surfaces $W = W_1$ and $W = W_2$ respectively, and take the line integral

$$J = \int_{P_1}^{P_2} p\, ds, \tag{1.3.5}$$

along the path s. J can readily be transformed into a time integral by making use of the relation

$$p\, ds = \frac{2K}{v}\, ds = 2K\, dt.$$

Thus

$$J = \int_{t_1}^{t_2} 2K\, dt \tag{1.3.6}$$

is the time integral of twice the kinetic energy of the particle between the time t_1 of departure from P_1 to the time t_2 of arrival at P_2, whose (constant) total energy is E and which is forced by some external constraint to follow the path s. The quantity (1.3.6) was called "*action*" by Maupertuis and the integral (1.3.5) is therefore referred to as the *action integral.*

Substituting from (1.3.1) into (1.3.5) one obtains

$$J = \int_{P_1}^{P_2} \frac{dW}{dn}\, ds. \tag{1.3.7}$$

It is seen from Fig. 1 that $ds/dn > 1$ for all paths s and that $ds/dn = 1$ only for the path σ. Hence the action integral has, for a fixed energy E, its *minimum value* if it is taken along the *actual* trajectory (without constraint) leading from P_1 to P_2. This theorem is known under the name of "*principle of least action*" or "*Maupertuis' principle*". This minimum value is

$$J_{\text{min}} = W_2 - W_1, \tag{1.3.8}$$

that is the difference between the values of the action function W at the two end points of the trajectory. This justifies the choice of the name given to this function.

In the case of a closed path the line integral (1.3.5) taken along the whole trajectory

$$J_0 = \oint p\, ds \tag{1.3.9}$$

is evidently always positive as p and ds have the same direction everywhere. This is, of course, not in contradiction to (1.3.8) because in deriving (1.3.8) we had assumed the trajectories outside S not to be closed.

The above outlined theory can be generalized for an arbitrary *conservative* dynamical system of n degrees of freedom in a potential field of force. Its movement can be represented in an n-dimensional "*configuration space*" with coordinates $q_1, \ldots, q_n$, and we can again, in generalization of (1.3.1), define an action function $W(q_1, \ldots, q_n)$ by

$$p_i = \frac{\partial W}{\partial q_i} \quad (i = 1, \ldots, n), \tag{1.3.10}$$

where the p_i are the generalized momenta (1.1.10). Substituting this into the energy equation (1.1.14) one obtains the general Hamilton–Jacobi equation in the form

$$H\left(q_1, \ldots, q_n; \frac{\partial W}{\partial q_1}, \ldots, \frac{\partial W}{\partial q_n}\right) = E. \tag{1.3.11}$$

This is again a partial differential equation of the first order for the action function W which can be solved if the shape of one of the "surfaces" $W = \text{const.}$ is given, which may be taken to be the locus of the starting points of the systems at $t = 0$ in configuration space. The "trajectories" are then, as before, the curves which are everywhere orthogonal to the surfaces $W = \text{const.}$ The dependence of the coordinates q_i on time is finally found by a second integration from the equations

$$\frac{d}{dt}\left(\frac{\partial W}{\partial q_i}\right) = -\frac{\partial H}{\partial q_i} \quad (i = 1, \ldots, n) \tag{1.3.12}$$

which follow from (1.1.12) and (1.3.10).

The just outlined method for solving problems of the dynamics of conservative systems can be extended to *non-conservative* systems, that is systems whose Hamiltonian function depends on time explicitly. It can indeed be shown that in such cases a *time-dependent* action function $\Omega(q_1, \ldots, q_n; t)$ can always be defined in such a way that, as before, the momenta are the partial derivatives of Ω with respect to the coordinates

$$p_i = \frac{\partial \Omega}{\partial q_i} \quad (i = 1, \ldots, n), \tag{1.3.13}$$

and the energy (which is now no longer a constant) is given by the relation

$$E = -\frac{\partial \Omega}{\partial t} \tag{1.3.14}$$

which evidently is dimensionally correct.

Substitution of (1.3.13) and (1.3.14) into the energy equation yields now the required Hamilton–Jacobi equation for non-conservative systems:

$$H\left(q_1, \ldots, q_n; \frac{\partial \Omega}{\partial q_1}, \ldots, \frac{\partial \Omega}{\partial q_n}; t\right) + \frac{\partial \Omega}{\partial t} = 0. \tag{1.3.15}$$

If this equation is solved under the initial condition that for $t = 0$, Ω is a given function of the coordinates, the dependence of the momenta and the energy on the coordinates and the time is obtained from the equations (1.3.13) and (1.3.14), and, finally, the dependence of the coordinates on time follows from the equations (1.3.12) when W is replaced by Ω.

The Hamilton–Jacobi equation (1.3.11) for conservative systems is seen to be a special case of the more general equation (1.3.15) when one puts

$$\Omega(q_1, \ldots, q_n; t) = W(q_1, \ldots, q_n) - Et, \tag{1.3.16}$$

where E is again the energy constant.

An action integral can again be defined by the time integral (1.3.6). If, as is usually the case, the kinetic energy K is a homogeneous function of second order in the momenta, one has

$$2K = \sum_i \frac{\partial K}{\partial p_i} p_i, \tag{1.3.17}$$

and thus the action integral takes the form

$$J = \sum_i \left(\int_{t_1}^{t_2} \frac{\partial K}{\partial p_i} p_i \, dt \right). \tag{1.3.18}$$

If this integral is taken over the actual trajectory connecting the points P_1 and P_2 in configuration space outside S one can make use of the first of the canonical equations (1.1.12)

$$\frac{\partial K}{\partial p_i} = \frac{\partial H}{\partial p_i} = \dot{q}_i$$

and transform the time integrals (1.3.18) into line integrals

$$J = \sum_i \left(\int_{P_1}^{P_2} p_i \, dq_i \right). \tag{1.3.19}$$

The integrals

$$J_i = \oint p_i \, dq_i \tag{1.3.20}$$

taken over a closed orbit (or a complete cycle of a periodic system) are called "*phase integrals*"; they played an important part in the earlier stages of development of quantum theory, as will be explained later in section 2.2.

Making use of the relations (1.3.10) one obtains from (1.3.19)

$$J = \int_{P_1}^{P_2} \sum_i \frac{\partial W}{\partial q_i} dq_i = W_2 - W_1, \tag{1.3.21}$$

that is again equation (1.3.8), showing the relationship between the action integral and the action function in the general case.

It will now be shown that the above explained procedure for constructing the paths of material particles in potential fields of force is completely analogous to the procedure used in *geometrical optics* for constructing the rays of light in an inhomogeneous medium.

Geometrical optics is essentially based on Huygens' principle. This maintains that if a *wave front* originally, say at $t = 0$, has the shape of a closed surface S, the wave front at a subsequent time t can be constructed as the enveloping surface of all the wavelets issuing from the various points P of S with velocities of propagation u, depending on the positions of P in space. In this way the whole set of gradually progressing wave fronts can be constructed step by step, and eventually the "light rays" traced as the orthogonal trajectories to these surfaces.

The wave fronts of a monochromatic light wave may be identified with the *surfaces of constant phase* $\Phi =$ const. The function $\Phi(\mathbf{r})$ is known as the "*eiconal function*". The vector $\nabla\Phi$ has everywhere the direction of the light ray and its magnitude is equal to the change of Φ along a unit length of the ray. As the phase changes by 2π over a length λ, the wavelength, of the ray one has

$$|\nabla\Phi|^2 = \left(\frac{2\pi}{\lambda}\right)^2 = 4\pi^2 k^2. \tag{1.3.22}$$

Written in Cartesian coordinates the equation (1.3.22) takes the form

$$\left(\frac{\partial\Phi}{\partial x}\right)^2 + \left(\frac{\partial\Phi}{\partial y}\right)^2 + \left(\frac{\partial\Phi}{\partial z}\right)^2 = 4\pi^2 k^2, \tag{1.3.23}$$

which is a first-order partial differential equation for $\Phi\,(x, y, z)$ called the "*eiconal equation*". For a given frequency $k = \nu/u$ is a known function $k(x, y, z)$ of the space coordinates, and $\Phi(x, y, z)$ can therefore be obtained by integrating the eiconal equation (1.3.23) under the boundary condition that the original wave front S be one of the surfaces Φ = const. The light rays emerging from S outwards can thus be constructed as the orthogonal trajectories to the surfaces Φ = const.

As can be seen, the Hamilton–Jacobi equation (1.3.3) has the same mathematical form as the eiconal equation (1.3.23) and the previously outlined procedure for the construction of the paths of particles in the mechanical problem is precisely analogous to the procedure for the construction of the light rays in geometrical optics. The analogy can even be carried further by considering the total time T taken by a "light signal" to travel from a point P_1 to a point P_2 along an arbitrary path s, connecting the two points (Fig. 1). T is given by a time integral which can readily be transformed into a line integral

$$T = \int_{t_1}^{t_2} dt = \int_{P_1}^{P_2} \frac{dt}{ds}\, ds = \int_{P_1}^{P_2} \frac{ds}{u}. \tag{1.3.24}$$

Making use of (1.3.22) one obtains

$$T = \frac{1}{2\pi\nu} \int_{P_1}^{P_2} \frac{d\Phi}{dn}\, ds. \tag{1.3.25}$$

Using the same argument as in the mechanical case, one sees from Fig. 1 (where now the surfaces W = const. are replaced by the surfaces Φ = const.) that T has a minimum for the *actual* ray σ passing through P_1 and P_2, and that the minimum value of T over the path σ is

$$T_{\min} = \frac{\Phi_2 - \Phi_1}{2\pi\nu}. \tag{1.3.26}$$

The above formulated minimum principle, which is a variational form of Huygens' principle, is known as "*Fermat's principle*" in geometrical optics.

It stipulates that a light signal always passes between two fixed points in an inhomogeneous medium along a line which will allow it to make the journey in the *shortest possible time.* Clearly Maupertuis' principle of least action is the mechanical analogue of Fermat's optical principle, and (1.3.26) is the analogue of (1.3.8).

To complete the developments of this section it is necessary to show under what conditions the eiconal equation (1.3.23) can take the place of the wave equation (1.3.2) in the process of propagation of waves. Let us again assume that the waves are harmonic with frequency ν and a space dependent wave velocity u. The wave equation then takes the form of the "amplitude equation" (1.2.4) in which now k is a function of space. The complex amplitude function ψ can be written in the form

$$\psi = a\, e^{i\Phi}, \tag{1.3.27}$$

where a is the *real* amplitude and Φ, as before, the phase. Substitution of (1.3.27) into (1.2.4) results in the complex equation

$$\nabla^2 a + 2i \nabla a \cdot \nabla \Phi + ia \nabla^2 \Phi - a\, |\nabla \Phi|^2 + 4\pi^2 k^2 a = 0.$$

If this equation is to be satisfied the real and the imaginary parts of its left-hand side must vanish separately. Hence one obtains

$$\left.\begin{aligned} \nabla^2 a - a\, |\nabla \Phi|^2 + 4\pi^2 k^2 a &= 0, \\ \nabla a \cdot \nabla \Phi + \frac{a}{2} \nabla^2 \Phi &= 0. \end{aligned}\right\} \tag{1.3.28}$$

If it is now assumed that the amplitude a changes sufficiently slowly in space, such that it remains practically constant over distances of the order of magnitude λ, the first term in the first of the equations (1.3.28) can be neglected against the two others and one obtains the eiconal equation (1.3.22). This, therefore, is the condition under which the methods of geometrical optics are valid.

Problems and exercises

P.1.1. Choose as generalized coordinates of a particle of mass m its spherical polar coordinates r (radius), θ (polar angle), and φ (azimuth angle). Derive the expressions for the kinetic energy K in terms of r, θ, φ, and hence the expressions for the generalized momenta p_r, p_θ, p_φ.

P.1.2. A particle of mass m is constrained to move on a spherical surface r = const. Using the results of P.1.1, set up the classical Hamiltonian equations of motion, in particular for the case where the potential energy U depends on θ alone. Show that in that case p_ϱ = const.

P.1.3. A system of particles of masses m_i, whose mutual distances are fixed, is equivalent to a "rigid body". If this system is constrained to rotation about a fixed axis as a whole (one-dimensional rigid rotator) the angle φ of rotation can be chosen as generalized coordinate. Derive the expression for the Hamiltonian in terms of φ and the corresponding "angular momentum" P and use the Hamiltonian equations for obtaining the equation of motion of the system under the action of an external couple M acting on it.

P. 1.4. Show that the Poisson equation (1.2.1) takes the form

$$\frac{1}{r}\,\frac{d^2(r,\Phi)}{dr^2} = -4\pi\varrho(r)$$

if the density distribution is spherically symmetric, i.e. if ϱ depends on the distance r from a fixed centre only. Solve this equation under the condition that ϱ = const between $r = 0$ and $r = R$ and $\varrho = 0$ for $r \geqslant R$, and the boundary condition $\Phi = 0$ for $r = \infty$.

P. 1.5. Derive the form of the wave equation (1.2.2) for the wave function Φ if it depends on the coordinate x only, i.e. for "plane waves". Show that the most general solution of this equation is

$$\Phi = f(x - ut) + g(x + ut),$$

where f and g are arbitrary but independent functions of their arguments. Discuss the meaning of this solution.

P. 1.6. Derive the form of the wave equation (1.2.2) if Φ is a function of the distance from a fixed centre only, i.e. for "spherical waves", making use of the solution of P. 1.4. Obtain the general solution of this equation with the help of the results of P. 1.5 and discuss the meaning of this solution.

P. 1.7. Verify that the functions (1.2.8) are indeed solutions of the amplitude equation (1.2.4).

P. 1.8. Use (1.3.11) for obtaining the Hamilton–Jacobi equation for a particle of mass m, moving in a plane under the influence of a potential $U = B/r$, in terms of the plane polar coordinates r, θ, and their associated momenta p_r and p_θ. Solve that equation by separation of variables, assuming that W can be split up in the form $W(r, \theta) = W_r(r) + W_\theta(\theta)$.

P. 1.9. A particle, starting from a point P_1 at a distance a_1 from a rigid wall, moves with velocity v in a direction making an angle α_1 with the normal to the wall. After reflection from the wall the particle is at some time found in a point P_2 at a distance a_2 from the wall, proceeding in a direction at an angle α_2 to the wall normal. Assuming that the kinetic energy of the particle is not changed by the impact with the wall, use the principle of least action to prove that $\alpha_1 = \alpha_2$.

P. 1.10. Suppose a certain region of space to be filled with a transparent medium whose refractive index μ is a linear function of y, but is independent of x and z. Set up the corresponding eiconal equation and solve it under the assumption that the eiconal function Φ is independent of z and can be expressed as the sum of a function Φ_x of x alone and a function Φ_y of y alone.

P. 1.11. Use Fermat's principle for proving Snell's law of refraction at the plane boundary between two media of refractive indices μ_1 and μ_2 respectively, folowing an argument similar to that used in P. 1.9.

P. 1.12. Prove by means of Fermat's principle that all light rays issuing from one of the focal points F_1 of a mirror in the form of a rotational ellipsoid must pass through the second focal point F_2.

CHAPTER 2

Short History of the Development of Quantum Theory

TOWARDS the end of the nineteenth century it had become possible to give a satisfactory account of most known atomic phenomena by applying the laws of classical particle physics to the movement of atoms and electrons, and those of classical continuum physics to the electromagnetic field. There remained, however, a number of phenomena which persistently defied classical explanation. In this chapter we shall first discuss these difficulties and then give a short history of the gradual development of quantum theory, which was designed to resolve these difficulties and which eventually led to he establishment of *quantum mechanics*, the main subject of Part I.

2.1. The difficulties of classical physics

The first of the phenomena to be considered here is the problem of the *black-body radiation* because of the decisive part it played in the foundation of quantum theory. It consists in determining the spectral distribution of intensity in the radiation which is in thermodynamic equilibrium with material bodies of temperature T, or which establishes itself in the interior of an "enclosure" of temperature T.

In order to explain the emission of electromagnetic radiation with a continuous spectrum from a material body by classical electromagnetic theory, one has to assume that the body contains a great number of *harmonic electronic oscillators* with a continuous range of frequencies ν. It follows from Maxwell's equations that the rate of energy emission from such an oscillator, whose electric moment P changes harmonically with time, is proportional to

$$\overline{(\ddot{P})^2} \simeq \nu^4(\overline{P^2}),$$

where the bars indicate time averages. The energy ε of the oscillator is proportional to

$$\varepsilon \simeq \frac{\nu^2(\overline{P^2})}{m},$$

m representing the total (mechanical plus electromagnetic) inertia of the oscillating system. Hence if there are N oscillators present in the narrow frequency range $\nu \cdots \nu + \delta\nu$, the total energy e, emitted per unit time by the body in this frequency range, is proportional to

$$e \simeq \frac{\nu^2 \varepsilon N \delta\nu}{m}. \tag{2.1.1}$$

On the other hand, an oscillator which is subjected to an oscillating electromagnetic field in the frequency range $\nu \cdots \nu + \delta\nu$ with intensity $I_\nu \delta\nu$ absorbs energy from that field at a rate proportional to $I_\nu \delta\nu/m$, and hence the total amount of energy a absorbed by the body per unit time is proportional to

$$a \simeq \frac{I_\nu N \delta\nu}{m}. \tag{2.1.2}$$

In thermodynamic equilibrium e must be equal to a, so that

$$I_\nu \simeq \nu^2 \varepsilon. \tag{2.1.3}$$

If the principle of continuity holds and consequently ε can assume any value within a continuous range of energies the *equipartition law* of statistical mechanics holds (see sections 13.3 and 13.4), according to which the mean energy of a harmonic oscillator at temperature T is equal to kT, irrespective of ν and m (k: Boltzmann's constant). Substituting this value for ε in (2.1.3) we obtain the required spectral energy distribution in the form of the relationship

$$I_\nu \simeq \nu^2 T, \tag{2.1.4}$$

which is known as "*Rayleigh's law*". This law, however, is not in agreement with experiment, except in the low frequency region of the spectrum. Moreover, the integral of the right-hand side of (2.1.4) over all frequencies diverges, and this means that the total energy density of black-body radiation in an enclosure would be infinitely large ("ultraviolet catastrophe") which is physically unacceptable.

Next we consider the problem of the *specific heat* of solids. A solid body consisting of N atoms is a system of $3N$ degrees of freedom for the translational movement of the atoms. The thermal movement of the atoms in a solid body consists in coupled oscillations of these atoms about their equilibrium positions, which constitute the "normal modes of vibration" and is therefore equivalent to the vibration of $3N$ "linear harmonic oscillators". Again, according to the classical equipartition law, the energy of each of these oscillators should be equal to kT so that the total energy of the thermal movement, which is equal to the "internal energy" U of the body, amounts to

$$U = 3NkT. \tag{2.1.5}$$

Hence the *heat capacity* of the body has the value

$$C = 3Nk, \tag{2.1.6}$$

which is independent of the nature of the constituent atoms and independent of temperature (Dulong–Petit's law).

Although this law is found to be in agreement with experiment at high temperatures, it fails at sufficiently low temperatures, C gradually decreasing to zero as the absolute zero point of temperature is approached. Moreover, the thermal kinetic energy of the *free electrons* in metals does not seem to contribute to the specific heat at all. These facts, together with those relating to the black-body radiation, seem to indicate that the classical law of equipartition of energy does not hold for very light particles, like electrons, and not for very low temperatures.

We now turn to the problem of the *photoelectric effect* which consists in the emission of electrons from metals when they are irradiated with electromagnetic radiation of sufficiently high frequency. Classically one would explain this phenomenon by the absorption of energy by the free electrons in the metal near its surface when they are made to vibrate in the electromagnetic radiation field; they will thus finally acquire sufficient kinetic energy to be able to escape from the metal. If this explanation were correct one would expect the kinetic energy of the emitted electrons to increase with increasing intensity of the radiation. In actual fact, however, this kinetic energy is *independent* of the radiation *intensity* and a function of its *frequency* ν only, linearly increasing with ν. It appears, therefore, that the photoelectric effect cannot be explained on the basis of the classical theory of electromagnetic waves.

We finally consider the problem of the *line spectra* produced in the emission of light from atoms or in the absorption of light by atoms. If the emission spectra originate in the periodic movement of electrons inside the atoms, one should expect these spectra to consist of lines whose frequencies coincide with the frequencies of all the harmonic vibrations of the electrons into which their periodic orbital movement can be resolved. This, however, would imply that the lines of any particular spectrum should have frequencies which are integer multiples of certain fundamental frequencies (higher harmonics); but this is not the case. Moreover, the continuous loss of energy of the electrons due to the emission of radiation would cause the electronic orbits to shrink gradually and the frequencies to increase during the emission process, which is incompatible with the steady nature of the structure of the spectra. In addition, this loss of energy would eventually (actually within a very short time) cause the electronic orbits to collapse completely, which is clearly incompatible with the apparent permanency of the atoms.

These and similar failures of the classical theory to account for the observed facts made it unavoidable to revise the classical ideas about the interaction between material particles and electromagnetic radiation.

2.2. The advent of quantum theory

Planck was the first to realize that the failure of classical theory to account for the phenomenon of black-body radiation must be due to the fact that the equipartition law was not applicable to electronic oscillators, and that this in turn implied that these oscillators were not capable of assuming a continuous range of energy values. On the basis of an ingenious argument he postulated in 1900 that the energy of a harmonic oscillator of frequency ν could only assume the *discrete* energy values

$$E_n = nh\nu \quad (n = 1, 2, 3, \ldots), \tag{2.2.1}$$

where h is a *universal constant* of the dimension of action, now generally known as "*Planck's constant*".

According to Boltzmann's statistical law (see section 14.2), the probability for the oscillator having an energy E_n within a body of temperature T is given by

$$W_n = \frac{\exp(-E_n/kT)}{\sum_{n=1}^{\infty} \exp(-E_n/kT)}, \tag{2.2.2}$$

and thus the mean energy $\bar{E}$ is, from (2.2.1) and (2.2.2),

$$\bar{E} = \sum_{n=1}^{\infty} E_n W_n = \frac{\sum_{n=1}^{\infty} n \exp(-nh\nu/kT)}{\sum_{n=1}^{\infty} \exp(-nh\nu/kT)} h\nu = \frac{h\nu}{e^{h\nu/kT} - 1}. \tag{2.2.3}$$

Equation (2.2.3) shows that for $h\nu/kT \ll 1$ the mean energy $\bar{E}$ is equal to its equipartition value kT. Thus the equipartition law will hold approximately for low frequencies and high temperatures, but it will break down for oscillators with high frequencies and at low temperatures as indicated in section 2.1.

If the expression (2.2.3) is substituted in (2.1.3), one obtains the formula

$$I_\nu \simeq \frac{\nu^3}{e^{h\nu/kT} - 1} \tag{2.2.4}$$

for the spectral intensity distribution of black-body radiation instead of the Rayleigh law (2.1.4). The integral over all ν of (2.2.4) now converges, giving a finite value for the energy density of the radiation, and the radiation law (2.2.4), known as "*Planck's radiation law*", is found to be in perfect agreement with experiment, Planck's constant having the value $h = 6{\cdot}62 \times 10^{-27}$ erg sec. A more rigorous derivation of Planck's law and a full discussion of its consequences will be given in section 15.3.

If Planck's quantum condition (2.2.1) for the harmonic oscillator is correct the emission and absorption of radiation by such oscillators must occur in the form of "packets" E_n of energy. On the basis of considerations concerning the statistical thermodynamics of black-body radiation, Einstein (1905a) proposed the even more radical hypothesis that radiation actually consisted of *light quanta*, nowaday usually called "*photons*", to which (1906) he ascribed energies $h\nu$. To this hypothesis he later (1909) added the postulate that a photon should also have a momentum p of magnitude $h\nu/c$ and should travel along a straight line with the vacuum velocity c of light. The argument leading to this conclusion will be discussed in section 10.3.

The quantum conditions for photons are therefore

$$E = h\nu, \quad p = h\nu/c = hk. \tag{2.2.5}$$

They give at once an explanation for the photoelectric effect. For the absorption of radiation by electrons now amounts to the transfer of the photon energy $h\nu$ to the electron, and thus its kinetic energy K on emerging from the emitter is determined by the relation

$$K = h\nu - w, \tag{2.2.6}$$

where w is the work the electron has to do in order to overcome the "potential barrier" at the metal surface. Thus K is indeed a linear function of ν, and emission can only occur if $\nu \geqslant w/h$.

A number of other experimental facts, like Stokes' law for incoherent light scattering, Duane's and Hunt's law for the continuous X-ray spectrum of the "*bremsstrahlung*", and the law governing photochemical reactions can be immediately explained on the basis of the theory of photons. On the other hand, the quantum conditions (2.2.5) seem to be incompatible with the continuum theory of radiation and incapable of explaining such phenomena as interference and diffraction of light which are so well accounted for by the wave theory. It will be shown later (section 3.2) how these apparent contradictions can be resolved.

In 1907 Einstein applied Planck's quantum condition (2.2.1) to the *atomic* oscillators whose vibrations constitute the thermal movement in solids as explained in section 2.1. For the sake of simplicity he assumed that the oscillations of the atoms in an element were not coupled so that each atom in one mole of substance performed a harmonic vibration with the same frequency ν (now usually called "*Einstein frequency*") which could be identified with the so-called "*rest-strahlen frequency*" of the substance in the far infrared. The mean thermal energy per atom is then three times the value (2.2.3), and hence the internal energy U per mole. containing N atoms,

$$U = 3N \frac{h\nu}{e^{h\nu/kT} - 1}, \tag{2.2.7}$$

from which follows for the molar specific heat c_v

$$c_v = \frac{dU}{dT} = 3R\varphi_E\left(\frac{h\nu}{kT}\right); \tag{2.2.8}$$

here $R = Nk$ is the universal gas constant and φ_E, called "*Einstein function*", is given by

$$\varphi_E(x) = \frac{x^2 e^x}{(e^x - 1)^2}. \tag{2.2.9}$$

For high temperatures (small x) $\varphi_E(x) \to 1$ and c_v assumes the Dulong–Petit value, but with decreasing temperature (increasing x) $\varphi_E(x)$ gradually decreases and tends to zero at $T = 0$, in agreement with experiment as mentioned in section 2.1. This was the first application of quantum concepts to the theory of the thermal properties of matter which will be dealt with systematically in Part III.

In 1913 Bohr attempted to develop a quantum theory of atomic spectra. He postulated that the *angular momentum* P of the orbital movement of an electron in an atom (in Rutherford's atomic model) should not be capable of assuming arbitrary values but be "quantized" according to the quantum condition

$$P = n\frac{h}{2\pi} \equiv n\hbar \quad (n = 1, 2, 3, \ldots) \tag{2.2.10}$$

and that emission and absorption of radiation should occur in the form of photons of frequency ν in sudden transitions between two quantum levels of energies E_n and $E_{n'}$ ("quantum jumps") according to the rule

$$E_n - E_{n'} = h\nu, \tag{2.2.11}$$

in contradiction to the classical ideas of interaction between charged particles and electromagnetic fields.

In the simplest case of a single electron of mass m and charge e, moving in a circular orbit of radius r with angular velocity ω about a fixed nucleus of charge Ze the condition of equilibrium between Coulomb attraction and centrifugal force on the electron is

$$m\omega^2 r^3 = e^2 Z$$

and the quantum condition (2.2.10) becomes

$$m\omega r^2 = n\hbar.$$

Elimination of ω from these two equations yields the expression

$$r_n = \frac{n^2\hbar^2}{me^2 Z} \quad (n = 1, 2, 3, \ldots) \tag{2.2.12}$$

for the radius of the orbit of the nth quantum state. The energy E_n of that state, the sum of the kinetic and the potential energies of the electron is

$$E_n = \frac{p_{\varphi_n}^2}{2mr_n^2} - \frac{e^2 Z}{r_n} = -\frac{me^4 Z^2}{2n^2\hbar^2}. \qquad (2.2.13)$$

The application of the relations (2.2.11) and (2.2.13) to the hydrogen atom ($Z = 1$) was the first step towards the explanation of atomic line spectra.

Bohr's quantum condition (2.2.10) was generalized by Sommerfeld in 1915 for general systems of N degrees of freedom. He postulated that of all the movements of the system, compatible with the laws of classical dynamics, only those should be permitted for which the phase integrals (1.3.20) were equal to integer multiples of h:

$$J_i = \oint p_i \, dq_i = n_i h \quad (i = 1, 2, \ldots, N). \qquad (2.2.14)$$

These equations constitute N quantum conditions with (in general) N different quantum numbers $n_1, n_2, \ldots, n_N$ for the system. It is easy to show that Planck's quantum condition (2.2.1) for the linear harmonic oscillator and Bohr's quantum condition (2.2.10) for a particle under the action of a central force are indeed special cases of (2.2.13).

In the case of the particle moving under the influence of a central force one can use the azimuth angle φ in the plane of the orbit as one of the generalized coordinates. The corresponding momentum is then the angular momentum P which remains constant during the movement. Hence from (2.2.14)

$$J \equiv \int_0^{2\pi} p_\varphi \, d\varphi = 2\pi P = nh, \qquad (2.2.15)$$

which is indeed identical with Bohr's quantum condition (2.2.10).

In the case of the harmonic oscillator we use the displacement q of the oscillating particle of mass m from its equilibrium position as generalized coordinate. Then

$$q = a \sin 2\pi\nu t \quad \text{and} \quad p = 2\pi\nu m a \cos 2\pi\nu t.$$

Eliminating t from these two equations we obtain a relation between q and p

$$\frac{p^2}{4\pi^2\nu^2 m^2 a^2} + \frac{q^2}{a^2} = 1,$$

which is the equation of an ellipse with semi-axes $2\pi\nu m a$ and a in the qp-plane. Hence the phase integral is equal to the area enclosed by the ellipse, and from (2.2.14) we obtain the quantum condition in the form

$$2\pi^2 \nu m a^2 = nh. \qquad (2.2.16)$$

On the other hand, the energy E of the oscillator has the constant value

$$E = \left(\frac{p^2}{2m}\right)_{t=0} = 2\pi^2\nu^2 m a^2. \qquad (2.2.17)$$

The combination of (2.2.16) and (2.2.17) results in the Planck condition (2.2.1).

An important addition to the Bohr–Sommerfeld theory of atomic systems is due to Uhlenbeck and Goudsmit who in 1925 suggested that an electron should perform not only a translational movement with 3 degrees of freedom but also a rotational movement about an internal axis with one further degree of freedom called the "*spin*". They were able to explain the multiplicity of certain spectral lines as a "splitting up" of the energy levels of the atom due to a quantization of the angular momentum of the spin according to Bohr's quantization rule (2.2.10) with half-integer quantum numbers $s = \pm\frac{1}{2}$.

Sommerfeld's theory offers no explanation as to why the movements of material particles should be restricted to those for which the quantum conditions (2.2.14) are satisfied. In 1925 De Broglie made the ingenious suggestion that this might be explained by attaching certain wave properties to moving material particles, which would create a similar situation to the one existing in optics where the light waves have certain particle properties as claimed by the photon theory. Consequently stationary systems of particles would only be capable of certain "normal modes", consisting of "standing waves", analogous to the normal modes of vibration of continuous wave fields, as discussed in section 1.2.

Following this idea de Broglie assigned a certain frequency ν to a particle of total energy E, satisfying the first of the Einstein relations (2.2.5) $E = h\nu$ for photons. By arguments based on the theory of relativity (as will be discussed in section 10.3) he concluded that to this relation must be added another relation $\mathbf{p} = h\mathbf{k}$ between the wave vector $\mathbf{k}$ of a wave and the momentum $\mathbf{p}$ of the particle, similar to the second of the Einstein relations (2.2.5). Thus the De Broglie relations are

$$E = h\nu, \quad \mathbf{p} = h\mathbf{k}. \tag{2.2.18}$$

It follows from (2.2.18) that the wavelength λ of this "*De Broglie wave*", called the "*De Broglie wavelength*" of the particle, is equal to

$$\lambda = \frac{1}{k} = \frac{h}{p}. \tag{2.2.19}$$

It is further seen that the "group velocity" of a De Broglie wave is, according to its definition, given by

$$\frac{d\nu}{dk} = \frac{\partial E}{\partial p} = \frac{p}{m} = v, \tag{2.2.20}$$

that is, it is equal to the particle velocity v.

Experimental evidence for the correctness of De Broglie's hypothesis is provided by the phenomenon of *diffraction* of beams of particles which allows their De Broglie wavelength to be measured and the relation (2.2.19) to be verified. This will be discussed in more detail in section 3.1.

Library
I.U.P.
Indiana, Pa.

530.1 F984f
c. 1

In order to show that the propagation of De Broglie waves gives rise to quantum conditions for stationary states of electronic orbits in atoms we consider the movement of a single electron along a closed path s. If this is to be equivalent to a stationary wave train along s the phase of the wave must change by an integer multiple of 2π over the circumference of s. Now the phase change over a length ds is from (2.2.19),

$$\frac{2\pi\, ds}{\lambda} = \frac{2\pi}{h}\, p\, ds,$$

and hence, integrating over the whole orbit one obtains the condition,

$$\oint p\, ds = nh \tag{2.2.21}$$

for stationarity. This is seen to be identical in form to the Sommerfeld conditions (2.2.14) if the arc along the orbit is chosen as one of the generalized coordinates.

The most decisive steps in creating a completely new and logically consistent quantum theory of mechanical processes were made independently of each other by Heisenberg and Dirac in 1925. In these theories the observable parameters are not, as in classical mechanics, represented by ordinary numbers but by other mathematical quantities which obey algebraic rules different from those valid for numbers. These theories were followed shortly, in 1926(a), by Schrödinger's theory which is based on the notion of De Broglie waves. The theories of Heisenberg and of Dirac became known as "*quantum mechanics*" and Schrödinger's theory as "*wave mechanics*"; but it was soon discovered by Schrödinger that both types of theories were in fact physically identical, using only different mathematical languages, a most remarkable fact, as they were developed from apparently completely different ideas.

In the following chapters we shall in the main proceed along the lines of Schrödinger's theory which links up more directly with the mathematical system of classical mechanics as described in Chapter 1.

Problems and exercises

P.2.1. Prove that the total intensity I of the radiation from a black body is, according to (2.2.4) proportional to T^4. Show further that the frequency ν_m at which I is a maximum at a given temperature T is proportional to T.

P.2.2. A particle is acted upon by "elastic" forces in the x- and y-directions, causing it to perform a motion in the xy-plane, which is a superposition of harmonic oscillations in the x- and y-directions with amplitudes a and b, frequencies ν_1 and ν_2, and phase angles φ_1 and φ_2. Discuss the character of this movement, applying the Sommerfeld quantum conditions (2.2.14). Show that the quantum levels are "degenerate", i.e. that more than one quantum state belongs to the same energy level if ν_1 and ν_2 are commensurate.

P.2.3. Calculate the De Broglie wavelength λ_n of an electron moving in the nth circular orbit of the simple atomic model treated in section 2.2, and show that it is n times the circumference of the first orbit.

CHAPTER 3

Principles of Schrödinger's Wave Mechanics

WE HAVE seen in section 1.3 that under certain circumstances one can use the methods of geometrical optics instead of those of wave optics for the treatment of optical phenomena with the exception of those which are essentially characteristic of waves, like interference and diffraction phenomena. In view of the Hamiltonian analogy between particle dynamics and geometrical optics it suggests itself to ascribe the failure of classical mechanics to account for atomic phenomena to the fact that the equations of classical dynamics are only approximations, valid under "macroscopic" conditions, which have to be replaced by "wave equations" because of the wave character of moving particles according to De Broglie's theory. In this chapter we shall show how Schrödinger, reasoning along these lines, succeeded in obtaining a quantum-mechanical wave equation, and discuss some of the basic features of his wave mechanics.

3.1. The wave equation

It was shown in section 1.3 how the eiconal equation (1.3.22), which governs geometrical optics, can be derived from the amplitude equation (1.2.4) of wave optics. Vice versa, it is possible to reconstruct the amplitude equation from the eiconal equation by the formal procedure of replacing the quantities on both sides of equation (1.3.22) by "operators". If the vector $\nabla\Phi$ is replaced by the "differential operator" $\frac{1}{i}\nabla$, the left-hand member is transformed into the iterated operator $\frac{1}{i}\nabla\left(\frac{1}{i}\nabla\right) = -\nabla^2$. The right-hand member simply becomes the "multiplication operator" $4\pi^2k^2$. Hence (1.3.22) is turned into the "operator equation"

$$\nabla^2 + 4\pi^2k^2 = 0, \tag{3.1.1}$$

and if this is made to operate on either a scalar function ψ or a vector function B of space the amplitude equation (1.2.4) is indeed obtained.

It seems plausible to apply the same procedure to the Hamilton–Jacobi equation (1.3.2) for a single particle in order to obtain a quantum mechanical "wave equation". But whereas the eiconal is a dimensionless quantity, W has the dimension of "action". Thus the operator to replace $\mathbf{p} = \nabla W$ will have to be of the form $\frac{\alpha}{i}\nabla$, where α is a universal factor of the dimension of action. The operator equation to replace (1.3.2) will therefore take the form

$$\alpha^2\nabla^2 + 2m(E - U) = 0, \tag{3.1.2}$$

and if this is made to operate on a scalar function ψ, one obtains the differential equation

$$\nabla^2\psi + \frac{2m}{\alpha^2}(E - U)\psi = 0. \tag{3.1.3}$$

In order to determine the value of the constant α one can take recourse to the De Broglie relationship (2.2.19) which may be written

$$2m(E - U) = h^2k^2. \tag{3.1.4}$$

Now, if (3.1.3) is to be the amplitude equation for a De Broglie wave with wave number k one must put $h^2/\alpha^2 = 4\pi^2$ or

$$\alpha = \hbar. \tag{3.1.5}$$

Thus the required quantum mechanical wave equation becomes

$$\nabla^2\psi + \frac{2m}{\hbar^2}(E - U)\psi = 0, \tag{3.1.6}$$

which is generally called the "*Schrödinger equation*".

According to the above outlined procedure this equation can also be formally derived from the energy equation for a single particle in a potential field (1.1.9) by the replacement of the linear momentum vector $\mathbf{p}$ by the "*momentum operator*"†

$$\mathscr{p} = \frac{\hbar}{i}\nabla. \tag{3.1.7}$$

The significance of the momentum operator in quantum mechanics will be more fully discussed later, in Chapter 7.

As the total energy E of the particle remains constant during its movement in a potential field the frequency ν of the corresponding De Broglie wave is

† Throughout this book operators will always be denoted by script lettering.

also a constant of the motion and given by the first of the relations (2.2.18)

$$\nu = \frac{E}{h} = \frac{E}{2\pi\hbar}. \tag{3.1.8}$$

According to (1.2.3) we can therefore assign a "wave function" $\varphi(x, y, z; t)$ to such a wave

$$\varphi(x, y, z; t) = \psi(x, y, z)\, e^{i\frac{E}{\hbar}t}. \tag{3.1.9}$$

The term "*wave function*" is, however, quite generally also used for the amplitude function ψ, and in the following we shall adhere to this convention as long as it is clear what is actually meant.

If $U =$ const., that is in a field-free unbounded space, $\mathbf{p}$ and hence the wave vector $\mathbf{k}$ also remain constant. The solution of the wave equation (3.1.6) has, therefore, in virtue of (2.2.18), the form of a "plane De Broglie wave" (1.2.10)

$$\varphi = \gamma\, e^{\frac{i}{\hbar}(Et - \mathbf{p}\cdot\mathbf{r})}. \tag{3.1.10}$$

The relationship between this wave and the straight classical path of the particle will be discussed in the following section 3.2.

If U is not constant, that is in the presence of a field of force, $\mathbf{k}$ does not remain constant and the solution of the wave equation cannot have the form of a plane wave. The De Broglie wave length λ and the wave number k will now, according to (3.1.4) for particles of a given energy E be functions of space and the field will therefore act as a kind of "inhomogeneous medium" for the propagation of the waves with a space-dependent "refractive index"

$$\mu(\mathbf{r}) \simeq k \simeq \sqrt{E - U(\mathbf{r})}. \tag{3.1.11}$$

Let us in particular consider the movement of electrons with charge $-e$ in an electrostatic field with potential V, emerging from a "point source" 0 in space with zero velocity. If we take the potential at 0 to be the zero of potential, then evidently $E = 0$ and $U = -eV$; hence (3.1.11) assumes the form

$$\mu \simeq \sqrt{V}. \tag{3.1.12}$$

Thus the paths of the electrons can be constructed according to the principles of geometrical optics as if they were rays of light, issuing from 0 in a medium with refractive index (3.1.12). This is the basis of "*electron optics*" in electrostatic fields.

The validity of electron optics, being analogous to geometrical light optics, is restricted to sufficiently weak fields, where the potential does not appreciably change over distances of the order of magnitude of the De Broglie wavelength of the electrons. If this condition is not satisfied, diffraction phenomena will occur, as, for example, when an electron beam passes through a thin layer of matter. These diffraction phenomena are thus essential quantum effects which cannot be explained on the basis of classical mechanics. Otherwise, however, the procedure for determining the paths of electrons in an electric field by the methods of electron optics must, of couıse, lead to the same results as the application of the classical laws of motion of material particles in fields of force.

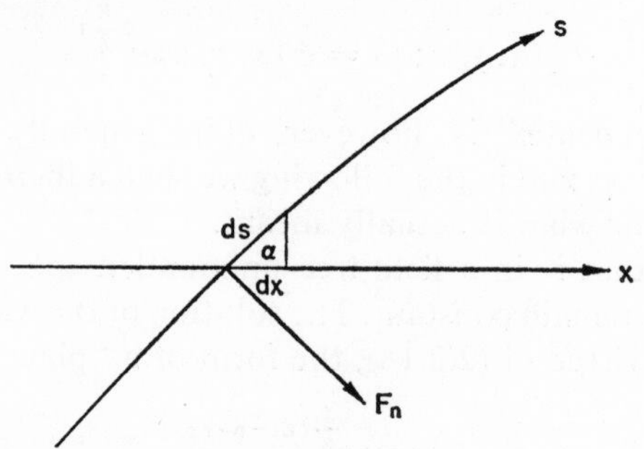

FIG. 2. Refraction of an electronic ray in an electric field of force.

This contention can be proved directly by means of the following argument. Let us first derive a relationship between the gradient of the refractive index in an inhomogeneous medium and the curvature of a ray of light. Let the direction of that gradient at a point P coincide with the x-axis of a coordinate system, and let α be the angle between a ray s, passing through P and the x-direction (Fig. 2). The relationship between α and μ then follows from Snell's law of refraction in its differential form

$$\frac{\sin(\alpha + d\alpha)}{\sin\alpha} = \frac{\mu}{\mu + d\mu}$$

or

$$\frac{\cos\alpha}{\sin\alpha}\, d\alpha = -\frac{d\mu}{\mu}.$$

As the line element ds of the ray is equal to $dx/\cos\alpha$, we have

$$\frac{1}{\sin\alpha}\frac{d\alpha}{ds} = -\frac{d(\log\mu)}{dx}. \tag{3.1.13}$$

The radius of curvature R of the ray at P is by definition

$$R = \frac{ds}{d\alpha}; \tag{3.1.14}$$

thus, from (3.1.13),

$$\frac{1}{R} = -\sin\alpha\,\frac{d(\log\mu)}{dx}. \tag{3.1.15}$$

If this formula is applied to electronic rays one obtains from (3.1.12)

$$\frac{1}{R} = -\frac{\sin\alpha}{2V}\frac{dV}{dx} = -\frac{\sin\alpha}{2Ve}F, \tag{3.1.16}$$

where F is the magnitude of the force of the electric field on the electron of charge $-e$. On the other hand, according to our convention about the zero point of V,

$$\frac{mv^2}{2} = eV,$$

and thus (3.1.16) can be put into the form

$$\frac{mv^2}{R} = -F\sin\alpha = -F_n. \tag{3.1.17}$$

The left-hand member of this equation is the centrifugal force on the electron and F_n is the component of F acting on the particle in the direction towards the centre of curvature of the path. Equation (3.1.17) is thus the condition for equilibrium between these forces, which is one way of expressing Newton's law of motion.

We now return to the quantum mechanical wave equation. We have seen how the Schrödinger equation (3.1.6) for a single particle can be derived from the corresponding energy equation by replacing the momentum vector **p** by the vector operator (3.1.7). In order to obtain the wave equation for a general conservative system of N degrees of freedom it suggests itself to follow a similar procedure in transforming the general energy equation (1.1.9) into an operator equation, namely to replace the generalized momenta $\mathbf{p}_k$ by the operators

$$\not{p}_k = \frac{\hbar}{i}\frac{\partial}{\partial q_k} \tag{3.1.18}$$

and to let them operate on a "wave function" ψ. The Hamiltonian function $H(q_1, \ldots, q_N; p_1, \ldots, p_N)$ will thus be transformed into the "*Hamiltonian operator*" $\mathcal{H}\left(q_1, \ldots, q_N; \frac{\hbar}{i}\frac{\partial}{\partial q_1}, \ldots, \frac{\hbar}{i}\frac{\partial}{\partial q_N}\right)$ and the wave equation takes the form

$$\mathcal{H}\left(q_1, \ldots, q_N; \frac{\hbar}{i}\frac{\partial}{\partial q_1}, \ldots, \frac{\hbar}{i}\frac{\partial}{\partial q_N}\right)\psi = E\psi, \tag{3.1.19}$$

which is a partial differential equation of the second order for the wave function $\psi(q_1, \ldots, q_N)$.

The Schrödinger equation (3.1.6) can further be generalized in a different direction for electrically charged particles subject to an *electromagnetic field* whose magnetic part cannot be derived from a scalar potential. For this purpose it is first necessary to obtain an expression for the Hamiltonian function. If V, as before, is the electric potential, and if we introduce the vector potential **A** for the magnetic field, we maintain that the Hamiltonian for a particle of mass m and charge e is given by

$$H = \frac{1}{2m}\left(\mathbf{p} - \frac{e\mathbf{A}}{c}\right)^2 + eV. \tag{3.1.20}$$

This can be proved as follows.

The Hamiltonian equations (1.1.7) are in this case

$$\dot{p}_x = -\frac{\partial H}{\partial x} = -e\frac{\partial V}{\partial x} + \frac{e}{mc}\left[\left(\mathbf{p} - \frac{e\mathbf{A}}{c}\right)\cdot\frac{\partial \mathbf{A}}{\partial x}\right] \tag{3.1.21}$$

(with similar equations for the y- and z-components of the momentum) and

$$\dot{\mathbf{r}} = \mathbf{v} = \nabla^p H = \frac{1}{m}\left(\mathbf{p} - \frac{e\mathbf{A}}{c}\right). \tag{3.1.22}$$

Substitution of (3.1.22) into (3.1.21) results in

$$\dot{p}_x = -e\frac{\partial V}{\partial x} + \frac{e}{c}\left(\mathbf{v}\cdot\frac{\partial \mathbf{A}}{\partial x}\right), \tag{3.1.23}$$

and differentiation of (3.1.22) with respect to time yields

$$m\dot{v}_x = \left[\dot{p}_x - \frac{e}{c}\left(\mathbf{v}\cdot\frac{\partial \mathbf{A}}{\partial x}\right)\right]. \tag{3.1.24}$$

Finally, substituting for $\dot{p}_x$ in (3.1.24) from (3.1.23), and doing the same for the other components, we obtain for the force **F** of the electromagnetic field on the particle

$$\mathbf{F} = -e\,\nabla V + \frac{e}{c}\,(\mathbf{v}\wedge \operatorname{curl}\mathbf{A}). \tag{3.1.25}$$

The first term here is clearly the electric force on the particle and, as curl **A** is the magnetic field strength, the second term represents the "Lorentz force" of the magnetic field on the moving particle (see section 9.3). Thus the expression (3.1.20) is indeed proved to be the required Hamiltonian.

In order to derive the Hamiltonian operator for the problem in question we first re-write the expression (3.1.20) in the symmetrical form

$$H = \frac{p^2}{2m} - \frac{e}{2mc}(\mathbf{p}\cdot\mathbf{A} + \mathbf{A}\cdot\mathbf{p}) + \frac{e^2}{2mc^2}A^2 + eV$$

and now replace **p** by the operator (3.1.7). This results in

$$\mathscr{H} = -\frac{\hbar^2}{2m}\nabla^2 + \frac{ie\hbar}{2mc}(\nabla\cdot\mathbf{A} + \mathbf{A}\cdot\nabla) + \frac{e^2}{2mc^2}A^2 + eV. \tag{3.1.26}$$

Finally, letting $\mathscr{H}$ operate on the wave function ψ and equating the result to $E\psi$, we obtain the wave equation

$$\nabla^2\psi - \frac{ie}{\hbar c}(\nabla \cdot \mathbf{A}\psi + \mathbf{A} \cdot \nabla\psi) - \frac{e^2}{\hbar^2 c^2} A^2\psi + \frac{2m}{\hbar^2}(E - eV)\psi = 0. \tag{3.1.27}$$

3.2. Statistical interpretation of the wave function

The dual nature of radiation exhibited by the simultaneous validity of the concepts of waves and photons was already briefly pointed out in section 2.2. One might think that this duality could be of a similar nature to the one exhibited by the phenomenon of sound propagation in matter which can be described as a continuous wave on a macroscopic scale but in fact consists of the vibrations of individual molecules of the medium. For if there is a large number n of photons of energies $h\nu$ present in the unit volume of a plane monochromatic light wave the energy density u in the wave is

$$u = nh\nu. \tag{3.2.1}$$

According to electromagnetic theory this is also equal to the flux density of electromagnetic momentum in the direction of the wave. On the other hand, this momentum flux should be made up of the momenta of all the photons crossing the unit area of a wave front per unit time, that is,

$$u = pnc. \tag{3.2.2}$$

The combination of this equation with (3.2.1) produces, indeed, the second of the Einstein relations (2.2.5). Furthermore, the electromagnetic radiation pressure will then appear to be the outcome of the transfer of momentum by the shower of photons to an absorbing surface on which they impinge.

However, diffraction experiments carried out with light of such small intensity that only one photon at a time enters the apparatus give, if extended over a sufficiently long period of time, exactly the same results as if the apparatus was exposed to strong light during a short period. This clearly indicates that *each individual photon* must be associated with a plane monochromatic wave according to the relations (2.2.5). These imply that the photon travels with the velocity c in the *direction* of the wave, but give no indication as to the actual *position* of the photon in space at any time, which thus remains arbitrary or subject to "chance".

The energy density u of a wave is proportional to the square of its amplitude; on the other hand, u is, according to (3.2.1), proportional to the number density n of the photons. This suggests a *statistical interpretation* of the amplitude function $\psi(\mathbf{r})$ of a light wave, namely the assumption that $|\psi|^2\,dV$ represents the *probability* for an individual photon to be found in the volume element dV at any particular time. $|\psi|^2$ is thus what is known as a "*probability density*" in statistical theory, and it must evidently satisfy the "normalizing condition"

$$\int\int\int |\psi|^2 \, dV = 1 \tag{3.2.3}$$

if the integral is extended over all space.

This statistical interpretation is also in keeping with the fact that, if a monochromatic light wave impinges on the boundary between two transparent media of different refractive index, it is split up into a penetrating and a reflected wave, both having the original frequency. This clearly indicates that the individual photons cannot themselves be split up into penetrating and reflected ones as then the *sum* of the frequencies of the latter would have to be equal to the frequency of the incoming photon. We are therefore led to believe that an individual photon can *either* penetrate through the boundary *or* be reflected by it, and as all photons are identical, it must be the result of chance whether the one or the other event happens. This idea is, by the way, remarkably close to the notions of Newton's particle theory of light, according to which the light particles should be subject to "fits" of easy reflection and easy penetration of boundaries.

A similar dual nature is exhibited in the movement of material particles because, according to De Broglie's theory, this movement can also be described as a propagation of De Broglie waves. Here, again a unique relationship between the direction and speed of the particle movement, on the one hand, and the direction and velocity of the corresponding wave is established by the equations (2.2.18), but the position of the particle is not determined at all.

In 1926 Born therefore suggested the statistical interpretation of the quantum mechanical wave function ψ, now almost generally accepted. According to this $|\psi|^2$ is the probability density for finding a particle at some specified point in space at any particular time, again provided that the normalization condition (3.2.3) is satisfied. This, of course, constitutes a break with the classical law of causation, as already indicated in the introduction, because in contrast to the situation in classical mechanics, the solution of the wave equation (3.1.6) does not determine the position $\mathbf{r}$ of the particle as a function of time but only allows the probabilities for the various values of $\mathbf{r}$ to be calculated. However, as will be explained later in detail in section 7.1, the "observable" quantities, which play the role of the classical parameters of physical systems, are in quantum mechanics represented by operators whose "expectation values" for sets of very many observations are defined in terms of the wave function. It is therefore possible to predict these expectation values precisely from the solution of the wave equation but not their actual values in single experiments.

The statistical interpretation of the wave function can be extended to general mechanical systems of N degrees of freedom. The modulus square of the solution of the general wave equation (3.1.19) $|\psi(q_1, \ldots, q_N)|^2$ can then again be interpreted as a probability density in the following sense: Let $w(q_1, q_2, \ldots, q_N)\, dq_1 dq_2 \cdots dq_N$ be the "*a priori* probability" for finding the coordinates of the system in the infinitesimal range $dq_1 dq_2 \cdots dq_N$, that is, the probability for finding them there in the *absence* of forces. $|\psi|^2 w\, dq_1 dq_2 \cdots dq_N$ will then represent the same probability in the *presence* of forces provided that ψ satisfies the normalizing condition

$$\underbrace{\int \cdots \int}_{N} |\psi|^2 w\, dq_1 \cdots dq_N = 1. \tag{3.2.4}$$

Imagine a "three-dimensional time exposure photograph" being taken of, say, an electron performing an orbital movement within an atom. The spatial density distribution in that photograph would then be proportional to the relative frequency of the various positions of the electron in space and hence proportional to the probability density of these positions. We may therefore imagine the mass m of the electron to be "smeared out" in the form of a continuous density ϱ within the atom such that

$$\varrho(\mathbf{r}) = m\, |\psi(\mathbf{r})|^2. \tag{3.2.5}$$

The space integral over the left-hand member of this equation must be equal to the mass of the particle, and thus equation (3.2.3) is satisfied. This is a very convenient way of visualizing the electronic wave function in this and similar cases. It is, however, strictly speaking restricted to one-particle systems, as in general the function $|\psi(q_1 \cdots q_N)|^2$ would represent a density distribution in an N-dimensional "configuration space" and not in ordinary space.

As explained before, the indeterminacy in the positions of photons and material particles and consequently the chance element in their movement is brought about by their wave character. This can be further clarified by a general consideration concerning "*wave packets*". This term is used for the superposition of waves in a continuous range of frequencies and wave vectors.

Let us first consider the superposition of harmonic oscillations in a continuous range of frequencies ν with amplitudes $f(\nu)$ in the form of the *Fourier integral*

$$\varphi(t) = \int_{-\infty}^{+\infty} f(\nu)\, e^{2\pi i \nu t}\, d\nu. \tag{3.2.6}$$

The function $\varphi(t)$ which is defined by this integral is known as the "*Fourier transform*" of the amplitude function $f(\nu)$. The so-called "inversion theorem" of the theory of Fourier transforms states that vice versa $f(\nu)$ is the "*inverse transform*" of $\varphi(t)$ according to

$$f(\nu) = \int_{-\infty}^{+\infty} \varphi(t)\, e^{-2\pi i \nu t}\, dt. \tag{3.2.7}$$

The time function $\varphi(t)$ is in communication theory called the "*signal*" because it may be made to represent a message to be sent from one station to another, and the function $f(\nu)$ is called its "*spectrum*".

An especially simple form of signal is the "square pulse" defined by

$$\left.\begin{aligned} \varphi(t) &= 0 \quad \text{for} \quad t < -\Delta t/2 \quad \text{and} \quad t > \Delta t/2, \\ \text{and} \qquad \varphi(t) &= A\, e^{2\pi i \nu_0 t} \quad \text{for} \quad -\Delta t/2 \leqslant t \leqslant \Delta t/2. \end{aligned}\right\} \tag{3.2.8}$$

which, for example, can represent a "Morse dash" on a carrier frequency ν_0. Substitution of (3.2.8) into (3.2.7) yields the expression

$$f(\nu) = A \int_{-\Delta t/2}^{+\Delta t/2} e^{2\pi i (\nu_0 - \nu)\, t}\, dt = \frac{A}{\pi\, (\nu_0 - \nu)} \sin\left[\pi\, (\nu_0 - \nu)\, \Delta t\right] \tag{3.2.9}$$

for the amplitude function, a diagram of which is shown in Fig. 3. The function has a maximum at $\nu = \nu_0$ and first zeros to both sides at distances $\pm 1/\Delta t$. The main contribution to the spectrum therefore comes from oscillations in a frequency band of width $\Delta\nu$ of the order of magnitude $1/\Delta t$. Thus the pulse width Δt and the band width $\Delta\nu$ are related by

$$\Delta\nu\, \Delta t \simeq 1. \tag{3.2.10}$$

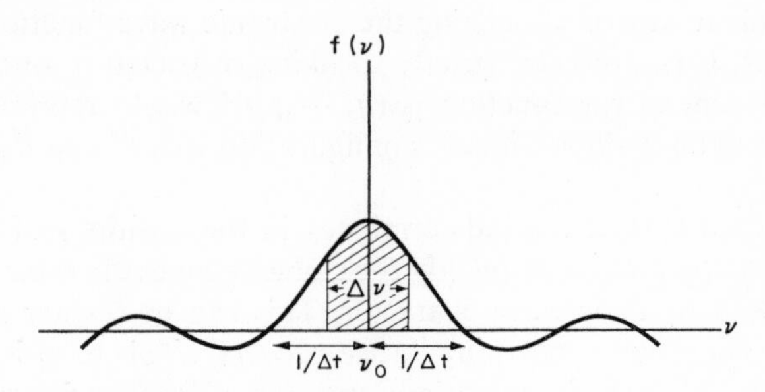

FIG. 3. Amplitude function of a square pulse and illustration of the optical uncertainty relation.

This relation indicates that, in order to produce a very sharp pulse, one has to superpose oscillations within a very wide range of frequencies, or in other words, in order to locate the signal precisely in time one has to be content with a very vague definition of signal frequency. Vice versa, if one wishes to measure the frequency of an oscillator very accurately the measurement has to be extended over a very long time interval. This shows that it is not possible to perform frequency *and* time measurements simultaneously with unlimited accuracy; the combined accuracy of the measurement is constant and expressed by formula (3.2.10).

We now turn to the consideration of a packet of plane waves travelling in the same direction s with the same velocity u. A pulse of duration Δt will then be transmitted in the form of a finite wave train with carrier frequency ν_0 and length $\Delta s = u\Delta t$. The wave numbers of the waves will further fill an interval $\Delta k = \Delta\nu/u$ around a carrier wave number ν_0/u. From (3.2.10) we obtain at once

$$\Delta k \, \Delta s \simeq 1, \tag{3.2.11}$$

a relation which expresses a limitation to the accuracy of the combined determination of the location of a wave packet in space and its wave number.

The "uncertainty relations" (3.2.10) and (3.2.11) are valid for any kind of wave and must therefore also hold for De Broglie waves. Combining them with the De Broglie relations (2.2.18) we obtain

$$\Delta E \, \Delta t \simeq h, \quad \Delta p \, \Delta s \simeq h, \tag{3.2.12}$$

where s is a space coordinate and p the corresponding momentum component. These are the "*Heisenberg uncertainty relations*" (1927) for the measurement of the energy of a material particle at a certain time, on the one hand, and the simultaneous measurement of its position and its momentum, on the other hand. They constitute a fundamental principle of quantum mechanics, namely that the simultaneous measurement of certain parameters of a physical system can only be performed with an accuracy of the order of magnitude of the characteristic universal constant h of quantum theory. This is, of course, in sharp contrast to classical mechanics which imposes no restriction on the measurement of these parameters at all. The smallness of h in comparison with the values of action involved in macroscopic phenomena explains at once why classical mechanics holds so extremely well in the realm of macroscopic physics. In atomic processes, however, the action integrals are of the order of h, as seen from the Sommerfeld quantum conditions (2.2.14), and therefore the Heisenberg uncertainty relations, are an essential feature of microscopic physics.

The uncertainty relations have here been derived from wave theoretical considerations. But they are in fact manifestations of the statistical character of the laws of quantum mechanics. For, as previously stated, precise values are predicted by them only for the *averages* of the observable parameters, but not for their *actual* values at individual experiments, and hence a non-reducible margin of error in such measurements is to be expected. This can also be understood if it is realized that any measuring process consists in an interaction between the object to be measured and the measuring device. Quantum theory maintains that these interactions can only occur in the form of *finite* exchanges of energy and momentum and that, moreover, the occurrence of these interactions is governed by chance. Consequently the results of repeated measuring processes must be subject to fluctuations which cannot be suppressed below a certain level dependent on the magnitude of h.

We finish this chapter by stating, for the moment without proof, that the uncertainty relations (3.2.12) can be generalized for mechanical systems of N

degrees of freedom in the form

$$\Delta E \, \Delta t \simeq h, \quad \Delta p_k \, \Delta q_k \simeq h \quad (k = 1, \ldots, N) \tag{3.2.13}$$

for any set of generalized coordinates q_k and their associated momenta p_k. We shall come back to this problem later in section 7.2.

Problems and exercises

P.3.1. Establish the law of "refraction" of a beam of electrons at a plane surface of discontinuity of electric potential V, making use of eq. (3.1.11). Show that the same formula follows from the treatment of this problem by classical mechanics.

P.3.2. Use eq. (3.1.19) for deriving the wave equation for the "one-dimensional rigid rotator", i.e. a rigid body which is constrained to a rotational movement about an axis fixed in space, making use of the results of P.1.3. Discuss its relation to the Schrödinger equation for a single particle.

P.3.3. Set up the wave equation for the wave function $\psi(r, \varphi)$ in plane-polar coordinates in the xy-plane for an electron moving in a homogeneous magnetic field of field strength $\mathbf{H}$ in the z-direction, making use of eqn. (3.1.27).

P.3.4. Calculate the "signal function" $\varphi(t)$ belonging to a Gaussian spectrum function $f(\nu) = Ce^{-\alpha\nu^2}$, using formula (3.2.6), and prove that $\varphi(t)$ is also Gaussian. Verify the inversion formula (3.2.7). Show that the product of $\Delta\nu = \sqrt{\overline{\nu^2}}$ and $\Delta t = \sqrt{\overline{t^2}}$ satisfies a relationship of the form (3.2.10).

P.3.5. Calculate the signal function $\varphi(t)$ belonging to a narrow "wave band" between $\nu_0 + \frac{\Delta\nu}{2}$ and $\nu_0 - \frac{\Delta\nu}{2}$, and show that the location of the signal in time is only possible with an accuracy $1/\Delta\nu$.

P.3.6. A beam of particles travelling in the z-direction impinges on a screen in the xy-plane with a slit of width Δx along the y-axis. Show that, after having penetrated the slit, the particles acquire momenta p_x of the order of magnitude Δp_x in the x-direction, due to the diffraction of the beam by the slit, and prove that $\Delta x \, \Delta p_x$ satisfies the second of the Heisenberg uncertainty relations (3.2.12).

CHAPTER 4

Eigenfunctions and Eigenvalues of the Wave Equation

In section 1.2 an outline of the boundary problem of the optical amplitude equation (1.2.4) was given. It was pointed out there that for a given boundary condition this equation had, in general, solutions only for certain discrete values of the parameter k, which were called the eigenvalues of the differential equation; the amplitude functions themselves, belonging to these eigenvalues, were called eigenfunctions. In this chapter we shall deal with the analogous boundary problem of the Schrödinger equation (3.1.6). The parameter here is the energy E of the system, and the fact that this in general will have discrete eigenvalues explains at once why the system is only capable of quantized energy levels. The corresponding eigenfunctions then provide the probability distributions for the positions of the particles in space.

4.1. Boundary conditions at discontinuities of potential

We shall first deal with the boundary conditions imposed by discontinuities of the potential energy function $U(r)$ at certain surfaces in space. Suppose that ψ is continuous and U finite on both sides of the discontinuity surface S (Fig. 4) and let a point P on S be enclosed in a flat box whose bottom and top are parallel to S. Then, according to the wave equation (3.1.6),

$$\nabla^2\psi + \frac{2m}{\hbar^2}(E - U)\psi = 0,$$

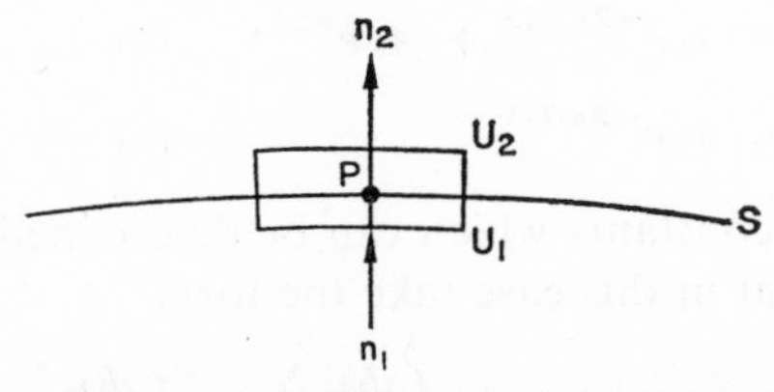

Fig. 4. Diagram illustrating the derivation of the boundary condition for the wave function at a discontinuity of potential.

$\nabla^2\psi$ must be finite everywhere inside the box. Now by Gauss's theorem the space integral of $\nabla^2\psi$ over the interior of the box must be equal to the integral of the normal component of the gradient of ψ over the surface of the box. In the limit of an infinitely flat box $\iiint \nabla^2\psi\, dV$ vanishes and therefore, in order that the surface integral be zero as well, the derivative $\partial\psi/\partial n$ at P must have the same value on both sides of S or, in other words, $\partial\psi/\partial n$ must be continuous on S. Consequently, ψ itself must also be continuous. The boundary conditions imposed on the wave function at a discontinuity surface of U are thus

$$\psi_1 = \psi_2, \quad \left(\frac{\partial\psi}{\partial n}\right)_1 = \left(\frac{\partial\psi}{\partial n}\right)_2. \tag{4.1.1}$$

In this section we shall be concerned with the quantum mechanics of particles moving in a field-free space with discontinuities of the potential energy function.

We shall first deal with the problem of the "*potential step*" which is defined by the following conditions for U:

$$\left.\begin{aligned} U &= 0 && \text{for} \quad x < 0, \\ U &= U_0 < E && \text{for} \quad x \geqslant 0. \end{aligned}\right\} \tag{4.1.2}$$

In this case evidently ψ must be a function of x alone and the wave equation takes the form

$$\left.\begin{aligned} \frac{d^2\psi}{dx^2} + 4\pi^2 k_1^2 \psi &= 0 \quad \text{for} \quad x < 0, \\ \frac{d^2\psi}{dx^2} + 4\pi^2 k_2^2 \psi &= 0 \quad \text{for} \quad x \geqslant 0, \end{aligned}\right\} \tag{4.1.3}$$

where

$$k_1^2 = \frac{2mE}{h^2}, \quad k_2^2 = \frac{2m(E - U_0)}{h^2}. \tag{4.1.4}$$

Assume now that a particle with kinetic energy E travels in the direction of the positive x-axis and impinges on the "step" at $x = 0$ from the left. The corresponding De Broglie wave will be partly reflected and will partly penetrate through the step; hence the solution of the wave equation is

$$\left.\begin{aligned} \psi_1 &= A_i\, e^{-2\pi i k_1 x} + A_r\, e^{2\pi i k_1 x} && \text{for} \quad x \leqslant 0, \\ \psi_2 &= A_t\, e^{-2\pi i k_2 x} && \text{for} \quad x \geqslant 0. \end{aligned}\right\} \tag{4.1.5}$$

Here A_i, A_r, A_t are constants which can be determined from the boundary conditions (4.1.1) that in this case take the form

$$\psi_1(0) = \psi_2(0), \quad \left(\frac{d\psi_1}{dx}\right)_0 = \left(\frac{d\psi_2}{dx}\right)_0. \tag{4.1.6}$$

From (4.1.5) and (4.1.6) one obtains

$$A_i + A_r = A_t, \quad k_1(A_i - A_r) = k_2 A_t,$$

leading to the relations

$$\frac{A_t}{A_i} = \frac{2k_1}{k_1 + k_2}, \quad \frac{A_r}{A_i} = \frac{k_1 - k_2}{k_1 + k_2}. \tag{4.1.7}$$

This result, if interpreted statistically, as discussed in section 3.2, indicates that there is a certain chance for the particle to be reflected back or to penetrate beyond the potential step, whereas one would expect from classical mechanics that all particles whose kinetic energies are larger than the "step height" ($E > U_0$) should penetrate and none be reflected.

In order to calculate the probabilities for reflection and penetration we first notice that, according to (4.1.4), the velocities v_i, v_r, v_t of the incoming, the reflected and the penetrating particles are

$$v_i = v_r = \frac{hk_1}{m}, \quad v_t = \frac{hk_2}{m}. \tag{4.1.8}$$

If the number density in a beam of impinging particles is n_i and the corresponding number densities in the reflected and transmitted beams are n_r and n_t, the squares of the amplitudes of the corresponding De Broglie waves are proportional to these number densities and one has

$$\frac{n_r}{n_i} = \left(\frac{A_r}{A_i}\right)^2, \quad \frac{n_t}{n_i} = \left(\frac{A_t}{A_i}\right)^2. \tag{4.1.9}$$

The numbers of particles crossing a unit area normal to the beam in unit time are given by

$$N_i = n_i v_i, \quad N_r = n_r v_r, \quad N_t = n_t v_t, \tag{4.1.10}$$

hence, from (4.1.7) and (4.1.10),

$$\frac{N_r}{N_i} = \left(\frac{A_r}{A_i}\right)^2 = \left(\frac{k_1 - k_2}{k_1 + k_2}\right)^2, \quad \frac{N_t}{N_i} = \left(\frac{A_t}{A_i}\right)^2 \frac{v_t}{v_i} = \frac{4k_1 k_2}{(k_1 + k_2)^2}. \tag{4.1.11}$$

These expressions also represent the probabilities for reflection and penetration of individual particles at the potential step, and one can see immediately that their sum is equal to 1, as it ought to be.

We now turn to the case where $U_0 > E$ so that classically the particle would be unable to penetrate the boundary and hence all the particles in a beam should be reflected. We therefore refer to such a boundary as a "*potential barrier*". The solution of the wave equation is then still given by the equations (4.1.3) except that the "wave number" k_2 in (4.1.4) will now be imaginary. We introduce a real quantity $\varkappa = ik_2$ and obtain thus, instead of the second of the equations (4.1.5)

$$\psi_2 = A_t e^{-2\pi\varkappa x} \quad x \geqslant 0. \tag{4.1.12}$$

This shows that beyond the barrier the wave function has no longer the form of a De Broglie wave but decreases exponentially with x, provided that $\varkappa > 0$. It appears that there is a finite probability for finding the particle *beyond* the potential barrier proportional to

$$|\psi_t|^2 = A_t^2 e^{-4\pi\varkappa x}, \quad x \geqslant 0, \tag{4.1.13}$$

which is again contrary to classical concepts but somehow similar to the situation in optics in cases of "total reflection" on a boundary between two transparent media. The "depth of penetration" is seen to be of the order of magnitude $1/\varkappa$ and thus inversely proportional to $\sqrt{U_0 - E}$.

The equations (4.1.7) are also valid in this case when $-i\varkappa$ is substituted for k_2. Hence

$$\frac{A_t}{A_i} = \frac{2k_1}{k_1 - i\varkappa}, \quad \frac{A_r}{A_i} = \frac{k_1 + i\varkappa}{k_1 - i\varkappa}. \tag{4.1.14}$$

The probability for reflection is thus $|A_r/A_i|^2 = 1$, i.e. as in classical physics, all particles are eventually reflected by the barrier, but only after having penetrated more or less deeply behind it. The number density n_t of particles of a beam with original density n_i just behind the barrier is from (4.1.9) and (4.1.14)

$$n_t = n_i \frac{4k_1^2}{k_1^2 + \varkappa^2} \tag{4.1.15}$$

which decreases with increasing $\varkappa$, that is increasing "barrier height".

Next we consider a potential energy function with two discontinuity surfaces according to the conditions

$$\left.\begin{array}{lll} U = 0 & \text{for} & x < 0 \;\text{ and }\; x > a, \\ U = U_0 > E & \text{for} & 0 \leqslant x \leqslant a. \end{array}\right\} \tag{4.1.16}$$

This may appropriately be called a "*potential wall*" of thickness a and height $U_0 - E$. Again, as in the previous case, a particle impinging on the wall from the left would classically not be able to penetrate through it. The solution of the wave equation in the three regions of x can now be written in the form

$$\left.\begin{array}{ll} \psi_1 = A_i e^{-2\pi ikx} + A_r e^{2\pi ikx} & x \leqslant 0, \\ \psi_2 = B_1 e^{-2\pi\varkappa x} + B_2 e^{2\pi\varkappa x} & 0 \leqslant x \leqslant a, \\ \psi_3 = A_t e^{-2\pi ikx} & x \geqslant a, \end{array}\right\} \tag{4.1.17}$$

with

$$k^2 = \frac{2mE}{h^2}, \quad \varkappa^2 = \frac{2m(U_0 - E)}{h^2}. \tag{4.1.18}$$

The boundary conditions (4.1.1) applied to the discontinuity planes at $x = 0$ and $x = a$ now yield the following four conditions for the five constants A_i, A_r, A_t, B_1, B_2:

$$\left.\begin{aligned} A_i + A_r &= B_1 + B_2, \quad i\frac{k}{\varkappa}(A_i - A_r) = B_1 - B_2, \\ A_t\, e^{-2\pi i k a} &= B_1\, e^{-2\pi\varkappa a} + B_2\, e^{2\pi k a}, \\ \frac{ik}{\varkappa} A_t\, e^{-2\pi i k a} &= B_1\, e^{-2\pi\varkappa a} - B_2\, e^{2\pi\varkappa a}, \end{aligned}\right\} \tag{4.1.19}$$

from which their ratios can be calculated. In particular one obtains

$$\frac{A_t}{A_i} = \frac{2e^{2\pi i k a}}{2\cosh(2\pi\varkappa a) - i\left(\dfrac{\varkappa}{k} - \dfrac{k}{\varkappa}\right)\sinh(2\pi\varkappa a)}, \tag{4.1.20}$$

and hence for the probability of penetration (because of $v_3 = v_1$) from (4.1.18) and (4.1.20)

$$\frac{N_t}{N_i} = \left|\frac{A_t}{A_i}\right|^2 = \left[1 + \frac{U_0^2 \sinh^2(2\pi\varkappa a)}{4E(U_0 - E)}\right]^{-1}. \tag{4.1.21}$$

This formula shows that, contrary to the predictions from classical mechanics, a certain finite number of particles in a beam impinging on the wall will penetrate through it. This typical quantum mechanical effect is called the "*tunnel effect*" because the particles behave as if they would tunnel their way through the wall over which they cannot climb. The relative number of these particles diminishes according to (4.1.21) with increasing height and increasing thickness of the wall.

In the three cases treated so far in this section no restriction was placed on the energy parameter by the boundary conditions. The reason for this is that a solution for the wave equation was required for the whole space and that the solutions (4.1.3) and (4.1.17) do not allow for a definite value of ψ to be fixed at $x = -\infty$. As a typical example of a "true" boundary problem we shall now consider a particle that is contained in a "box" from which it cannot escape. In this case the wave function must be zero everywhere outside the box and hence, because of its continuity, also at the surface of the box. Physically the escape can be made impossible by putting up infinitely high potential barriers at the box surface. This, however, makes the second of the boundary conditions (4.1.1) invalid, for whose derivation an infinity of U was explicitly excluded. It would, indeed, be impossible to obtain a solution of the wave equation if two independent boundary conditions were imposed on a surface enveloping the whole system.

For the sake of simplicity we shall assume the "box" simply to consist of two parallel infinite walls normal to the x-axis at $x = \pm l/2$ and the particle to be confined to the space between these planes. According to what has been said before the boundary conditions are here

$$\psi = 0 \quad \text{for} \quad x = \pm l/2, \tag{4.1.22}$$

and we shall assume that there is no field within this space so that we can put $U = 0$ everywhere. ψ is evidently a function of x only so that the wave equation has the form

$$\frac{d^2\psi}{dx^2} + \frac{2m}{\hbar^2} E\psi = 0. \tag{4.1.23}$$

The complete solution of this differential equation is

$$\psi = A\, e^{2\pi ikx} + B\, e^{-2\pi ikx}, \tag{4.1.24}$$

where k has the value (4.1.18) and A and B are constants to be determined from the condition (4.1.22). For symmetry reasons one must have $A = \pm B$ and hence (4.1.24) assumes one of the two forms

$$\psi = C \sin 2\pi kx \quad \text{or} \quad \psi = C \cos 2\pi kx, \tag{4.1.25}$$

the first of which is an odd function of x and the second an even function. If these solutions are to satisfy the conditions (4.1.22) k must have one of the values

$$k_n = n/2l \quad (n = 1, 2, 3, \ldots) \tag{4.1.26}$$

with an even n for an odd wave function and an odd n for an even wave function. The corresponding wave functions are therefore given by the equations

$$\left.\begin{aligned} \psi_n &= C_n \sin \frac{n\pi x}{l} \quad (n = 2, 4, 6, \ldots), \\ \psi_n &= C_n \cos \frac{n\pi x}{l} \quad (n = 1, 3, 5, \ldots). \end{aligned}\right\} \tag{4.1.27}$$

These evidently represent standing De Broglie waves with "nodes" at $x = \pm l/2$. The "fundamental", with $n = 1$, has no other nodes than these, the others are "harmonics" with a total number $n + 1$ of nodes, spaced at equal distances between $x = l/2$ and $x = -l/2$.

From (4.1.18) and (4.1.26) one now obtains the eigenvalues of the particle energy, or its allowed energy levels

$$E_n = \frac{n^2h^2}{8ml^2} \quad (n = 1, 2, 3, \ldots), \tag{4.1.28}$$

and the functions (4.1.27) are the corresponding eigenfunctions. The coefficients C_n can be determined from the normalization condition (3.2.3) which here takes the form†

$$\int_{-l/2}^{+l/2} \psi_n \psi_n^* \, dx = 1$$

from which follows

$$|C_n|^2 = 2/l. \tag{4.1.29}$$

We notice here that all integrals involving two different eigenfunctions vanish:

$$\int_{-l/2}^{+l/2} \psi_n \psi_{n'}^* \, dx = 0 \quad \text{for} \quad n \gtrless n'. \tag{4.1.30}$$

It is customary to refer to the integrals of the form

$$\left.\begin{aligned} (f \cdot g) &\equiv \int f^*(x)\, g(x)\, dx \\ (g \cdot f) &\equiv \int g^*(x) f(x)\, dx = (f \cdot g)^* \end{aligned}\right\} \tag{4.1.31}$$

of two complex functions f and g, taken over the whole range of the definition of these functions, as the "*inner product*" of the functions. They are said to be "orthogonal" to each other if their inner product vanishes (in analogy to the fact that the "inner" or "scalar" product of two orthogonal vectors is zero). The inner product of a function with itself is called its "*norm*". Two functions whose norms are equal to unity and whose inner product is zero are called "*orthonormal*". The eigenfunctions of the present boundary problem therefore form an infinite set of orthonormal functions.

From (4.1.27) and (4.1.29) one obtains for the probabilities for finding the particle at a certain place within the box the density functions

$$\left.\begin{aligned} |\psi_n(x)|^2 &= \frac{4}{l^2} \sin^2 \frac{n\pi x}{l} \quad (n = 2, 4, 6, \ldots), \\ |\psi_n(x)|^2 &= \frac{4}{l^2} \cos^2 \frac{n\pi x}{l} \quad (n = 1, 3, 5, \ldots), \end{aligned}\right\} \tag{4.1.32}$$

which show that there is maximum probability for finding the particle midway between two nodes and zero probability for finding it at a node. This is a characteristic feature of the wave nature of matter, falling in the same category as the previously mentioned diffraction phenomena. With increasing quantum number n, that is with increasing energy of the particle, the distances between the maxima and minima become smaller and smaller, and eventually, when they have become smaller than the particle size, the probability tends to become uniform within the whole interior of the box, as one would expect from classical probability considerations.

This problem also provides an opportunity to demonstrate the Heisenberg uncertainty principle in a special case. If nothing more is known about the

† Throughout this book an asterisk indicates a conjugate complex quantity.

particle than that it is in the "box", then its location in space is only known with an accuracy $\Delta x \simeq l$. On the other hand, as we do not know in which direction it is just moving its momentum p_n in the nth quantum state is only known with an accuracy $\Delta p_n = h\,\Delta k_n = \frac{hn}{l}$, and hence the smallest inaccuracy with which one can measure p is $\Delta p_x \simeq \frac{h}{l}$. Thus $\Delta x\, \Delta p_x \simeq h$, in accordance with the second of the equations (3.2.12).

As a last example in this section we consider the following conditions for the potential energy function U:

$$\begin{aligned} U &= 0 && \text{for} \quad -l/2 < x < l/2 \\ U &= U_0 > E && \text{for} \quad x \leqslant -l/2 \quad \text{and} \quad x \geqslant l/2. \end{aligned} \tag{4.1.33}$$

We shall refer to this field as a "*rectangular potential well*" because a particle inside the well could classically never get out by itself. (The well becomes a box in the previously defined sense if U_0 is infinitely large.) The solution ψ_1 of the wave equation inside the well is again given by (4.1.25) and outside the well ψ_2 has an exponential form as in (4.1.12) to both sides of the well as it must not become infinite anywhere. Hence as in the previous example, there are two sets of solutions, an odd and an even one, namely:

$$\begin{aligned} \psi_1 &= C \sin 2\pi k x && \text{for} \quad -\frac{l}{2} \leqslant x \leqslant \frac{l}{2} \\ \psi_2 &= D\, e^{-2\pi\varkappa x} && \text{for} \quad x \geqslant \frac{l}{2} \\ \psi_2 &= -D\, e^{2\pi\varkappa x} && \text{for} \quad x \leqslant -\frac{l}{2} \end{aligned}$$

and

$$\begin{aligned} \psi_1 &= C \cos 2\pi k x && \text{for} \quad -\frac{l}{2} \leqslant x \leqslant \frac{l}{2} \\ \psi_2 &= D\, e^{-2\pi\varkappa x} && \text{for} \quad x \geqslant \frac{l}{2} \\ \psi_2 &= D\, e^{2\pi\varkappa x} && \text{for} \quad x \leqslant -\frac{l}{2} \end{aligned} \tag{4.1.34}$$

with

$$k = \frac{\sqrt{2mE}}{h}, \quad \varkappa = \frac{\sqrt{2m\,(U_0 - E)}}{h}.$$

These solutions must satisfy the boundary conditions (4.1.1)

$$\psi_1 = \psi_2 \quad \text{and} \quad \frac{d\psi_1}{dx} = \frac{d\psi_2}{dx} \quad \text{at} \quad x = \pm\frac{l}{2}.$$

For an odd wave function these conditions demand that

$$\frac{D}{C} = e^{\pi \varkappa l} \sin \pi k l \quad \text{and} \quad \frac{D}{C}\frac{\varkappa}{k} = -e^{\pi \varkappa l} \cos \pi k l. \tag{4.1.35}$$

These two equations for D/C are only compatible if the transcendental equation

$$\frac{k}{\varkappa} = -\tan(\pi k l)$$

or

$$\sqrt{\frac{E}{U_0 - E}} = -\tan\left(\frac{\pi l}{h}\sqrt{2m\,E}\right) \tag{4.1.36}$$

is satisfied. For a given "well width" l and "well depth" U_0, this equation yields an infinite number of eigenvalues of energy which may be computed by a numerical or graphical method. We refer to them as E_2, E_4, E_6 ....

By a completely analogous procedure one obtains for an even wave function the conditions

$$\frac{D}{C} = e^{\pi \varkappa l} \cos \pi k l \quad \text{and} \quad \frac{D}{C}\frac{\varkappa}{k} = e^{\pi \varkappa l} \sin \pi k l, \tag{4.1.37}$$

from which follows

$$\frac{k}{\varkappa} = \cot(\pi k l) \tag{4.1.38}$$

leading to the equation

$$\sqrt{\frac{E}{U_0 - E}} = \cot\left(\frac{\pi l}{h}\sqrt{2m\,E}\right) \tag{4.1.39}$$

for the energy eigenvalues to be denoted by E_1, E_3, E_5 ....

It is easy to show that the sequence of energy levels E_1, E_2, E_3, ..., increases monotonically, and that, as in the previous examples, the even quantum numbers belong to odd eigenfunctions and the odd quantum numbers

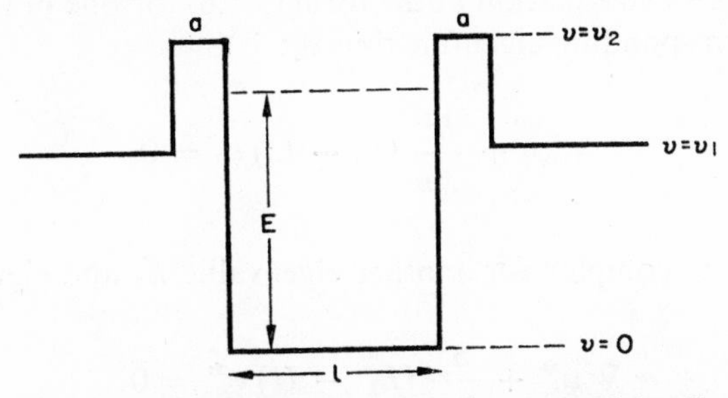

FIG. 5. Potential well surrounded by a potential wall.

to even eigenfunctions. The eigenfunctions themselves are obtained from (4.1.34) when the values of the constants from (4.1.35) and (4.1.37) respectively are inserted.

We conclude this section by indicating another boundary problem consisting of a combination of a "potential well" with a "potential wall", as indicated in Fig. 5. Here again the particle could, with the energy at its disposal, never escape from the well by "jumping" over the wall, but quantum mechanically it has a finite chance to escape by tunnelling its way through the wall. This model is the basis for Gamow's theory of the mechanism by which "nucleons" can escape from the interior of a nucleus in the process of radioactive decay.

4.2. The eigenvalue problem in continuous fields

In this section we shall be concerned with the eigenvalue problem of the wave equation for the movement of a particle in a field of force with no discontinuities of the potential energy function $U(r)$. We shall confine ourselves to such cases where the particle remains permanently attached to a certain system during its movement, that is in cases to which in the older quantum theory the Sommerfeld quantum conditions would apply. We shall assume that the field vanishes at infinity or the potential function approaches a certain constant value at infinity, which we shall choose as its zero point. Thus U will have to be *negative* everywhere else.

Under these circumstances the wave function must satisfy the condition that $\psi(\mathbf{r})$ and its derivatives with respect to the coordinates must be continuous functions of space everywhere. Furthermore, in view of the statistical meaning of the wave function, $\psi(\mathbf{r})$ must also be regular and univalued and, in order to satisfy the normalization condition (3.2.3), it must also tend to zero to a sufficiently high order at infinity. These conditions replace the explicit boundary conditions of section 4.1, and it will be shown that they are, indeed, sufficient to enforce discrete energy eigenvalues of the Schrödinger equation.

It will first be shown that these eigenvalues are always *real*. For this purpose we use the wave equation in the form (3.1.6) for one of the eigenvalues E_i and the corresponding eigenfunction ψ_i:

$$\nabla^2\psi_i + \frac{\hbar^2}{2m}(E_i - U)\,\psi_i = 0 \tag{4.2.1}$$

and its conjugate complex for another eigenvalue E_k and eigenfunction ψ_k:

$$\nabla^2\psi_k^* + \frac{\hbar^2}{2m}(E_k^* - U)\,\psi_k^* = 0. \tag{4.2.2}$$

Multiplying (4.2.2) by ψ_i and (4.2.1) by ψ_k^*, subtracting and integrating over all space we obtain

$$\iiint (\psi_i \nabla^2 \psi_k^* - \psi_k^* \nabla^2 \psi_i)\, dV = \frac{2m}{\hbar^2} \iiint (E_i - E_k^*)\, \psi_i \psi_k^*\, dV. \quad (4.2.3)$$

The volume integral on the left-hand side of this equation, extended over a finite region, can be transformed into an integral over the surface of that region by means of Green's theorem

$$\iiint (\psi_i \nabla^2 \psi_k^* - \psi_k^* \nabla^2 \psi_i)\, dV = \iint \left(\psi_i \frac{\partial \psi_k^*}{\partial n} - \psi_k^* \frac{\partial \psi_i}{\partial n} \right) da.$$

If now the region is extended to cover all space the surface integral must vanish as a result of the above stated conditions for the wave function at infinity, and hence the left-hand member of equation (4.2.3) is zero. Thus

$$(E_i - E_k^*) \iiint \psi_i \psi_k^*\, dV = 0. \quad (4.2.4)$$

Assume first that $i = k$ and hence $\psi_k^* = \psi_i^*$. In this case the integral in (4.2.4) is finite and therefore we must have

$$E_i = E_i^* \quad (4.2.5)$$

for all eigenvalues, which is only possible if they are all real. This, of course, is a necessary prerequisite for the whole theory being applicable to physical systems since the energy must necessarily be a real quantity.

Assume next that $i \gtrless k$. Making use of the fact, just proved, that $E_k = E_k^*$, one obtains from (4.2.4)

$$(E_i - E_k) \iiint \psi_i \psi_k^*\, dV = 0. \quad (4.2.6)$$

Hence for $E_i \gtrless E_k$ the integrals must vanish:

$$\iiint \psi_i \psi_k^*\, dV = 0 \quad (i \gtrless k). \quad (4.2.7)$$

This means that any two eigenfunctions belonging to different eigenvalues of the wave equation are *orthogonal* to each other.

If $E_i = E_k$ but $\psi_i \gtrless \psi_k$ the energy state is "degenerate" according to the nomenclature introduced in 1.2. We say that the state is "*n-fold degenerate*" if there exist n linearly independent eigenfunctions belonging to the same eigenvalue. One sees directly from (4.2.1) that any linear combination of these eigenfunctions must also satisfy the wave equation, that is, it must also be an eigenfunction. There is therefore in this case an n-fold infinity of eigenfunctions from which one can always choose n independent ones which will satisfy the condition (4.2.7). With this proviso it can then be stated that *all* the eigenfunctions ψ_i of the Schrödinger equation form a set of mutually orthogonal functions. We have already come across a special example for this in section 4.1 when we dealt with the problem of the particle in the box.

Because of the linearity of the wave equation (4.2.1), any solution of that equation can, of course, be multiplied by an arbitrary constant. By an appropriate choice of this constant we can therefore always "normalize" the eigenfunctions so that they fulfill the condition (3.2.3). The eigenfunctions of the Schrödinger equation can thus be made into a *set of "orthonormal" functions* in the sense of the definition given in section 4.1.

An important conclusion concerning the *shape* of the eigenfunctions can be drawn from the one-dimensional form of the wave equation

$$\frac{d^2\psi}{dx^2} + \frac{2m}{\hbar^2}(E - U)\,\psi = 0.$$

$E - U$, the kinetic energy of the particle, is evidently an essentially positive quantity. Hence it follows that ψ and $\frac{d^2\psi}{dx^2}$ must always have opposite signs, that is, the curve representing $\psi(x)$ must be convex (as seen from above) everywhere where ψ is positive and concave where it is negative in those regions of x into which the particle can penetrate "classically". This fact, together with the fact that ψ must approach zero asymptotically at both ends,

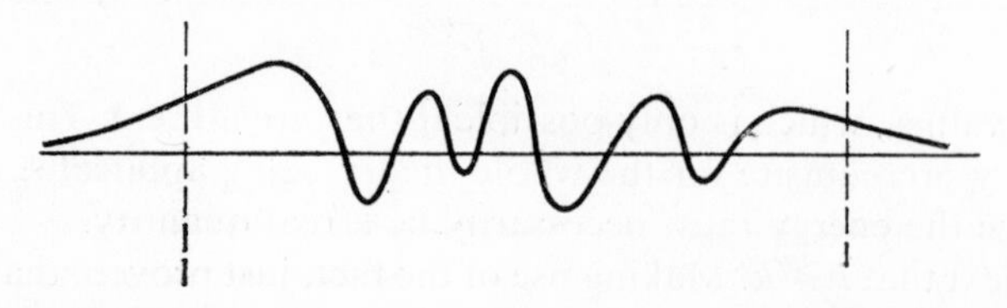

FIG. 6. Characteristic form of a one-dimensional wave function ψ, obeying the condition $\psi = 0$ at infinity, showing loops and nodes.

demands that the curve must have a "wavy" form with maxima and minima and zero's in between, as shown in Fig. 6 in the region between the two vertical dotted lines to which the particle is classically confined. Thus the wave function has "loops" and "nodes", like a standing wave, and the different eigenfunctions are characterized by their shapes and, in particular, by the number of nodes which must be in close relationship to the quantum number, as was already pointed out in the special case of the particle in the box.

Similarly, we have to expect that in a two-dimensional problem there will be "nodal lines" and in a three-dimensional problem "nodal surfaces" where the wave function is zero, apart from infinity, and that the number of these nodal lines or nodal surfaces for a certain eigenfunction will be related to the corresponding quantum numbers. We shall discuss this in more detail in connection with some special problems in sections 5.2. and 5.3.

We finish this section by stating the following important theorem: *Any function $f(\mathbf{r})$, that satisfies some very general conditions, can be expanded into a* *finite or infinite series of orthonormal functions* ψ_i, *which may, for example, be the eigenfunctions of a certain differential equation*

$$f(\mathbf{r}) = \sum_i a_i \psi_i(\mathbf{r}). \tag{4.2.8}$$

Multiplying this equation on both sides by ψ_k^* and integrating over all space we obtain, in virtue of (3.2.3) and (4.2.7),

$$\iiint f\psi_k^* \, dV = \sum_i a_i \iiint \psi_k^* \, \psi_i \, dV = a_k. \tag{4.2.9}$$

Using the symbol for the "inner product" of two functions, defined by the equations (4.1.31), we can write the expressions for the coefficients in the expansion (4.2.8) in the form

$$a_i = (\psi_i f). \tag{4.2.10}$$

From these formulae the coefficients can be evaluated uniquely, which proves the validity of the expansion theorem. It is, in fact, a generalization of Fourier's famous theorem concerning the expansion of a function into a series of harmonic functions (sin- and cos-functions), and the relations (4.2.10) are the generalizations of the well-known formulae for calculating the Fourier coefficients. In the following we shall make frequent use of the expansion theorem.

4.3. Matrices belonging to eigenfunctions

At the end of section 2.2 it was pointed out that in the version of quantum mechanics originally founded by Heisenberg and Dirac, certain mathematical quantities played an essential part which were different from ordinary numbers and obeyed different algebraical rules. It was discovered by Born that those quantities, which were introduced by Heisenberg in his theory, were in actual fact identical with the so-called "*matrices*" known and used for a long time in pure mathematics. In this section we shall therefore give a brief outline of the algebra of matrices and, following Schrödinger, prove that there exists an intimate connection between the eigenvalue problem in quantum mechanics and the theory of matrices.

A matrix is a two-dimensional array of numbers which we denote by a_{ik}, the subscript i indicating the number of the row and k that of the column in which the "element" a_{ik} is situated. For simplicity we shall assume first that the array is quadratic and that i and k run from 1 to n. We shall denote the matrix symbolically by $\mathfrak{A}$.† Thus

$$\mathfrak{A} = \begin{pmatrix} a_{11} & a_{12} & a_{13} & \cdots & a_{1n} \\ a_{21} & a_{22} & a_{23} & \cdots & a_{2n} \\ \cdot & \cdot & \cdot & \cdots & \cdot \\ \cdot & \cdot & \cdot & \cdots & \cdot \\ a_{n1} & a_{n2} & a_{n3} & \cdots & a_{nn} \end{pmatrix}. \tag{4.3.1}$$

† Throughout this book matrix quantities will always be denoted by gothic lettering.

One defines as the *sum* (respectively *difference*) of two matrices $\mathfrak{A}$ and $\mathfrak{B}$ a matrix whose elements are $a_{ik} \pm b_{ik}$, and as the product of a matrix with a number s a matrix with elements sa_{ik}. The *product* $\mathfrak{C} = \mathfrak{A}\mathfrak{B}$ of two matrices $\mathfrak{A}$ and $\mathfrak{B}$ is obtained by combining the rows of $\mathfrak{A}$ with the columns of $\mathfrak{B}$ according to the following rule:

$$c_{kj} = \sum_{i=1}^{n} a_{ki} b_{ij}. \tag{4.3.2}$$

This implies that, in general, $\mathfrak{A}\mathfrak{B}$ is not the same as $\mathfrak{B}\mathfrak{A}$, in other words matrices do not obey the commutative law of multiplication of ordinary number algebra.

A matrix in which all the elements except the diagonal ones with $i = k$ are zero is called a "*diagonal matrix*", and in particular if all these elements are equal to 1, it is called a "*unit matrix*", which we shall symbolically denote by $\mathfrak{J}$. It follows from the multiplication rule (4.3.2) that

$$\mathfrak{A}\mathfrak{J} = \mathfrak{J}\mathfrak{A} = \mathfrak{A}, \tag{4.3.3}$$

which justifies the chosen nomenclature.

As "*reciprocal*" to a matrix $\mathfrak{A}$ is defined a matrix $\mathfrak{A}^{-1}$ which satisfies the equation

$$\mathfrak{A}\mathfrak{A}^{-1} = \mathfrak{J}. \tag{4.3.4}$$

If both sides of this equation are multiplied by $\mathfrak{A}$ from the right one obtains

$$\mathfrak{A}(\mathfrak{A}^{-1}\mathfrak{A}) = \mathfrak{J}\mathfrak{A} = \mathfrak{A},$$

and hence

$$\mathfrak{A}^{-1}\mathfrak{A} = \mathfrak{J} \tag{4.3.5}$$

is also satisfied. According to this rule one can *divide* a matrix $\mathfrak{A}$ by another matrix $\mathfrak{B}$ by multiplying it by $\mathfrak{B}^{-1}$.

We finally introduce the useful notion of the "*adjoined matrix*" $\mathfrak{A}^{+}$ to a matrix $\mathfrak{A}$; this is obtained by "transposing" $\mathfrak{A}$, that is interchanging its rows and columns, and by replacing the elements by their conjugate complex values. Thus

$$a_{ik}^{+} = a_{ki}^{*}. \tag{4.3.6}$$

A matrix is called "*self-adjoined*" or "*Hermitian*" if $\mathfrak{A} \equiv \mathfrak{A}^{+}$ or if, according to (4.3.6),

$$a_{ik} = a_{ki}^{*}. \tag{4.3.7}$$

Thus in an Hermitian matrix elements which are situated symmetrically to the main diagonal are conjugate complex to each other and the diagonal elements a_{ii} must be real.

These notions can readily be extended to *infinite matrices* with an infinite number of rows and columns, provided that the sums (4.3.2) converge when $n \to \infty$.

According to Schrödinger one can associate with any real function $f(\mathbf{r})$ of the coordinates a matrix $\mathfrak{F}$ in terms of an orthonormal set of discrete eigenfunctions of the wave equation by defining its elements f_{ik} by the relations

$$f_{ik} = \iiint \psi_i^*(r) f(\mathbf{r}) \psi_k(\mathbf{r})\, dV. \tag{4.3.8}$$

As the number of eigenfunctions is infinite $\mathfrak{F}$ is an infinite matrix with an enumerably infinite number of elements, whereas the function $f(\mathbf{r})$ itself has a continuously infinite number of values.

In order to prove that the so defined matrices obey the above formulated laws of matrix algebra we consider the matrix $\mathfrak{H}$ that belongs to the product of two functions $f(\mathbf{r})$ and $g(\mathbf{r})$, whose elements are given by

$$h_{ik} = \iiint \psi_i^* f g \psi_k\, dV. \tag{4.3.}$$

The function $g\psi_k$ may be expanded in a series of eigenfunctions ψ_j according to the theorem stated at the end of section 4.2:

$$g\psi_k = \sum_j c_j \psi_j. \tag{4.3.10}$$

The coefficients c_j of this expansion are, according to (4.2.9) and (4.3.8), given by

$$c_j = \iiint g\psi_k \psi_j^*\, dV = g_{jk}. \tag{4.3.11}$$

Substituting from (4.3.10) and (4.3.11) into (4.3.9) we obtain with the help of (4.3.8)

$$h_{ik} = \sum_j c_j \iiint \psi_i^* f \psi_j\, dV = \sum_j f_{ij} g_{jk}, \tag{4.3.12}$$

which on comparison with (4.3.2) shows that $\mathfrak{H}$ is indeed the product of $\mathfrak{F}$ and $\mathfrak{G}$.

It is further easy to prove that the matrices of the form (4.3.8) are self-adjoined. For taking the conjugate complex values on both sides of the equation (4.3.8) one gets

$$f_{ik}^* = \iiint \psi_k^* f(\mathbf{r}) \psi_i\, dV = f_{ki} \tag{4.3.13}$$

which is identical with (4.3.7).

So far the matrices (4.3.8) have been introduced on a purely formal basis. We shall now have to see what their physical significance is. The diagonal elements are defined by

$$f_{ii} = \iiint \psi_i^*(\mathbf{r}) f(\mathbf{r}) \psi_i(\mathbf{r})\, dV. \tag{4.3.14}$$

According to the statistical interpretation of the wave function the quantity $\psi_i \psi_i^*\, dV$ is the probability for finding the particle in the volume element dV at any specific time. f_{ii} is therefore the so-called "*expectation value*" of $f(\mathbf{r})$ for repeated observations of a system which is known to be in the quantum state with quantum number i. This can also be expressed in a slightly different way by saying that f_{ii} is the *average* value of f within a system in the quantum state i.

It is plausible to assign an analogous meaning to the off-diagonal elements of the matrix (4.3.8) for the non-stationary transition process from the quantum state i to the state k or vice versa, by interpreting the arithmetic mean $(f_{ik} + f_{ki})/2$ $(i \gtrless k)$ as the average value of $f(\mathbf{r})$ during such a transition. Because of the condition (4.3.13) this will always be a *real* quantity (like f_{ii}), which is, of course, a necessary condition for this interpretation to be valid. We thus see that the Hermitian character of the matrices (4.3.8) is a necessary prerequisite for their interpretation as observable physical quantities.

A particularly important and simple case arises when $f(\mathbf{r})$ is just one of the coordinates x, y, z, themselves. The corresponding matrices are called "*coordinate matrices*"; in the Born–Heisenberg–Jordan version of quantum mechanics (1925–6) they play the part of the coordinates of classical mechanics. In vector form they can be written

$$\mathbf{r}_{ik} = \int\int\int \psi_i^*(\mathbf{r})\, \mathbf{r}\psi_k(\mathbf{r})\, dV. \tag{4.3.15}$$

For $i = k$ this equation enables one to calculate the expectation values of the position vector $\mathbf{r}$ of the particle in the various quantum states. If, as indicated in section 3.2, the particle mass is imagined to be "smeared out" continuously in space with a mass density ϱ_i, one has, from (3.2.5),

$$m\mathbf{r}_{ii} = \int\int\int \mathbf{r}\varrho_i(\mathbf{r})\, dV \tag{4.3.16}$$

which means that $\mathbf{r}_{ii}$ is the position vector of the "mass centre" of this distribution. Similarly, for $i \gtrless k$ one can visualize the quantity $(\mathbf{r}_{ik} + \mathbf{r}_{ki})/2$ as the average position vector of the mass centre in the intermediate state between the states i and k.

Problems and exercises

P.4.1. Treat the problem of the "potential wall", outlined in section 4.1, under the assumption that $U_0 < E$, when classically the particles would be able to "jump" over the wall. Show that quantum theoretically only a fraction N_t/N_i of the impinging particles will be transmitted and the rest reflected ("forward and backward scattering" by the wall). Discuss the dependence of N_t/N_i on $E - U_0$ and comment on it.

P.4.2. Show how eqns. (4.1.36) and (4.1.39) for the energy levels in the rectangular well problem, treated in section 4.1, can be solved graphically by plotting the functions

$$f_1 = \sqrt{\frac{U_0 - E}{E}}, \quad f_2 = \sqrt{\frac{E}{U_0 - E}} \quad \text{and} \quad f_3 = \tan\left(\frac{\pi l}{h}\sqrt{2m\,E}\right)$$

against E, and determining the intersection points of f_1 and f_3, and f_2 and f_3 respectively. Use this graph further for proving that the sequence of the levels E_n is monotonic and that even quantum numbers n belong to odd eigenfunctions and odd numbers n to even eigenfunctions.

P. 4.3. A beam of particles, moving in the positive x-direction in a field-free space with energy E, encounters a "trap" in the form of a rectangular potential well of depth U_0. Classically all particles falling into the trap must come out again at the opposite side so that there is perfect transmission. Show, by an argument similar to that used in P. 4.1, that quantum theoretically the trap acts as a scattering centre and only a fraction N_t/N_i of the impinging particles is let through and the rest is reflected.

P. 4.4. Solve the problem indicating in the last paragraph of section 4.1 (and Fig. 5), following the same procedure as used in the problems of the "potential wall" and the "potential well". Show that the energy of a particle inside the well is quantized and the quantum levels given by eqns. (4.1.36) and (4.1.39). Show further that the ratio of the number densities of the particles outside and inside the wall is given by the expression

$$\frac{1 - U_1/U_2}{1 - U_1/E} \exp\left[-\frac{4\pi a}{h}\sqrt{2m\,(U_2 - E)}\right].$$

Comment on this result.

P. 4.5. An analytical function $f(x)$ is defined in the interval between $x = -l/2$ and $x = l/2$, vanishing at these points. Use the expansion theorem (4.2.8) and formula (4.2.10) for the development of $f(x)$ in a series of the eigenfunctions $\psi_n(x)$ of the "one-dimensional box" problem of section 4.1. Assume in particular that $f(x)$ has the triangular shape

$$f(x) = -\frac{2a}{l}x + a \quad \text{for} \quad x \geqslant 0, \quad f(x) = \frac{2a}{l}x + a \quad \text{for} \quad x \leqslant 0.$$

P. 4.6. Prove that the distributive and the associative laws of ordinary algebra are also valid in matrix algebra, i.e. that

$$\mathfrak{A}\,(\mathfrak{B} + \mathfrak{C}) = \mathfrak{A}\mathfrak{B} + \mathfrak{A}\mathfrak{C} \quad \text{and} \quad \mathfrak{A}\,(\mathfrak{B}\mathfrak{C}) = (\mathfrak{A}\mathfrak{B})\,\mathfrak{C}.$$

P. 4.7. Show that the matrices

$$\mathfrak{A} = \begin{pmatrix} 0 & 0 & 0 & 1 \\ 0 & 0 & -1 & 0 \\ 0 & -1 & 0 & 0 \\ 1 & 0 & 0 & 0 \end{pmatrix} \quad \text{and} \quad \mathfrak{B} = \begin{pmatrix} 1 & 0 & 0 & 0 \\ 0 & 1 & 0 & 0 \\ 0 & 0 & -1 & 0 \\ 0 & 0 & 0 & -1 \end{pmatrix}$$

obey the relations

$$(\mathfrak{A}\mathfrak{A}) = \mathfrak{J}, \quad (\mathfrak{B}\mathfrak{B}) = \mathfrak{J}, \quad \mathfrak{A}\mathfrak{B} + \mathfrak{B}\mathfrak{A} = 0.$$

P. 4.8. Calculate the reciprocal matrix $\mathfrak{A}^{-1}$ to the matrix $\mathfrak{A} = \begin{pmatrix} a_{11} & a_{12} \\ a_{21} & a_{22} \end{pmatrix}$ from the definition (4.3.4).

P. 4.9. Evaluate the coordinate matrix x_{ik} belonging to the eigenfunctions (4.1.27), (4.1.29) of the "particle in the box" problem according to the definition (4.3.15) and show that it is Hermitian. Prove that the diagonal elements are zero and give the statistical explanation of this fact. Derive an expression for the magnitude of the non-vanishing elements.

$$(\mathfrak{A}\mathfrak{A})=\mathfrak{I}, \quad (\mathfrak{B}\mathfrak{B})=\mathfrak{I}, \quad \mathfrak{A}\mathfrak{B}+\mathfrak{B}\mathfrak{A}=0.$$

CHAPTER 5

Application of Quantum Mechanics to Simple Problems of Atomic Physics

ALTHOUGH Part I of this book is primarily meant to give an introduction to the *fundamental principles* of quantum mechanics, it was considered useful to include a chapter on the *applications* of the quantum mechanical methods to some simple problems of atomic physics. We shall restrict ourselves to *one-particle* problems, namely to the solution of the eigenvalue problems of a single particle under the influence of a potential field of force. The rotational spin movement of the particle (see section 2.2) will be neglected.

5.1. The harmonic oscillator

The simplest example of a system in which a particle would classically perform a periodic movement in space is the *harmonic oscillator* consisting of a particle of mass m under the influence of an *elastic force* from a fixed centre, say the origin of coordinates, that is a force directed towards that centre and proportional to the distance from it. The potential energy U of the particle is then proportional to the square of that distance

$$U(\mathbf{r}) = \tfrac{1}{2}\alpha r^2, \quad \alpha > 0, \tag{5.1.1}$$

and it follows from elementary classical mechanics that the particle performs an elliptic orbit with frequency

$$\nu = \frac{1}{2\pi}\sqrt{\frac{\alpha}{m}}. \tag{5.1.2}$$

We shall first consider the one-dimensional problem where the movement of the particle takes place along a straight line, say the x-axis. We shall refer to this system as a "*linear harmonic oscillator*". The wave equation (3.1.6) then takes the form

$$\frac{d^2\psi}{dx^2} + \frac{2m}{\hbar^2}\left(E - \frac{\alpha x^2}{2}\right)\psi = 0 \tag{5.1.3}$$

which has to be solved under the conditions specified in section 4.2.

Introducing the two new constants

$$\gamma = \sqrt{\frac{2\pi\nu m}{\hbar}}, \quad \lambda = \frac{E}{\pi\hbar\nu} = \frac{2E}{h\nu}, \tag{5.1.4}$$

and using, instead of x, the dimensionless variable

$$\xi = \gamma x \tag{5.1.5}$$

we obtain easily from (5.1.3) the differential equation

$$\frac{d^2\psi}{d\xi^2} + (\lambda - \xi^2)\,\psi = 0 \tag{5.1.6}$$

for $\psi(\xi)$.

According to 4.2 we have to seek a solution of (5.1.6) which is univalued everywhere and vanishes at $x = \pm\infty$ to a sufficiently high order. These conditions are evidently satisfied if we set

$$\psi(\xi) = e^{-\beta\xi^2} f(\xi), \tag{5.1.7}$$

where β is a positive constant and $f(\xi)$ is a polynomial of finite order n in ξ. Substituting the function (5.1.7) into the differential equation (5.1.6), we find that it will, indeed, be a solution of this equation provided that $f(\xi)$ satisfies the differential equation

$$\frac{d^2f}{d\xi^2} - 4\beta\xi\,\frac{df}{d\xi} + (\lambda - 2\beta)\,f + (4\beta^2 - 1)\,\xi^2 f = 0. \tag{5.1.8}$$

The first term in (5.1.8) is of order $n - 2$ in ξ, the second and third terms are of order n, and the last term is of order $n + 2$. Thus for large ξ the last term will predominate and the equation can only be satisfied if

$$\beta = \tfrac{1}{2}. \tag{5.1.9}$$

Consequently (5.1.7) becomes

$$\psi(\xi) = e^{-\xi^2/2} f(\xi), \tag{5.1.10}$$

and (5.1.8) simplifies to

$$\frac{d^2f}{d\xi^2} - 2\xi\,\frac{df}{d\xi} + (\lambda - 1)\,f = 0. \tag{5.1.11}$$

The solutions of this differential equation are supposed to have the form

$$f_n(\xi) = \sum_{k=0}^{n} A_k \xi^k, \tag{5.1.12}$$

where the A_k are constants. Let us first again assume ξ to be large. Substitution of (5.1.12) into (5.1.11) then yields the asymptotic equation

$$A_n\,(-2n + \lambda_n - 1)\,\xi^n = 0 \quad (\xi \text{ large})$$

which can evidently only be satisfied if

$$\lambda_n = 2n + 1 \quad (n = 0, 1, 2, \ldots). \tag{5.1.13}$$

The λ_n, determined by this condition, are the *eigenvalues* of the differential equation (5.1.11). It follows at once from the second of the equations (5.1.4) that the energy E of the system is restricted to the *energy levels*

$$E_n = \frac{h\nu}{2}\lambda_n = \left(n + \frac{1}{2}\right) h\nu \quad (n = 0, 1, 2, \ldots). \tag{5.1.14}$$

(5.1.14) is thus the quantization condition for the linear harmonic oscillator and the n are the *quantum numbers.* As in Planck's original theory, neighbouring energy levels are separated by steps of magnitude $h\nu$ but, in contrast to Planck's quantization rule (2.2.1), the energy levels are *half-integer multiples* of $h\nu$ and the lowest possible energy level is now $E_0 = h\nu/2$ instead of zero. This means that according to the concepts of quantum mechanics the particle can never be at rest. The energy E_0 is usually referred to as "*zero-point energy*" for reasons which will be discussed later in sections 15.1 and 15.2.

Substituting $\lambda = \lambda_n$ into (5.1.11) we now obtain the differential equations

$$\frac{d^2 f_n}{d\xi^2} - 2\xi \frac{df_n}{d\xi} + 2nf_n = 0 \quad (n = 0, 1, 2, \ldots), \tag{5.1.15}$$

which are well known to mathematicians. Their eigenfunctions $f_n(\xi)$ are called "*Hermitian polynomials*" and are usually denoted by H_n. On substituting the expressions (5.1.12) and their derivatives

$$\left.\begin{aligned} \frac{df_n}{d\xi} &= \sum_{k=1}^{n} kA_k\xi^{k-1} \\ \frac{d^2 f_n}{d\xi^2} &= \sum_{k=2}^{n} k(k-1) A_k\xi^{k-2} = \sum_{k=0}^{n-2} (k+2)(k+1) A_{k+2}\xi^k \end{aligned}\right\} \tag{5.1.16}$$

into the differential equation (5.1.15) one obtains algebraic equations for ξ which can only be satisfied identically if all the coefficients of the various powers of ξ vanish separately. This procedure leads to the recurrence equations

$$(k+2)(k+1) A_{k+2} = 2(k-n) A_k \quad (k = 0, 1, 2, \ldots, n; \quad n = 0, 1, 2, \ldots), \tag{5.1.17}$$

for the coefficients of the Hermitian polynomials.

Since $A_{n+2} = A_{n+1} = 0$ by definition the equations (5.1.17) are certainly satisfied for $k = n$. For $k = n - 1$ one obtains $A_{n-1} = 0$ and hence, putting $k = n - 3$, one gets $A_{n-3} = 0$, etc. That is, for *even* n all the *odd* coefficients vanish and the polynomials are therefore even in ξ, and similarly for *odd n* they are *odd* in ξ. Keeping this in mind one obtains easily from (5.1.17)

$$\left.\begin{aligned} A_2 &= -nA_0, \quad A_3 = \frac{1-n}{3} A_1, \\ A_4 &= \frac{n(n-2)}{6} A_0, \quad A_5 = \frac{(n-1)(n-3)}{30} A_1, \ldots \end{aligned}\right\} \tag{5.1.18}$$

and thus the following expression for the first five Hermitian polynomials:

$$\left.\begin{aligned} &H_0 = A_0, \quad H_1 = A_1\xi, \\ &H_2 = A_0(1 - 2\xi^2), \quad H_3 = A_1(\xi - \tfrac{2}{3}\xi^3), \\ &H_4 = A_0(1 - 4\xi^2 + \tfrac{4}{3}\xi^4), \quad H_5 = A_1(\xi - \tfrac{4}{3}\xi^3 + \tfrac{4}{15}\xi^5). \end{aligned}\right\} \tag{5.1.19}$$

The coefficients A_0 and A_1 remain, of course, undetermined owing to the fact that the equations (5.1.15) are homogeneous.

The eigenfunctions ψ_n of the wave equation (5.1.6) now follow from (5.1.10) and (5.1.19) in terms of the variable ξ. As the first factor in (5.1.10) is even in ξ the ψ_n are even functions of ξ for even n and odd functions for odd n. All the $|\psi_n|^2$ are therefore symmetric in ξ (or x) as is indeed required by the symmetry of the system. It can further be verified easily that the functions ψ_n are orthogonal to each other as demanded by the general theorem (4.2.7). The constants A_0 and A_1 can also always be chosen in such a way that the normalizing conditions (3.2.3) are fulfilled.

In particular for the "ground state" of the linear harmonic oscillator, corresponding to the lowest quantum number $n = 0$, one gets from (5.1.10)

$$\psi_0(\xi) = \text{const } e^{-\xi^2/2}. \tag{5.1.20}$$

The classical amplitude a of the oscillation of the particle is connected with its energy by the relation

$$E = 2\pi^2 m a^2 \nu^2.$$

For $E_0 = \dfrac{h\nu}{2} = \pi\hbar\nu$ we have therefore from (5.1.4) and (5.1.5)

$$\left.\begin{aligned} \gamma &= \sqrt{\frac{2\pi\nu m}{\hbar}} = \frac{1}{a}, \\ x &= a\xi. \end{aligned}\right\} \tag{5.1.21}$$

Substituting this into (5.1.20) and normalizing, we obtain finally for the eigenfunction of the ground state the "*Gaussian distribution*"

$$\psi_0(x) = \frac{1}{\left(a\sqrt{\pi}\right)^{1/2}} e^{-x^2/2a^2}. \tag{5.1.22}$$

This formula illustrates clearly the fundamental difference between the classical and the quantum mechanical aspects of the mechanics of the linear harmonic oscillator. Whereas classically the movement of the particle is restricted to the region between a and $-a$, quantum theoretically there is a finite probability density $|\psi_0|^2$ for finding it at any particular time outside this region and a plays only the part of the "standard deviation" of the probability distribution $|\psi_0|^2$.

The shape of the eigenfunctions for the higher energy states can similarly be obtained from the equations derived above. A few of the corresponding curves are shown in Fig. 7. In particular one notices that they are indeed

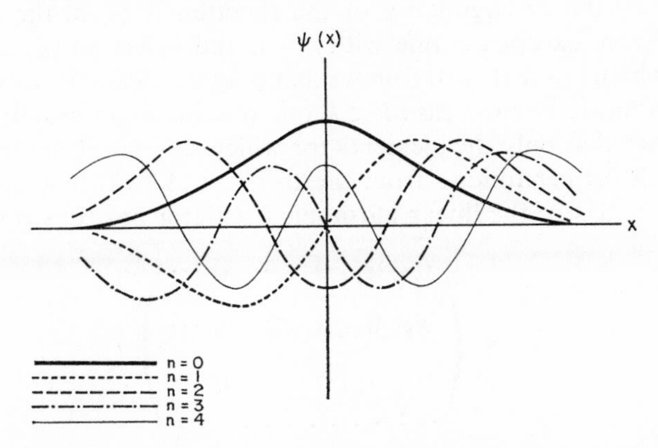

FIG. 7. Eigenfunctions of the one-dimensional harmonic oscillator.

symmetric about $x = 0$ (even) for even quantum numbers n and antisymmetric (odd) for odd n. Further, the number of nodes (apart from those at infinity) is equal to $(n-1)$, as in the case of the particle in the box (section 4.1).

Having solved the eigenvalue problem for the linear harmonic oscillator we can now proceed to calculate its coordinate matrix according to its definition (4.3.15) which, in view of (5.1.10), here reduces to

$$x_{ik} = \int_{-\infty}^{+\infty} \psi_i^*(x)\, x\psi_k(x)\, dx = \text{const} \int_{-\infty}^{+\infty} \xi\, e^{-\xi^2} f_i(\xi)\, f_j(\xi)\, d\xi\,. \quad (5.1.23)$$

For $i = k$ the integrand in this formula is an odd function of ξ and therefore the integral vanishes. Thus the diagonal elements of the coordinate matrix are all zero. According to the statistical interpretation given in section 4.3 this means that the mean value of x is zero for all quantum states, which is, of course, obvious on account of the symmetry properties of the system.

For $i \lessgtr k$ let us first assume that $i > k$. ξf_k is then a polynomial of order $k + 1 \leqslant i$, and by making use of the expansion theorem (4.2.8) we can expand it in terms of the eigenfunctions f_j of the differential equation (5.1.11) in a finite series of the form

$$\xi f_k(\xi) = \sum_{j=0}^{k+1} \alpha_j f_j(\xi). \tag{5.1.24}$$

Substitution of this into (5.1.23) with interchange of the order of summation and integration gives the result

$$x_{ij} = \text{const} \sum_{j=0}^{k+1} \alpha_j \int_{-\infty}^{+\infty} f_j(\xi) f_i(\xi)\, e^{-\xi^2} d\xi. \tag{5.1.25}$$

In virtue of the orthogonality of the functions $f(\xi)$ all the integrals in (5.1.25) are zero except the one with $j = i$, and hence all the elements x_{ik} vanish for which $i > k + 1$. It follows that only the elements for which $i - k = 1$ remain finite. For the case $i < k$ one obtains in an exactly analogous way the result that only the elements for which $k - i = 1$ are finite. As the x_{ij} are real it further follows from the theorem (4.3.13) that $x_{ij} = x_{ji}$; the coordinate matrix of the linear harmonic oscillator has thus the form

$$x_{ij} = \begin{pmatrix} 0 & x_{12} & 0 & 0 & 0 & \cdots \\ x_{12} & 0 & x_{13} & 0 & 0 & \cdots \\ 0 & x_{13} & 0 & x_{14} & 0 & \cdots \\ \cdot & \cdot & \cdot & \cdot & \cdot & \cdot \end{pmatrix}. \tag{5.1.26}$$

From the statistical interpretation of the coordinate matrix in section 4.3 we can now conclude that only those transitions from one quantum state to another have finite probability to occur in which the quantum number n changes by unity or the energy of the system by $h\nu$; all other transitions are said to be "*forbidden*". We shall discuss some of the implications of this "selection rule" later in section 6.2.

We now go over to the treatment of the eigenvalue problem for a harmonic oscillator in more than one dimensions. Let us first deal with the "*two-dimensional harmonic oscillator*" which is supposed to consist again of a particle of mass m under the action of a central force from the origin of coordinates whose movement is now restricted to the xy-plane; the potential energy function is still given by (5.1.1). The wave equation is therefore

$$\frac{\partial^2 \psi}{\partial x^2} + \frac{\partial^2 \psi}{\partial y^2} + \frac{2m}{\hbar^2}\left[E - \frac{\alpha}{2}(x^2 + y^2)\right]\psi = 0. \tag{5.1.27}$$

In this and in many other cases it is possible to obtain a solution of the partial differential equation (5.1.27) in the form of a product of a function ψ' of x alone and a function ψ'' of y alone:

$$\psi(x, y) = \psi'(x)\, \psi''(y). \tag{5.1.28}$$

If in addition E is split into two parts:

$$E = E' + E''; \tag{5.1.29}$$

the wave equation (5.1.27) assumes the form

$$\frac{1}{\psi'}\frac{d^2\psi'}{dx^2} + \frac{1}{\psi''}\frac{d^2\psi''}{dy^2} + \frac{2m}{\hbar^2}\left[\left(E' - \frac{\alpha}{2}x^2\right) + \left(E'' - \frac{\alpha}{2}y^2\right)\right] = 0.$$

This equation is equivalent to the two ordinary differential equations

$$\left.\begin{aligned} \frac{d^2\psi'}{dx^2} + \frac{2m}{\hbar^2}\left(A + E' - \frac{\alpha}{2}x^2\right)\psi' = 0, \\ \frac{d^2\psi''}{dy^2} + \frac{2m}{\hbar^2}\left(B + E'' - \frac{\alpha}{2}y^2\right)\psi'' = 0, \end{aligned}\right\} \tag{5.1.30}$$

in which A and B are constants which are subject to the condition

$$A + B = 0. \tag{5.1.31}$$

Each of the equations (5.1.30) has the form of the equation (5.1.3) for the linear harmonic oscillator, and it therefore follows at once that the eigenvalues must satisfy the equations

$$\left.\begin{aligned} E_i' + A &= (i + \tfrac{1}{2})\,h\nu, \\ E_j'' + B &= (j + \tfrac{1}{2})\,h\nu \quad (i, j = 0, 1, 2, \ldots). \end{aligned}\right\} \tag{5.1.32}$$

From (5.1.29), (5.1.31), and (5.1.32) one now obtains for the energy eigenvalues of the two-dimensional harmonic oscillator

$$E_n = (n + 1)\,h\nu \quad (n = i + j; \quad i, j = 0, 1, 2, \ldots). \tag{5.1.33}$$

Thus the energy levels here are *integer* multiples of $h\nu$ and they are *one-fold degenerate.*

The eigenfunctions ψ_n belonging to the energy eigenvalues E_n are obtained from those of the linear harmonic oscillator by combining linearly functions of the form (5.1.28)

$$\psi_n(x, y) = \sum_{i,j}^{(i+j=n)} c_{ij}\psi_i'(x)\,\psi_j''(y) \tag{5.1.34}$$

where the coefficients are arbitrary and only restricted by the normalizing condition (3.2.3) which, in this case, has the form

$$\iint_{-\infty}^{+\infty} |\psi_n(x, y)|^2\,dx\,dy = 1. \tag{5.1.35}$$

The *three-dimensional harmonic oscillator* can be treated in exactly the same way when the wave function is assumed to be of the form

$$\psi\,(x, y, z) = \psi'(x)\,\psi''(y)\,\psi'''(z), \tag{5.1.36}$$

satisfying the wave equation

$$\frac{\partial^2\psi}{\partial x^2} + \frac{\partial^2\psi}{\partial y^2} + \frac{\partial^2\psi}{\partial z^2} + \frac{2m}{\hbar^2}\left[E - \frac{\alpha}{2}(x^2 + y^2 + z^2)\right]\psi = 0. \quad (5.1.37)$$

As a result one obtains instead of (5.1.33) now the quantization relation

$$E_n = (n + \tfrac{3}{2})\,h\nu \quad (n = i + j + k, \quad i, j, k = 0, 1, 2, \ldots), \quad (5.1.38)$$

which shows that the energy levels are again *half-integer* multiples of $h\nu$, as for the linear harmonic oscillator, and that they are *twofold degenerate.* The corresponding eigenfunctions are now

$$\psi_n(x, y, z) = \sum_{i,j,k}^{(i+j+k=n)} c_{ijk}\,\psi'_i(x)\,\psi''_j(y)\,\psi'''_k(z) \quad (5.1.39)$$

with arbitrary coefficients c_{ijk}, subjected to the condition that

$$\iiint_{-\infty}^{+\infty} \psi_n(x, y, z)\,dx\,dy\,dz = 1. \quad (5.1.40)$$

5.2. Particle moving at a fixed distance from a centre

In this section we shall deal with the eigenvalue problem of another simple mechanical system which, although it is not directly related to any real atomic system, like the harmonic oscillator, is very useful for demonstrating the application of certain mathematical methods which are frequently used for the solution of problems of atomic physics. It may also be regarded as a preliminary to the following section.

The system in question consists again of a single particle of mass m which, by some appropriate mechanism, is constrained to move on the surface of a sphere of radius r in an otherwise field-free space. In this and similar cases it is appropriate to use, instead of Cartesian coordinates x, y, z as before, spherical polar coordinates r, θ, φ (θ polar angle; φ azimuth angle). The relationship between the two coordinate systems is expressed by the transformation equations

$$\left.\begin{aligned} x &= r\sin\theta\cos\varphi, \\ y &= r\sin\theta\sin\varphi, \\ z &= r\cos\theta. \end{aligned}\right\} \quad (5.2.1)$$

Using these equations it is easy to verify the following expression for the "*Laplace operator*" ∇^2 in polar coordinates:

$$\nabla^2 = \frac{1}{r^2}\frac{\partial}{\partial r}\left(r^2\frac{\partial}{\partial r}\right) + \frac{1}{r^2\sin\theta}\frac{\partial}{\partial\theta}\left(\sin\theta\frac{\partial}{\partial\theta}\right) + \frac{1}{r^2\sin^2\theta}\frac{\partial^2}{\partial\varphi^2}, \quad (5.2.2)$$

and the wave equation (3.1.6) therefore takes the form

$$\sin^2\theta \frac{\partial}{\partial r}\left(r^2 \frac{\partial \psi}{\partial r}\right) + \sin\theta \frac{\partial}{\partial \theta}\left(\sin\theta \frac{\partial \psi}{\partial \theta}\right) + \frac{\partial^2 \psi}{\partial \varphi^2}$$

$$+ \frac{2mr^2 \sin^2\theta}{\hbar^2} [E - U(r, \theta, \varphi)]\, \psi = 0. \tag{5.2.3}$$

Let us first assume that r and θ are fixed, so that the particle is not only forced to remain on the surface of the sphere $r =$ const. but has also to move along the parallel circle with radius $a = r \sin\theta$. We shall refer to a system of this type as a "*one-dimensional rotator*".

As the space is supposed to be field-free we have $K = E - U =$ const. Hence the partial differential equation (5.2.3) now reduces to the ordinary differential equation

$$\frac{d^2\psi}{d\varphi^2} + \frac{2ma^2}{\hbar^2} K\psi = 0. \tag{5.2.4}$$

The general solution of this equation is

$$\psi(\varphi) = A\, e^{iu\varphi} + B\, e^{-iu\varphi}, \tag{5.2.5}$$

where A and B are arbitrary, in general complex, constants, and

$$u = \frac{a}{\hbar} \sqrt{2mK}\,. \tag{5.2.6}$$

This solution must satisfy the condition (see section 4.2) that φ has to be a univalued function of its argument, which implies that it must assume the same value when φ is increased by an integer multiple of 2π. Consequently u must be restricted to *integer* values, from which follows that, according to (5.2.6), the kinetic energy K is restricted to the quantum levels

$$K_u = \frac{\hbar^2}{2ma^2} u^2 \quad (u = 0, 1, 2, \ldots). \tag{5.2.7}$$

The quantization rule (5.2.7) for the kinetic energy of the system can be transformed into one for its angular momentum P about the polar axis which is connected with K by the relation

$$K = P^2/2ma^2.$$

This leads to

$$P_u = u\hbar \quad (u = 0, 1, 2, \ldots), \tag{5.2.8}$$

which is seen to be identical with Bohr's quantization rule (2.2.10); this rule thus *follows* from the quantum mechanical wave equation.

The eigenfunctions (5.2.5) have the character of standing waves along the circumference of the circle of radius a. This is in keeping with De Broglie's original ideas as explained in section 2.2. If these functions are to be normalized the constants A and B have to satisfy the relation

$$\int_0^{2\pi} \psi\psi^* d\varphi = 2\pi (AA^* + BB^*) = 1, \tag{5.2.9}$$

which shows that the quantum levels of the one-dimensional rotator are *one-fold degenerate.*

It was pointed out in section 4.2 that it is always possible to select from the infinite number of eigenfunctions belonging to one and the same eigenvalue just one in such a way that these eigenfunctions form an orthonormal set. This can evidently be achieved by setting either $A = 0$ or $B = 0$. On account of (5.2.9) we then obtain the following set of eigenfunctions:

$$\left.\begin{aligned} \psi_u(\varphi) &= A\, e^{iu\varphi} \quad (u = 0, \pm 1, \pm 2, \ldots), \\ AA^* &= 1/2\pi, \end{aligned}\right\} \tag{5.2.10}$$

which indeed satisfy the conditions

$$\int_0^{2\pi} \psi_u \psi_{u'}^{\prime *}\, d\varphi \begin{array}{l} = 0 \quad \text{for} \quad u \gtrless u', \\ = 1 \quad \text{for} \quad u = u'. \end{array}$$

The elements of the coordinate matrix for φ are obtained from (4.3.8) and (5.2.10):

$$\varphi_{uu'} = \frac{1}{2\pi} \int_0^{2\pi} e^{i(u-u')\varphi}\, \varphi\, d\varphi \begin{array}{ll} = 1/i\,(u - u'), & \text{for} \quad u \gtrless u' \\ = \pi & \text{for} \quad u = u' \end{array} \tag{5.2.11}$$

These equations show that the coordinate matrix of the one-dimensional rotator is indeed Hermitian as demanded by the general theorem (4.3.13). There are no "forbidden" transitions here as all matrix elements are different from zero. The expectation value of φ is, according to 4.3, equal to π for all quantum states and zero for all transitional states. The former result follows, of course, directly from the fact that the probability density for all angles φ is equal ($|\psi_0|^2 = 1/2\pi$).

We now return to the general wave equation (5.2.3) and assume that only r has a fixed value and that φ and θ can vary freely. We shall call such a system a "*two-dimensional rotator*". Under these conditions the wave equation becomes

$$\sin\theta \frac{\partial}{\partial\theta}\left(\sin\theta \frac{\partial\psi}{\partial\theta}\right) + \frac{\partial^2\psi}{\partial\varphi^2} + \frac{2mr^2 \sin^2\theta}{\hbar^2} K\psi = 0. \tag{5.2.12}$$

It suggests itself to try the same method for the solution of this equation which we used in the case of the two-dimensional harmonic oscillator, namely to set

$$\psi(\varphi, \theta) = \psi'(\varphi)\, \psi''(\theta). \tag{5.2.13}$$

If this is substituted in (5.2.12) the wave equation takes the form

$$\frac{1}{\psi'}\frac{d^2\psi'}{d\varphi^2}+\frac{\sin\theta}{\psi''}\frac{d}{d\theta}\left(\sin\theta\frac{d\psi''}{d\theta}\right)+\frac{2mr^2\sin^2\theta}{\hbar^2}K=0. \qquad (5.2.14)$$

Here the first term depends on φ only and the two others depend on θ only. Hence if the equation (5.2.14) is to be satisfied for all values of φ and θ the two ordinary differential equations

$$\frac{1}{\psi'}\frac{d^2\psi'}{d\varphi^2}+u^2=0 \qquad (5.2.15)$$

and

$$\frac{\sin\theta}{\psi''}\frac{d}{d\theta}\left(\sin\theta\frac{d\psi''}{d\theta}\right)+\frac{2mr^2\sin^2\theta}{\hbar^2}K-u^2=0 \qquad (5.2.16)$$

with an arbitrary constant u must be valid separately.

Equation (5.2.15) is identical with (5.2.4) in combination with (5.2.6), and hence the eigenvalues of u are *integers*. For what is to follow it is more convenient here to select, instead of the set (5.2.10), another set of *real* orthogonal functions to represent the eigenfunctions of (5.2.15), namely the functions

$$\left.\begin{aligned}\psi'_u&=\frac{1}{\sqrt{\pi}}\cos u\varphi\\ \text{and}\qquad \psi'_u&=\frac{1}{\sqrt{\pi}}\sin u\varphi\end{aligned}\right\}\quad (u=0,1,2,\ldots), \qquad (5.2.17)$$

the first of which are even in φ and the second odd in φ.

In order to solve (5.2.16) we introduce a new independent variable $\zeta = \cos\theta$, as a result of which the equation now becomes

$$(1-\zeta^2)\frac{d}{d\zeta}\left[(1-\zeta^2)\frac{d^2\psi''}{d\zeta^2}\right]+[\lambda(1-\zeta^2)-u^2]\psi''=0, \qquad (5.2.18)$$

where we have put

$$\lambda=\frac{2mr^2K}{\hbar^2}. \qquad (5.2.19)$$

If ψ'' is to be a regular function of θ in the interval $-1\leqslant\zeta\leqslant+1$ it must, near the ends of this interval, evidently be of the form

$$\psi''=(1-\zeta^2)^\alpha,\quad \alpha>0\quad(\zeta\to\pm1).$$

On substituting this into (5.2.18) one obtains in the limit $|\zeta|=1: \alpha=u/2$. Thus a function of the type

$$\psi''(\zeta)=(1-\zeta^2)^{u/2}f(\zeta), \qquad (5.2.20)$$

where $f(\zeta)$ is a polynomial of finite order s in ζ of the form (5.1.12) (with ξ replaced by ζ) will conform to the required conditions of regularity and uniqueness over the whole range of ζ.

If now (5.2.20) is to be a solution of the differential equation (5.2.18) $f(\zeta)$ has to satisfy the equation

$$(1 - \zeta^2)\frac{d^2 f}{d\zeta^2} - 2(u+1)\zeta\frac{df}{d\zeta} + (\lambda - u - u^2)f = 0. \quad (5.2.21)$$

Following the same procedure as in 5.1 and making use of the relations (5.1.12) and (5.1.16) one obtains the recurrence equations

$$(k+1)(k+2)A_{k+2} = [k(k-1) + 2(u+1)k - (\lambda - u - u^2)]A_k \quad (5.2.22)$$

for the coefficients of the polynomials f.

Let us first put $k = s$. As f is by definition of order s it follows that $A_{k+2} = 0$ but $A_k \gtrless 0$, and hence the expression in square brackets in (5.2.22) must vanish; the resulting equation

$$\lambda = s(s-1) + 2s(u+1) + u(u+1) = (s+u)(s+u+1), \quad (5.2.23)$$

for λ, together with the condition that u should be an integer, constitutes the quantization rule for the two-dimensional rotator. As u and s are both integers we can introduce a new quantum number $l = s + u$, and use l and u (instead of s and u) for characterizing the various quantum states of the system; this implies that $u \leqslant l$.

From (5.2.19) and (5.2.23) we obtain now for the energy levels of the two-dimensional rotator the relation

$$K_l = \frac{\hbar^2}{2mr^2}\, l(l+1) \quad (l = 0, 1, 2, \ldots), \quad (5.2.24)$$

which shows that they depend on the quantum number l only (and not on u). This relation is similar to the corresponding relation (5.2.7) for the one-dimensional rotator, except for the fact that the square of the quantum number u is now replaced by the product $l(l+1)$. By introducting the magnitude P of the angular momentum of the particle about the origin, which is related to K by $K = P^2/2mr^2$, we can transform the quantum condition (5.2.24) for the kinetic energy into one for the angular momentum, namely

$$P_l^2 = \hbar^2 l(l+1) \quad (l = 0, 1, 2, \ldots), \quad (5.2.25)$$

which is analogous to (5.2.8), but where again u^2 is replaced by $l(l+1)$. Because the quantum number l determines the total angular momentum of the system it is appropriately called the "*angular momentum quantum number*".

The eigenfunctions of the differential equation (5.2.18) were well known to mathematicians long before the advent of quantum theory. They are called "*Legendre functions*" or "*Legendre polynomials*" for $u = 0$ and "*associated Legendre functions*" for $u > 0$, and are usually denoted by $P_l(\xi)$ and $P_l^u(\xi)$ respectively. Their explicit expressions can easily be obtained from their defining equations (5.2.20) with the help of the recurrence relations (5.2.22) for the coefficients of the polynomials $f(\zeta)$, apart from an arbitrary factor.

The complete eigenfunctions of the wave equation (5.2.12) are now, from (5.2.13) and (5.2.17),

$$\psi_{l,u}(\varphi, \theta) = \text{const} \begin{Bmatrix} \cos u\varphi \\ \sin u\varphi \end{Bmatrix} P_l^u(\cos\theta), \quad (l, u = 0, 1, 2, 3, \ldots; u \leqslant l). \tag{5.2.26}$$

They are known as "*surface harmonics*" and are widely used for the expansion of arbitrary functions of φ and θ on the surface of a sphere by a procedure similar to the one referred to at the end of section 4.2.

It follows from (5.2.26) that to each energy level K_l belong $2l + 1$ different eigenfunctions $\psi_{l,u}$ as u runs from 0 to l and to each value of $u \gtrless 0$ belong two different eigenfunctions, a symmetric and an antisymmetric one.

The loci of the points on the surface of the sphere where $\psi_{l,u} = 0$ are the "nodal lines" referred to in section 4.2, and it can be shown that the number of nodal lines is always equal to l.

Table 1 contains the Legendre functions for $l = 0, 1, 2$, the corresponding surface harmonics according to (5.2.26) (both arbitrarily normalized), and the equations of the nodal lines; the latter are also shown in Fig. 8.

It now remains to calculate the coordinate matrix for the two-dimensional rotator. The subscripts i, k of the matrix elements have here to be replaced by two *pairs* of quantum numbers, say u, l and u', l'. From the definition (4.3.15) we obtain, using (5.2.1),

$$\left.\begin{aligned} x_{ul,u'l'} &= \text{const} \int\!\!\int \sin\theta\cos\varphi\,\psi_{lu}(\varphi,\theta)\,\psi_{l'u'}(\varphi,\theta)\,da, \\ y_{ul,u'l'} &= \text{const} \int\!\!\int \sin\theta\sin\varphi\,\psi_{lu}(\varphi,\theta)\,\psi_{l'u'}(\varphi,\theta)\,da, \\ z_{ul,u'l'} &= \text{const} \int\!\!\int \cos\theta\,\psi_{lu}(\varphi,\theta)\,\psi_{l'u'}(\varphi,\theta)\,da, \end{aligned}\right\} \tag{5.2.27}$$

where dV is now replaced by the surface element da on the sphere of radius r, and the integration is to be extended over the whole surface, that is over φ from 0 to 2π, and over θ from 0 to π.

Substitution of the expressions (5.2.26) for the surface harmonics into (5.2.27) shows immediately that the integrals over φ in the first two of these relations must vanish except when $u - u' = \pm 1$ and the corresponding integral in the third relation is zero except when $u = u'$. It can further be proved by exactly the same method which we used before in the case of the linear harmonic oscillator in section 5.1, that because of the orthogonality of the Legendre functions P_l^u all the integrals over θ in (5.2.27) vanish except

TABLE 1

Legendre functions	Surface harmonics	Nodal lines
$P_0^0 = 1$	1	none
$P_1^0 = \zeta$	$\cos\theta$	$\theta = \pi/2$
$P_1^1 = (1-\zeta^2)^{1/2}$	$\sin\theta \begin{Bmatrix}\cos\varphi \\ \sin\varphi\end{Bmatrix}$	$\varphi = \pi/2$ or $\varphi = 0$
$P_2^0 = 1 - 3\zeta^2$	$1 - 3\cos^2\theta$	$\cos\theta = \pm 1/\sqrt{3}$
$P_2^1 = \zeta(1-\zeta^2)^{1/2}$	$\sin 2\theta \begin{Bmatrix}\cos\varphi \\ \sin\varphi\end{Bmatrix}$	$\varphi = \pi/2$ or $\varphi = 0,\ \theta = \pi/2$
$P_2^2 = 1 - \zeta^2$	$\sin^2\theta \begin{Bmatrix}\cos 2\varphi \\ \sin 2\varphi\end{Bmatrix}$	$\varphi = \pm\pi/4$ or $\varphi = \pm\pi/2$

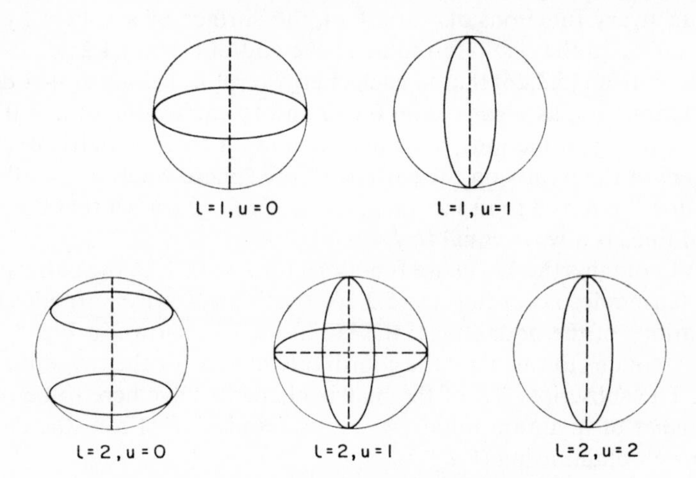

FIG. 8. Nodal lines of the two-dimensional rigid rotator.

when $l - l' = \pm 1$. It follows that all transitions between two quantum states of the two-dimensional rotator, characterized by the pairs of quantum numbers u, l and u', l' respectively, are "forbidden" where u and l would change by more than one unit.

5.3. Particle moving under the action of a central force

A problem of particular importance for atomic physics is presented by the quantum mechanical treatment of a system consisting of a single particle in a field of force in which the potential energy U is a function of the distance r of the particle from a fixed centre only (which we may choose for the origin of our coordinate system). This means that the particle is under the action of a force directed towards this centre with a magnitude depending on r only. Such a force is usually called a "*central force*". The potential energy function (5.1.1) for the harmonic oscillator is a special type of a function of this type.

As in section 5.2, we shall use polar coordinates so that the wave equation has the form (5.2.3) where ψ is now dependent on all three variables φ, θ, and r. Again, as before, we try to obtain solutions of this equation of the form

$$\psi(r, \theta, \varphi) = \chi(r)\,\psi'(\varphi)\,\psi''(\theta). \tag{5.3.1}$$

Substitution of (5.3.1) into (5.2.3) results in the differential equation

$$\frac{1}{\chi r^2}\frac{d}{dr}\left(r^2\frac{d\chi}{dr}\right) + \frac{1}{r^2\sin^2\theta}\left\{\frac{1}{\psi'}\frac{d^2\psi'}{d\varphi^2} + \frac{\sin\theta}{\psi''}\frac{d}{d\theta}\left(\sin\theta\frac{d\psi''}{d\theta}\right) + \frac{2mr^2\sin^2\theta}{\hbar^2}K\right\} + \frac{2m}{\hbar^2}[E - U(r) - K] = 0. \tag{5.3.2}$$

Here the expression in curly brackets is identical with the left-hand side of equation (5.2.14) and equation (5.3.2) will therefore be identically satisfied if (5.2.14) holds and simultaneously the function $\chi(r)$ is a solution of the ordinary differential equation

$$\frac{1}{r^2}\frac{d}{dr}\left(r^2\frac{d\chi}{dr}\right) + \frac{2m}{\hbar^2}[E - U(r) - K]\,\chi = 0. \tag{5.3.3}$$

$\chi(r)$ is called a "*radial wave function*" and the equation (5.3.3) a "*radial wave equation*".

It follows that the eigenfunctions of our present system are the products of the eigenfunctions (5.2.26) of the two-dimensional rotator and the eigenfunctions of the radial wave equation (5.3.3) in which K must be one of the eigenvalues (5.2.24). We thus have the task of finding the eigenvalues and eigenfunctions of the equation

$$\frac{1}{r^2}\frac{d}{dr}\left(r^2\frac{d\chi}{dr}\right) + \left\{\frac{2m}{\hbar^2}[E - U(r)] - \frac{l(l+1)}{r^2}\right\}\chi = 0 \quad (l = 0, 1, 2, \ldots). \tag{5.3.4}$$

It is mathematically more convenient to introduce a new dependent variable

$$g(r) = r\chi(r) \tag{5.3.5}$$

instead of χ. It is easily verified that the function $g(r)$ must satisfy the differential equation

$$\frac{d^2g}{dr^2} + \left\{\frac{2m}{\hbar^2}[E - U(r)] - \frac{l(l+1)}{r^2}\right\}g = 0. \tag{5.3.6}$$

A certain amount of information concerning the general shape of the function $g(r)$ can be obtained by direct inspection of (5.3.6). In the region to which the particle would "classically" be confined the kinetic energy $E - U$ must be positive. Thus for not too small values of r the coefficient of g will be positive from which it follows, by the same reasoning as used in section 4.2, that $g(r)$ must have the form of a wavy line with maxima and minima and a certain number k of nodes in between. For larger values of r, outside the classical region, the coefficient of g will be negative and, as g must vanish at infinity (according to the general requirements concerning the wave function formulated in section 4.2) $g(r)$ must approach zero asymptotically outside the classical region. For small values of r the coefficient of g becomes again negative and g itself becomes zero at $r = 0$. The general appearance of the function $g(r)$ will therefore be represented by a curve of the shape shown in Fig. 9.

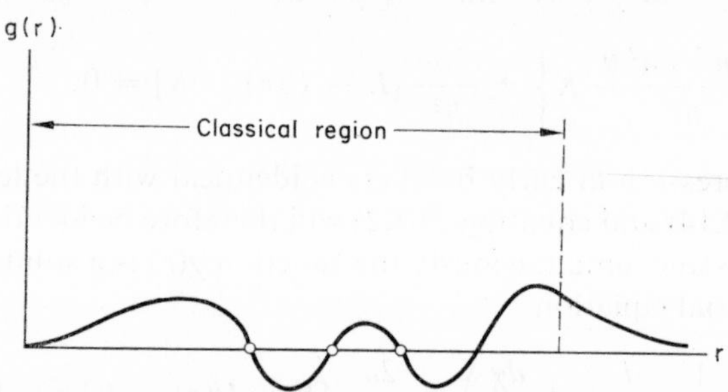

FIG. 9. Characteristic shape of the radial wave function $\chi(r)$ of a particle. Shown is the function $g(r) = r\chi(r)$.

The eigenfunctions of our present problem are characterized by the *three* quantum numbers k, l, u corresponding to the 3 degrees of freedom of the system. It is, however, customary to use instead of k another quantum number $n = l + k + 1$, called "*principal quantum number*" for reasons which will emerge later on in this section. The eigenfunctions will therefore be denoted by $\psi_{n,l,u}$ where, as before, $l, u = 0, 1, 2, 3, \ldots$ $(u \leqslant l)$ and $n = 1, 2, 3, \ldots$ $(l \leqslant n - 1)$.

To each principal quantum number n belong n different quantum numbers l, and to each l belong $2l + 1$ different quantum states. Thus to each value of n belong altogether

$$\sum_{l=0}^{n-1} (2l + 1) = n^2$$

different quantum states. The various energy levels depend, however, on the quantum numbers n and l only. It is customary to denote them by symbols consisting of the figure for n followed by a letter denoting l, the letters $s, p, d, f, \ldots,$ standing for $l = 0, 1, 2, 3, \ldots$ respectively. Thus, for example, the symbol $2p$ denotes the quantum level belonging to the quantum numbers $n = 2$ and $l = 1$.

The various eigenfunctions, defined in 3-dimensional space, are now characterized by *nodal surfaces* with equations

$$\psi_{n,l,u}(r, \varphi, \theta) = 0.$$

It follows from (5.3.1) that one set of nodal surfaces is given by the equations $\chi_n(r) = 0$ and another set by the equations $\psi_{l,u}(\varphi, \theta) = 0$. The former are concentric spheres, k in number, the latter are partly cones with their apexes at the origin and partly planes through the origin, corresponding to the nodal lines of the two-dimensional rotator. The total number of nodal surfaces is therefore equal to $k + l = n - 1$.

We shall now investigate more closely one particularly important case where the potential energy function has the form

$$U(r) = -C/r \quad (C > 0). \tag{5.3.7}$$

This represents an attractive force on the particle which is inversely proportional to the square of the distance from the attracting centre and is characteristic of a *gravitational* or an *electrostatic* interaction. We shall refer to this problem as the "*Kepler problem*" because it is evidently the quantum mechanical analogue of the classical problem of determining the orbits of the planets about the sun.

Substituting for $U(r)$ from (5.3.7) into the radial wave equation (5.3.6) we obtain

$$\frac{d^2g}{dr^2} + \left\{\frac{2m}{\hbar^2}\left(E - \frac{C}{r}\right) - \frac{l(l+1)}{r^2}\right\} g = 0. \tag{5.3.8}$$

This equation can be simplified by the use of a new independent dimensionless variable ϱ instead of r, defined by

$$\varrho^2 = -\frac{2mE}{\hbar^2} r^2, \tag{5.3.9}$$

where the negative sign is used on account of the fact that E is negative (see the remark at the end of the first paragraph in section 4.2). The wave equation thus takes the form

$$\frac{d^2g}{d\varrho^2} + \left\{\frac{\alpha}{\varrho} - 1 - \frac{l(l+1)}{\varrho^2}\right\} g = 0, \tag{5.3.10}$$

where we have put

$$\alpha^2 = -\frac{2mC^2}{E\hbar^2}. \tag{5.3.11}$$

As in the case of the linear harmonic oscillator, we can enforce the necessary conditions the wave function $g(\varrho)$ has to fulfil in the whole range of the variable ϱ from 0 to ∞ by setting

$$g = e^{-\beta\varrho} f(\varrho) \quad (\beta > 0), \tag{5.3.12}$$

where $f(\varrho)$ is again a polynomial of finite order n whose lowest term is of order s in ϱ. Substituting (5.3.12) into (5.3.10) we get the differential equation

$$\frac{d^2f}{d\varrho^2} - 2\beta\frac{df}{d\varrho} + \left[\beta^2 - 1 + \frac{\alpha}{\varrho} - \frac{l(l+1)}{\varrho^2}\right]f = 0 \qquad (5.3.13)$$

for $f(\varrho)$.

For large values of ϱ this equation is only satisfied if $\beta = 1$. In the limit of small ϱ the polynomial f reduces to its first term, proportional to ϱ^s, and all other members on the left-hand side of equation (5.3.13) become negligibly small compared with the first and the last. Hence it follows that

$$s(s-1) = l(l+1).$$

The only positive solution of this equation is $s = l + 1$, and (5.3.12) thus takes the form

$$g(\varrho) = e^{-\varrho}\,\varrho^{l+1}\sum_{i=0}^{k} A_i\varrho^i. \qquad (5.3.14)$$

We now follow the same procedure as on previous occasions, namely substituting the expression (5.3.14) for g into (5.3.10) and putting the coefficients of the various powers of ϱ equal to zero. This results in the following recurrence formula for the coefficients A_i:

$$[(i+l+1)(i+l) - l(l+1)]\,A_i = [2(i+l) - \alpha]\,A_{i-1}. \qquad (5.3.15)$$

Put first $i = k + 1$. In this case $A_i = 0$ and $A_{i-1} \gtrless 0$ as the polynomial in (5.2.14) is supposed to be of order k, and hence one must have

$$n = k + 1 + l = \alpha/2. \qquad (5.3.16)$$

(5.3.16) combined with (5.3.11) constitutes the quantization rule for the energy eigenvalues E_n, namely

$$E_n = -\frac{mC^2}{2n^2\hbar^2} \quad (n = 1, 2, 3, \ldots). \qquad (5.3.17)$$

One sees that they depend on the principal quantum number n only which is the reason for the nomenclature chosen.

This result can immediately be applied to the "hydrogen-like atom", consisting of an electron of charge $-e$ moving about a nucleus of charge Ze, a system which was first treated on quantum theoretical grounds by Bohr (see section 2.2). In this case we have $C = Ze^2$, and the relation (5.3.17) becomes identical with the formula (2.2.13) derived by Bohr. With the help of the expression

$$a = \frac{\hbar^2}{me^2Z} = \frac{0{\cdot}529}{Z}\,\text{Å} \qquad (5.3.18)$$

for the "Bohr radius" of the lowest quantum orbit, which follows from (2.2.12), we may further put the relationship between the variables ϱ and r for the nth eigenstate into the simple form

$$\varrho = \frac{r}{na}. \tag{5.3.19}$$

We now turn to the determination of the eigenfunctions for the Kepler problem. Using the expression (5.3.16) for α we first transform the recurrence equations (5.3.15) for the coefficients of the polynomial in (5.3.14) into

$$i\,(2l + i + 1)\,A_i = 2\,(i - k - 1)\,A_{i-1}. \tag{5.3.20}$$

These polynomials are known under the name "*associated Laguerre polynomials*"; we shall denote them by L_k^l, the subscript indicating their order. From (5.3.20) one deduces easily the following expressions for the first three Laguerre polynomials, putting arbitrarily $A_0 = 1$:

$$\left.\begin{aligned} L_1^l &= 1 - \frac{1}{l+1}\varrho, \quad L_2^l = 1 - \frac{2}{l+1}\varrho + \frac{2}{(l+1)(2l+3)}\varrho^2, \\ L_5^l &= 1 - \frac{3}{l+1}\varrho + \frac{6}{(l+1)(2l+3)}\varrho^2 - \frac{2}{(l+1)(l+2)(2l+3)}\varrho^3. \end{aligned}\right\} \tag{5.3.21}$$

For the radial eigenfunctions χ_{nl} one has now, from (5.3.5), (5.3.14), and (5.3.19),

$$\chi_{nl} = \text{const}\, e^{-r/na}\, r^l L_{n-1-l}^l\left(\frac{r}{na}\right) \qquad (l \leqslant n - 1). \tag{5.3.22}$$

In particular for the lowest quantum levels $n = 1, 2, 3$ one obtains from this formula the arbitrarily normalized expressions

$$\left.\begin{aligned} \chi_1^0 &= e^{-r/a}, \quad \chi_2^0 = e^{-r/2a}\left(1 - \frac{r}{2a}\right), \quad \chi_2^1 = e^{-r/2a}\, r, \\ \chi_3^0 &= e^{-r/3a}\left[1 - \frac{2}{3}\frac{r}{a} + \frac{2}{27}\left(\frac{r}{a}\right)^2\right], \\ \chi_3^1 &= e^{-r/3a}\, r\left(1 - \frac{1}{6}\frac{r}{a}\right), \\ \chi_3^2 &= e^{-r/3a}\, r^2. \end{aligned}\right\} \tag{5.3.23}$$

Diagrams of these functions are shown in Fig. 10, from which it is seen that, as previously stated in this section, the number of nodes of the radial wave function is equal to $k = n - l - 1$.

The complete eigenfunctions $\psi_{n,l,u}(r, \varphi, \theta)$ for the Kepler problem are, according to (5.3.1), the products of the eigenfunctions (5.2.26) of the two-dimensional rotator and the radial eigenfunctions (5.3.22) with an arbitrary factor which is determined by the normalizing condition

$$\int_0^\infty \int_0^{2\pi} \int_0^\pi \psi_{n,l,u}^2 (r, \theta, \varphi)\, r^2 \sin\theta \, dr\, d\varphi\, d\theta = 1. \tag{5.3.24}$$

In particular for the "ground state" of the system ($n = 1, l = 0, u = 0$) one has

$$\psi_{1,0,0} = \frac{1}{\sqrt{\pi a^3}} e^{-r/a}. \tag{5.3.25}$$

It is well known that the classical solution of the Kepler problem, obtained for the first time by Newton, consists in elliptical orbits of the particle about the centre of attraction so that it remains within a fixed range of distances from the centre. The above obtained quantum mechanical solution consists only of a finite probability density ψ^2 for the particle to be found anywhere in space, but also in regions into which it could not penetrate classically with the energy available.

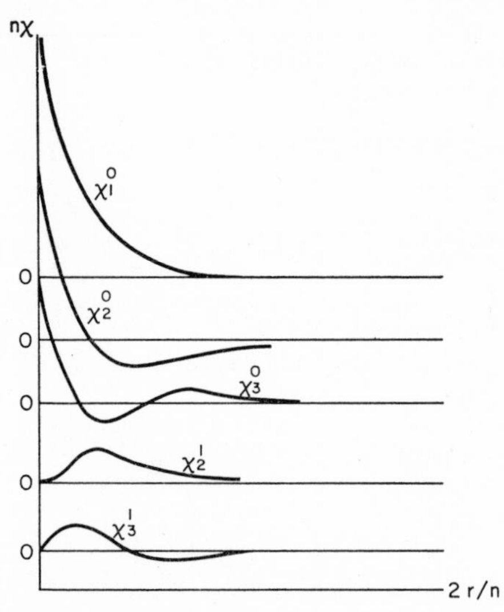

FIG. 10. Radial wave function for an electron in a hydrogen-like atom.

We may also interpret our results in a different manner, as previously pointed out in section 3.2, by visualizing the function ψ^2 for an electron in a hydrogen-like atom as representing the density of an "electronic cloud" surrounding the nucleus with a continuous density proportional to ψ^2; we shall discuss this aspect of the solution more closely in section 6.2.

We shall not endeavour to calculate the coordinate matrix for the Kepler problem in this context. May it suffice to state that none of the integrals over r involved vanishes, so that all transitions between states with different principal quantum numbers n are allowed in principle. The only transition rules valid for the Kepler problem are thus the ones holding for the two-dimensional rotator which we derived in section 5.2, namely that the quantum numbers l and u must not change by more than one unit. Thus an "s-state" of a hydrogen-like atom can only change into a "p-state", a "p-state" only into an "s-state" or a "d-state", etc. This has a very important bearing on the structure of the emission and absorption spectra of such atoms, a problem to which we shall come back later in section 6.2.

Problems and exercises

P.5.1. Extend the calculation of the coefficients (5.1.18) of the Hermitian polynomials and write them in such a way that a general formula for A_k can be guessed by induction.

P.5.2. In extension to (5.1.22) calculate the first five normalized eigenfunctions $\psi_i(x)$ of the linear harmonic oscillator, and hence the probability density for finding the particle at a certain point x in the quantum state i.

P.5.3. Write down the explicit expressions for the eigenfunctions $\psi_0, \psi_1, \psi_2, \psi_3$ of the two-dimensional harmonic oscillator problem, using the equations (5.1.10), (5.1.19), and (5.1.34). Show that they imply the existence of "nodal lines" in the xy-plane where $\psi(x, y) = 0$, and derive the equations for these nodal lines. Draw qualitative sketches of these lines.

P.5.4. The expression for the Laplace operator in general orthogonal curvilinear coordinates q_1, q_2, q_3 is

$$\nabla^2 = \frac{1}{g_1 g_2 g_3}\left\{\frac{\partial}{\partial q_1}\left(\frac{g_2 g_3}{g_1}\frac{\partial}{\partial q_1}\right) + \frac{\partial}{\partial q_2}\left(\frac{g_3 g_1}{g_2}\frac{\partial}{\partial q_2}\right) + \frac{\partial}{\partial q_3}\left(\frac{g_1 g_2}{g_3}\frac{\partial}{\partial q_3}\right)\right\},$$

where g_1^2, g_2^2, g_3^2 are the coefficients in the expression for the line element

$$ds^2 = g_1^2\, dq_1^2 + g_2^2\, dq_2^2 + g_3^2\, dq_3^2.$$

Use the above formula for deriving the expression (5.2.2) for ∇^2 in the spherical polar coordinates r, θ, φ.

P.5.5. Describe in detail how the data in Table 1 on p. 70 are obtained. Extend the calculations to the quantum number $l = 3$, and draw diagrams of the corresponding nodal lines of the surface harmonics.

P.5.6. Normalize the surface harmonics $\psi_{l,u}(\theta, \varphi)$ up to $l = 3$ and $u = 3$ so that the integrals of $|\psi|^2$ over the whole surface of the sphere are equal to unity. Give the explicit formulae for the probability density of the particle for a given position θ, φ on the sphere.

P.5.7. Derive the general expression for the coefficient A_i of ϱ^i in the Laguere polynomials L_k^l from the recurrence formula (5.3.20), putting $A_0 = 1$ as in (5.3.21).

P.5.8. Determine the distances r from the centre at which the radial wave functions χ_n^l of the "Kepler problem" are zero, and hence the probability density for finding the particle at any of these distances is also zero. Also determine at what distance this probability density is a maximum (a) generally for $n - l = 1$ (from 5.3.22), commenting on the obtained results, and (b) for χ_2^0 and χ_3^1.

P.5.9. Give a table of the complete (non-normalized) eigenfunctions $\psi_{n,l,u}$ of the Kepler problem for the first three principal quantum numbers $n = 1, 2, 3$.

P.5.10. Obtain from (5.3.6) the radial wave equation for a "spherically symmetric" harmonic oscillator whose potential energy has the form (5.1.1). Transform this equation in analogy to the procedure adopted for the Kepler problem, and show that its solution has the form

$$g = e^{-(\lambda/2)\varrho^2}\varrho^{l+1}\sum_{i=0}^{n} A_i\varrho^i,$$

where

$$\varrho^2 = \frac{2nE}{\hbar^2}r^2 \quad \text{and} \quad \lambda^2 = \frac{\alpha\hbar^2}{4mE^2}.$$

Derive the quantization rule for this problem in analogy to the treatment of the linear harmonic oscillator, and compare the result obtained with that for the three-dimensional harmonic oscillator (5.1.38).

CHAPTER 6

Non-stationary Phenomena

So FAR we have only dealt with the principles of quantum mechanics of conservative systems whose Hamiltonian function does not explicitly depend on time. We have seen that such systems can only exist in certain quantum states with energies which are the eigenvalues of the wave equation (3.1.6) or, more generally, the equation (3.1.19), and that the probabilities of the various configurations of the system in these states are determined by the time-independent amplitude function ψ of the coordinates. These phenomena are therefore called "*stationary*". In this chapter we shall develop the basic quantum-mechanical theory of "*non-stationary*" phenomena occurring in systems which are under the action of a time-dependent potential field whose Hamiltonian therefore depends on time explicitly. These systems are consequently non-conservative and will not be confined to definite quantum states.

6.1. The time-dependent wave equation

It was shown in section 3.1 how the general wave equation for a conservative mechanical system may be derived from its classical Hamilton–Jacobi equation (1.3.11) by the replacement of the quantities $\frac{\partial W}{\partial q_k}$ by the operators $\frac{\hbar}{i}\frac{\partial}{\partial q_k}$; this transforms it into an operator equation which is made to operate on the amplitude function $\psi(q_1, q_2, \ldots, q_n)$.

This suggests using a similar procedure for deriving the wave equation for non-stationary phenomena by starting from the Hamilton–Jacobi equation (1.3.15) for non-conservative systems and replacing the quantities $\frac{\partial \Omega}{\partial q_k}$ and $\frac{\partial \Omega}{\partial t}$ by the operators $\frac{\hbar}{i}\frac{\partial}{\partial q_k}$ and $\frac{\hbar}{i}\frac{\partial}{\partial t}$ respectively. If the ensuing operator equation is made to operate on the time-dependent wave function $\varphi(q_1, \ldots, q_n; t)$ the time-dependent wave equation

$$\mathscr{H}\left(q_1, \ldots, q_N;\ \frac{\hbar}{i}\frac{\partial}{\partial q_1}, \ldots, \frac{\hbar}{i}\frac{\partial}{\partial q_N};\ t\right)\varphi = i\hbar\frac{\partial \varphi}{\partial t} \qquad (6.1.1)$$

is obtained. Alternatively (6.1.1) may also be derived from the wave equation (3.1.19) by replacing the energy E by an "energy operator"

$$\mathscr{E} = i\hbar \frac{\partial}{\partial t} \tag{6.1.2}$$

and using φ instead of ψ.

It can be readily verified that a wave function of the form

$$\varphi(q_1, \ldots, q_N; t) = e^{-\frac{iE}{\hbar}t} \psi(q_1, \ldots, q_N) \tag{6.1.3}$$

always satisfies the wave equation (6.1.1) provided that ψ satisfies the time-independent amplitude equation (3.1.19). This is indeed a necessary condition for the wave equation (6.1.1) to be correct, for (6.1.3) is the expression for the wave function of a conservative system with constant energy content E, whose time dependence has the character of a harmonic oscillation with the De Broglie frequency (3.1.8)

$$\nu = \frac{E}{2\pi\hbar} = \frac{E}{h}.$$

In the special case of a system consisting of a single particle of mass m (6.1.1) reduces to the "*time-dependent Schrödinger equation*"

$$\nabla^2\varphi - \frac{2m}{\hbar^2} U\varphi = \frac{2m}{i\hbar} \frac{\partial\varphi}{\partial t} \tag{6.1.4}$$

for the wave function $\varphi(x, y, z; t)$. This again reduces to the time-independent Schrödinger equation (3.1.6) if φ has the form

$$\varphi(x, y, z; t) = e^{-\frac{iE}{\hbar}t} \psi(x, y, z) \tag{6.1.5}$$

which, apart from the insignificant change of sign in the exponent, is identical with (3.1.9).

One might have expected the differential equation for φ to have the form of the classical wave equation (1.2.2) which is a partial differential equation of the second order in the coordinates *and* the time. Instead, the equation (6.1.4) is only of the second order in x, y, z but of the *first* order in t. (It is therefore not really correct to call it a "wave equation".) We shall discuss the implications of this characteristic feature in section 6.3. The boundary problem of the solution of the quantum mechanical wave equation is, however, essentially the same as that for the classical wave equation which was formulated in section 1.2. If the values of φ are given at the boundary of a finite region in ordinary or in configuration space for all time and its values at $t > 0$ in the interior of the domain the solution of the equation (6.1.1) or (6.1.4) yields φ as a unique function of the coordinates and the time in the interior of the domain.

It follows from (6.1.3) and (6.1.5) that for any of the stationary states of a conservative system $\varphi\varphi^* = \psi\psi^*$. In keeping with the statistical interpretation of the quantum mechanical wave function (section 3.2) $|\varphi|^2$ therefore represents the probability density for the various configuration of the system in such a state. It suggests itself to extend this interpretation to the general case of a non-stationary system by interpreting $|\varphi|^2$ as the probability density for its configuration at time t. A necessary condition for this interpretation to be valid is that it must be possible to normalize φ in such a way that the normalization remains in force in the course of time. It will be shown in section 7.2 that this is, indeed, the case, but will at the moment be proved here only for a system consisting of a single particle whose wave function satisfies (6.1.4).

The conjugate complex of the wave equation (6.1.4) which, of course, must be equally valid, is

$$\nabla^2\varphi^* - \frac{2m}{\hbar^2}\,U\varphi^* = -\frac{2m}{i\hbar}\,\frac{\partial\varphi^*}{\partial t}. \tag{6.1.6}$$

Multiplying this equation by φ on both sides, multiplying (6.1.4) by φ^* on both sides and subtracting, one obtains

$$\frac{\partial\,(\varphi\varphi^*)}{\partial t} = \frac{\hbar}{2mi}\,(\varphi\,\nabla^2\varphi^* - \varphi^*\,\nabla^2\varphi). \tag{6.1.7}$$

If this equation is integrated over the interior of a closed surface S, the sequence of integration and time derivation interchanged on the left-hand side, and the right-hand member is transformed into a surface integral over the surface S by means of Green's theorem, one obtains, finally,

$$\frac{\partial}{\partial t}\iiint \varphi\varphi^*\,dV = \frac{\hbar}{2mi}\iint\left(\varphi\,\frac{\partial\varphi^*}{\partial n} - \varphi^*\,\frac{\partial\varphi}{\partial n}\right)da. \tag{6.1.8}$$

Now it was pointed out in section 4.2 that the wave function has to vanish at infinity to a sufficiently high order if there are no explicit boundary conditions to be fulfilled. Hence the surface integral in (6.1.8) must tend to zero if S is extended to infinity and thus we have

$$\frac{d}{dt}\iiint \varphi\varphi^*\,dV = 0, \tag{6.1.9}$$

where the integral is extended over all space. This means that if at any time the normalizing condition

$$\iiint \varphi\varphi^*\,dV = 1 \tag{6.1.10}$$

is imposed (which one can always do owing to the linearity of the wave equation (6.1.4)), the condition will automatically hold for all subsequent times.

On the basis of this statistical interpretation of the wave function φ we can now calculate *time-dependent expectation values* of any function f of the coordinates by means of

$$\bar{f}(t) = \iiint \varphi^* f \varphi \, dV \tag{6.1.11}$$

and further, in generalization of the remark at the end of section 4.3, we can define the position vector $\mathbf{R}_{kj}$ with components X_{kj}, Y_{kj}, Z_{kj} for the centre of the continuous mass distribution in the intermediate state between the quantum states k and j by

$$\mathbf{R}_{kj} = \tfrac{1}{2}\left[\iiint \varphi_k^* \mathbf{r} \varphi_j \, dV + \iiint \varphi_j^* \mathbf{r} \varphi_k \, dV\right]. \tag{6.1.12}$$

Making use of the relations (6.1.5) and (4.3.15) we can transform this into the simpler form

$$\mathbf{R}_{kj} = \tfrac{1}{2}\left[e^{i/\hbar\,(E_k - E_j)} \mathbf{r}_{kj} + e^{i/\hbar\,(E_j - E_k)} \mathbf{r}_{jk}\right], \tag{6.1.13}$$

where $\mathbf{r}_{kj}$ are the elements of the coordinate matrix. In view of the fact that this matrix is Hermitian we may further write (6.1.13) in component form as follows:

$$X_{kj} = |x_{kj}| \cos\left(2\pi\nu_{kj} + \gamma_{kj}\right), \tag{6.1.14}$$

with analogous expressions for Y_{kj} and Z_{kj}, where

$$\nu_{kj} = \frac{|E_k - E_j|}{2\pi\hbar} = \frac{|(E_k - E_j)|}{h} \tag{6.1.15}$$

and γ_{kj} are phase constants. This shows that the mass centre performs a harmonic oscillation with frequency ν_{kj} and amplitude components x_{kj}, y_{kj}, z_{kj} in the x-, y-, and z-directions respectively.

Let us now consider wave functions which are linear combinations of the wave functions φ_k belonging to the eigenvalues E_k of a stationary system:

$$\varphi(x, y, z; t) = \sum_k c_k \varphi_k = \sum_k c_k \, e^{-\frac{iE_k}{\hbar}t} \, \psi_k(x, y, z). \tag{6.1.16}$$

As the eigenfunctions ψ_k are orthogonal to each other the φ_k also form an orthonormal system of functions and hence, according to (4.2.9) the coefficients c_k of the expansion (6.1.16) are given by

$$c_k = \iiint \varphi \varphi_k^* \, dV. \tag{6.1.17}$$

From (6.1.10), (6.1.16), and (6.1.17) we obtain now

$$\iiint \varphi\varphi^* \, dV = \iiint \varphi \sum_k c_k^* \varphi_k^* \, dV = \sum_k c_k c_k^* = 1. \tag{6.1.18}$$

This suggests the following statistical interpretation of a wave function of the form (6.1.16). If $|\varphi_k|^2$ is a "probability density", then $|\varphi_k|^2 \, dV$ measures the probability for a particle in a conservative system in the kth quantum state to be found in the volume element dV at time t. Alternatively, we can imagine a multitude of very many identical systems, a so-called "*ensemble*" of such systems,† all in the quantum state k. $|\varphi_k|^2\, dV$ then measures the *relative number* of these systems for which the particle is within the volume element dV at time t. Similarly, we may assume the wave function (6.1.16) to represent an ensemble of identical systems which, however, are in *different* quantum states. The fact that the sum of the quantities $|c_k|^2$ is always equal to unity then leads one to assume that $|c_k|^2$ represents the relative number of those systems in the ensemble which are in the quantum state k, or the probability for finding a randomly selected member of the ensemble in the quantum state k, for the sum of these probabilities for all possible values of k must evidently be equal to 1.

Let us now assume that all the systems of which the ensemble consists are at $t = 0$ in the same quantum state k, but that they are subsequently subjected to external forces derivable from a time-dependent potential energy function $U(x, y, z; t)$ for a time t. The wave function at t can then, at least in principle, be calculated by solving the wave equation (6.1.4) under the initial condition $\varphi = \varphi_k$ for $t = 0$. According to the expansion theorem (4.2.8) φ can then always be expanded in a series of the form (6.1.16) with coefficients c_{kj} which will depend on time. In the language of the just stated statistical interpretation this means that after the cessation of the external disturbance the members of the ensemble will in general no longer be in the initial quantum state k but will be found in other states j with relative numbers $|c_{kj}|^2$. The quantities $|c_{kj}(t)|^2$ are therefore the "*transition probabilities*" for the transition of the system from the quantum state k into the state j under the influence of the external disturbance during the time interval t; the matrix c_{kj} is therefore called the "*transition probability matrix*".

Problems of this type are frequently met in atomic physics when one wishes to calculate the probabilities for the transition of the electronic system of an atom from one quantum state to another under the influence of an external electromagnetic field, in particular an electromagnetic wave. The corresponding probabilities are called "*induced transition probabilities*". Such transitions may also be caused by the variable electromagnetic field of the atomic electrons themselves, and these "spontaneous" transitions are governed by "*spontaneous transition probabilities*". It can be shown that they are proportional to t for short periods of time so that there exist constant "*transition probability rates*" w_{kj}, that is that there exist time independent probabilities $w_{kj}\, dt$ for the system to go over from a state k to a state j in time dt. We shall return to these problems in the following section 6.2 and in Part III.

† A more elaborate definition of the notion of ensembles will be found in Chapter 13.

6.2. Charge and current density

In section 3.2 it was shown how one may visualize the wave function of a particle in a stationary state as a "smeared out" continuous mass density distribution (3.2.5). Similarly, one may imagine the charge e of an electron to be smeared out in the form of a continuous distribution of *charge density* ϱ_e, according to the relation

$$\varrho_e(\mathbf{r}, t) = e\,|\varphi\,(\mathbf{r}, t)|^2 = e\varphi\varphi^*, \tag{6.2.1}$$

which, in virtue of (6.1.10), satisfies the necessary condition that the total charge in this "electronic cloud" must be equal to e for all time:

$$\iiint \varrho_e dV = e \iiint \varphi\varphi^* dV = e. \tag{6.2.2}$$

In particular in a stationary state with quantum number k where $\varphi\varphi^* = \psi\psi^*$ one has then

$$\varrho_e(\mathbf{r}) = e\psi_k\psi_k^*, \tag{6.2.3}$$

which is evidently time independent.

Schrödinger, in accepting the idea of continuous De Broglie waves, originally believed that the relation (6.2.1) actually meant that in a closed system, instead of a movement of discrete charged particles, there existed a continuous charge distribution. This immediately provided an ingenious explanation for the fact that atoms in their stationary states do not emit electromagnetic radiation. For a necessary prerequisite for this is that the electric field due to the charges must be *static*. We shall show later that the other necessary condition, that the magnetic field must also be static, is satisfied as well in a stationary state.

We now consider an ensemble of non-stationary electronic systems, described by a wave function of the type (6.1.16). If we again define the charge density distribution in these systems by means of equation (6.2.1) we obtain

$$\varrho_e(\mathbf{r}, t) = e \sum_{kj} c_k^* c_j\, e^{\frac{i(E_k - E_j)t}{\hbar}}\, \psi_k^* \psi_j. \tag{6.2.4}$$

The time dependence of ϱ_e has thus the form of a superposition of harmonic oscillations with frequencies (6.1.15) which are identical with the Bohr frequencies (2.2.11). The laws of classical electrodynamics then predict the *emission* of radiation by the systems in the form of a *line spectrum* with frequencies which are determined by the second of Bohr's postulates (section 2.2).

The task of evaluating the intensities of the spectral lines can be greatly simplified if one assumes the space charge of the electronic cloud to be concentrated in its electric "centre of gravity" which evidently coincides with the mass centre with position vector **R** (X, Y, Z). The electronic charge, together with an equal and opposite charge, fixed in the atomic nucleus, then constitutes an "*electric dipole*". Its *moment* $\mathbf{\Pi}$ is given by

$$\mathbf{\Pi} = e\mathbf{R} \tag{6.2.5}$$

with components

$$\Pi_x = eX, \quad \Pi_y = eY, \quad \Pi_z = eZ, \tag{6.2.6}$$

provided the nucleus has been chosen as origin of coordinates.

According to (6.1.14) the charge distribution belonging to the transition from a quantum state k to a quantum state j (or vice versa) is then equivalent to an oscillating electric dipole with x-component

$$\Pi_x = eX_{kj} = e\,|x_{kj}|\cos(2\pi\nu_{kj} + \gamma_{kj}) \tag{6.2.7}$$

and similar expression for the other two components.

Now it is a well-known result of Maxwell's theory that the total amount of electromagnetic energy emitted per unit time by an electric dipole oscillator with frequency ν and amplitude Π_0 is equal to

$$I = \frac{(2\pi\nu)^4}{3c^3}\,\Pi_0^2. \tag{6.2.8}$$

Hence, if we assume this formula to be applicable in spite of the fact that, according to the principles of quantum theory (as explained in section 2.2) the emission of radiation should actually occur discontinuously in the form of photons, we obtain from (6.2.7) and (6.2.8) the following expression for the intensity per atom of a line of frequency ν_{kj} in an emission spectrum:

$$I_{kj} = \frac{e^2}{3c^3}(2\pi\nu_{kj})^4([|x_{kj}|^2 + |y_{kj}|^2 + |z_{kj}|^2). \tag{6.2.9}$$

In particular a line will be missing if all the components of the corresponding element of the coordinate matrix are zero.

Formula (6.2.9) allows not only the intensities but also the state of polarization of the spectral lines to be predicted, for the intensities of the components of the radiation polarized in the x-, y-, and z-directions respectively will on the basis of the same argument, be in the ratios of the quantities $|x_{kj}|^2$, $|y_{kj}|^2$, $|z_{kj}|^2$.

In an analogous manner the phenomenon of *absorption* of electromagnetic waves by an atomic system can be treated by applying the classical electromagnetic relations which govern the transfer of energy from an electromagnetic wave to a harmonic dipole oscillator. It is well known that this transfer occurs with maximum intensity when the frequency of the oscillator is equal to that of the wave; this explains the fact that the *absorption spectra* of atoms are also line spectra and the frequencies of the absorption lines coincide with those of corresponding emission lines.

We have already stated that this treatment of the emission and absorption processes of atomic systems was actually incompatible with the basic quantum mechanical idea of discontinuous interaction between matter and radiation. As pointed out at the end of section 6.1, there exist instead certain time-independent "spontaneous transition probability rates" w_{kj}^s for the transition of an atom from the state k to a state j. In every such transition a photon of energy $h\nu_{kj}$ is emitted, and hence the intensity of the radiation, which is equal to the energy emitted per unit time, is equal to

$$I_{kj} = w_{kj}^s h\nu_{kj}. \tag{6.2.10}$$

The "spontaneous" transition probability rate for the "dipole radiation" of an atom, resulting in the emission of a photon polarized in the x-direction, can be calculated by the method indicated in section 6.1. The result of this calculation is

$$w_{kj}^s = \frac{16\pi^4 e^2}{3c^3 h} \nu_{kj}^3 \, |x_{kj}|^2. \tag{6.2.11a}$$

If this expression is substituted in (6.2.10) one obtains indeed the previously derived formula (6.2.9) with $y_{kj} = z_{kj} = 0$. This justifies, *a posteriori*, the application of the much simpler method adopted in this section.

A similar procedure can be adopted for calculating the "induced" transition probability rate w_{kj}^i under the influence of an electromagnetic radiation field with "spectral density function" $u(\nu)$, $u(\nu)\,\Delta\nu$ being the energy density of the field in the narrow spectral region between ν and $\nu + \Delta\nu$. The result of this calculation is

$$w_{kj}^i = \frac{2\pi^3 e^2}{3h^2} |x_{kj}|^2 \, u(\nu_{kj}). \tag{6.2.11b}$$

The relations (6.2.11) reveal yet another statistical feature of quantum mechanics. They show, namely, that the squares of the elements of the coordinate matrix are proportional to the transition probability rates between the various quantum states of a system. In particular it now becomes clear why the transition between two states is "forbidden" if the corresponding coordinate matrix elements are zero: it means that the probability for the transition between these two states in a finite time interval is zero.

This also provides the clue for the understanding of the fact that the emission spectra of the atoms consist of *spectral series*, each of which contains all the lines emitted in the spontaneous transition between two states in which the quantum number l changes by ± 1, as outlined at the end of section 5.3, because all other transitions have vanishing transition probabilities. Thus for example the "*hydrogen-like*" spectra consist of the *principal series* due to the transition from a "p-state" to an "s-state", the *sharp series* due to transitions from an "s-state" to a "p-state", the *diffuse series* due to transitions from a "d-state" to a "p-state", etc.

The lines of the absorption spectra are caused by "induced" transitions from the ground state of the atom, which is always an s-state, to a p-state, the only ones for which the transition probabilities are finite. They therefore also form a principal series which coincides with the corresponding principal emission series.

Following Schrödinger we have shown, at the beginning of this section, how one can assign to a charged particle a continuous *charge density* distribution ϱ_e which is connected to the wave function φ by equation (6.2.1). We shall now show that we can also associate with the movement of such a particle a continuous *current* distribution with *current density* $\mathbf{j}$.

Making use of the well-known formula of vector analysis,

$$\operatorname{div}(a\mathbf{A}) = a \operatorname{div} \mathbf{A} + \mathbf{A} \cdot \nabla a,$$

where a is a scalar and $\mathbf{A}$ a vector function of space, we can transform (6.1.7) into

$$\frac{\partial(\varphi\varphi^*)}{\partial t} = \frac{\hbar}{2mi} \operatorname{div}(\varphi \nabla\varphi^* - \varphi^* \nabla\varphi). \tag{6.2.12}$$

If we now define $\mathbf{j}$ by the expression

$$\mathbf{j}(\mathbf{r}, t) = \frac{\hbar e}{2mi} (\varphi^* \nabla\varphi - \varphi \nabla\varphi^*), \tag{6.2.13}$$

we obtain from (6.2.1) and (6.2.12)

$$\frac{\partial \varrho_e}{\partial t} + \operatorname{div} \mathbf{j} = 0, \tag{6.2.14}$$

which is the *continuity equation* of classical electrodynamics. The definition (6.2.13) for the current density is therefore logically consistent with the definition (6.2.1) for the charge density, and as the expression in brackets in (6.2.13) is purely imaginary it appears that the so defined quantity $\mathbf{j}$ is real, which is, of course, a necessary condition for it to represent an observable physical quantity.

For a stationary state with quantum number k the wave function φ has the form (6.1.5) and the expression (6.2.13) then becomes

$$\mathbf{j}_k(\mathbf{r}) = \frac{\hbar e}{2mi} (\psi_k^* \nabla\psi_k - \psi_k \nabla\psi_k^*). \tag{6.2.15}$$

In this case the current density is independent of time and the movement of the electrically charged particle is equivalent to a *stationary current*. Since the magnetic field of a stationary current is *static* we see that the second necessary condition for absence of electromagnetic radiation from such a system is indeed satisfied, as already mentioned previously.

It is further seen from (6.2.15) that the current density is zero if ψ_k is real. We have seen that this is, for example, the case for the harmonic oscillator (section 5.1), and it is immediately understandable on account of the fact that this system is completely symmetrical. This is in marked contrast to the classical behaviour of such an oscillator which constitutes an oscillating charge and hence an oscillating electric current.

For a plane De Broglie wave of the form (3.1.10) the amplitude function is equal to

$$\psi = A\, e^{\frac{i}{\hbar}\mathbf{pr}}, \tag{6.2.16}$$

where $\mathbf{p}$ is the momentum vector of the corresponding particle. If this particle carries a charge e the wave will, according to (6.2.15) carry with it an electric current of density

$$\mathbf{j} = \frac{e}{m} AA^*\mathbf{p} = eAA^*\mathbf{v} = e\psi\psi^*\mathbf{v} = \varrho_e\mathbf{v}. \tag{6.2.17}$$

This constitutes a "convection current" produced by the displacement as a whole of the charge density ϱ_e, associated with the particle, with the classical velocity $\mathbf{v}$ of the particle.

The classical theory of electrons maintains that a moving electron is surrounded by an electromagnetic field which is carried along with the particle. We shall deal with this aspect of the theory of electrons in detail in sections 9.3 and 10.4 from the point of view of the theory of relativity. According to the quantum mechanical principles laid down in the present section, however, electrons moving within an atom, say, in a stationary quantum state, are equivalent to a stationary distribution of charge and current. In keeping with these principles we are therefore justified in assuming that such a system will be surrounded by a *stationary electromagnetic field* as determined by Maxwell's relations between charge and current density, on the one hand, and electric and magnetic field strength, on the other hand.

It was already assumed by Ampère that the atoms behaved like tiny magnets owing to the presence of closed electric currents in their interior. According to our modern ideas such currents do indeed exist in atoms in which the electrons perform orbital movements, and we can expect such atoms to be equivalent to small magnets with *magnetic moments* **M**. We shall now show how these moments may be evaluated by carrying out this calculation for a "hydrogen-like" atomic eigenfunction as obtained in section 5.3.

The radial wave function χ_{nl} (5.3.22) is here real and the electronic current density has therefore no component in a radial direction, which is obvious for symmetry considerations. Similarly, as the part P_l^u (cos θ) of the wave function [see (5.2.26)], which depends on the polar angle θ, is also real, there will be no component of electric current in a meridional direction; this again is easy to understand from reasons of symmetry. The only part of the wave function $\psi_{n,l,u}$ which is complex is thus the function (5.2.10)

$$\psi_u(\varphi) = A\, e^{iu\varphi} \tag{6.2.18}$$

which depends on the azimuth angle θ of the electron; thus there will be a closed current about the polar axis as predicted by Ampère.

In order to calculate the corresponding current density j_u from (6.2.15) we have first to evaluate the quantities $\nabla^\varphi \psi_u$ and $\nabla^\varphi \psi_u^*$. Now for a particle rotating about the polar axis along a circle of radius b the line element ds is equal to $b\, d\varphi$ and hence

$$\nabla^\varphi = \frac{1}{b}\frac{\partial}{\partial\varphi}. \tag{6.2.19}$$

Applying (6.2.19) to the function (6.2.18) and its conjugate complex we have

$$\nabla^\varphi \psi_u = \frac{iu}{b}\psi_u, \quad \nabla^\varphi \psi_u^* = -\frac{iu}{b}\psi_u^*, \tag{6.2.20}$$

and thus from (6.2.15)

$$j_k = \frac{\hbar e u}{mb}\psi_u \psi_u^*. \tag{6.2.21}$$

The current dI_u, flowing through a cross-sectional area da of a tube surrounding the circle, is equal to $j_u\, da$, and according to the elementary classical formula for the magnetic moment of a current loop the corresponding moment dM_u is

$$dM_u = \frac{dI_u\, b^2 \pi}{c} = \frac{\hbar e u b \pi}{mc}\psi_u\psi_u^*\, da = \frac{\hbar e u}{2mc}\psi_u\psi_u^*\, dV, \tag{6.2.22}$$

where dV is the volume of the tube.

The total magnetic moment M_u associated with the electron is obtained from (6.2.22) by integrating over all space which, on account of the condition (5.2.9), results in

$$M_u = \frac{\hbar e}{2mc}u. \tag{6.2.23}$$

This shows that the considered type of atom has, in a quantum state with quantum number u, a magnetic moment which is an integer multiple of the universal moment

$$M_0 = \frac{\hbar e}{2mc} = \frac{he}{4\pi mc}, \tag{6.2.24}$$

called the "*Bohr magneton*". The quantum number u is therefore usually referred to as "*magnetic quantum number*".

It is believed that the "spin" movement of an elementary particle of charge e likewise produces a magnetic moment equal to one Bohr magneton.

The expression (6.2.24) for the magneton is identical with the one originally derived by Bohr from his quantum condition (2.2.10) for an electron moving in a circular orbit with radius b. For replacing n by u in this formula we obtain for the angular velocity ω of the electron the expression

$$\omega = \frac{\hbar u}{mb^2}. \tag{6.2.25}$$

On the other hand the electric charge e of the electron crosses any point on its orbit $\omega/2\pi$ times per second and should therefore generate a magnetic moment perpendicular to the plane of the orbit equal to

$$M_u = \frac{b^2\pi}{c}\frac{\omega}{2\pi}. \tag{6.2.26}$$

Combination of (6.2.25) and (6.2.26) results, indeed, in the expression (6.2.23). The derivation of this formula from the general quantum mechanical principles is, however, much more satisfactory and holds quite generally.

6.3. The connection between the wave equation and the Newtonian equations of motion

As shown in section 6.1, the time-dependent wave equation (6.1.1) of quantum mechanics is based on the classical Hamilton–Jacobi equation and contains as an essential element the Hamiltonian operator $\mathscr{H}$ which is derived from the classical Hamiltonian function. However, the solution of a quantum mechanical problem consists, as repeatedly emphasized before, of certain *probabilities* for the change of the configuration of the system in time, whereas in classical mechanics the configuration at any time is *uniquely* determined by the initial condition of the system. We shall now, following the procedure of Ehrenfest (1927), show that nevertheless the *average* behaviour of a quantum mechanical system is still governed by the Newtonian equations of motion.

The average position vector **r** of a particle is, according to (6.1.11). Given by the expression

$$\bar{\mathbf{r}}(t) = \int\int\int \mathbf{r}\varphi\varphi^* \, dV. \tag{6.3.1}$$

We define as the *average velocity* **v** of the particle the derivative of **r** with respect to time, that is the quantity

$$\bar{\mathbf{v}}(t) = \frac{d\bar{\mathbf{r}}}{dt} = \frac{\partial}{\partial t}\int\int\int \mathbf{r}\varphi\varphi^* \, dV. \tag{6.3.2}$$

If the operations integration and differentiation with respect to time are interchanged in (6.3.2) this expression can, by means of the relation (6.1.7), be transformed so that the x-component of the average velocity $\bar{v}_x$ now becomes:

$$\bar{v}_x = \frac{\hbar}{2mi} \iiint x\,(\varphi\,\nabla^2\varphi^* - \varphi^*\,\nabla^2\varphi)\,dV, \tag{6.3.3}$$

and similar expressions hold for the other two components $\bar{v}_y$ and $\bar{v}_z$.

Now it is easily verified that

$$\nabla^2\,(x\varphi) = x\,\nabla^2\varphi + 2\frac{\partial\varphi}{\partial x},$$

and hence the expression (6.3.3) may be re-written in the form

$$\bar{v}_x = \frac{\hbar}{2mi}\left\{\iiint [(x\varphi)\,\nabla^2\varphi^* - \varphi^*\,\nabla^2(x\varphi)]\,dV + 2\iiint \varphi^*\frac{\partial\varphi}{\partial x}\,dV\right\}.$$

The first of the two integrals here can be transformed into a surface integral by means of Green's theorem and, by the same argument as applied to (6.1.7), one finds that this integral is zero if extended over all space. Hence we finally obtain

$$\bar{v}_x = \frac{\hbar}{mi}\iiint \varphi^*\frac{\partial\varphi}{\partial x}\,dV, \tag{6.3.4}$$

and analogous expressions for $\bar{v}_y$ and $\bar{v}_z$. They can be combined into the vector equation

$$\bar{\mathbf{v}} = \frac{\hbar}{mi}\iiint \varphi^*\,\nabla\varphi\,dV, \tag{6.3.5}$$

which does not depend on time explicitly.

As $\bar{\mathbf{v}}$ must necessarily be real (which is also immediately evident from (6.3.2)) it is equal to its conjugate complex value, and we therefore can write (6.3.5) in the alternative form

$$\bar{\mathbf{v}} = -\frac{\hbar}{mi}\iiint \varphi\,\nabla\varphi^*\,dV. \tag{6.3.6}$$

We may finally put the expression for $\bar{\mathbf{v}}$ in a symmetrical form by taking the arithmetic mean of the two expressions (6.3.5) and (6.3.6):

$$\bar{\mathbf{v}} = \frac{\hbar}{2mi}\iiint (\varphi^*\,\nabla\varphi - \varphi\,\nabla\varphi^*)\,dV. \tag{6.3.7}$$

Comparing (6.3.7) with (6.2.13) we notice that

$$e\bar{\mathbf{v}} = \iiint \mathbf{j}\,dV, \tag{6.3.8}$$

which shows that the total current $\mathbf{J}$ associated with a charged particle is equal to the product of its charge e and its mean velocity $\bar{\mathbf{v}}$. This relationship is analogous to the classical relationship for an electronic current (see section 9.3), which proves that our definitions of mean velocity and current density are logically consistent.

Following the classical procedure we define as *average momentum* of a particle of mass m the product

$$\bar{\mathbf{p}} = m\bar{\mathbf{v}}. \tag{6.3.9}$$

If we substitute here the expression (6.3.5) for the average velocity vector we obtain the following alternative definition for the average momentum:

$$\bar{\mathbf{p}} = \iiint \varphi^* \left(\frac{\hbar}{i} \nabla\right) \varphi \, dV. \tag{6.3.10}$$

This is of particular interest as it has the form of the general definition (6.1.11) for the averages of ordinary functions, but extended to operators. If this definition is to be unique the order of the factors in the product under the integral must not be changed. In the present case the operator concerned is the previously defined "momentum operator" (3.1.7), and formula (6.3.10) expresses a very important relationship between the average of an observable quantity and its associated operator. This subject will be dealt with in detail in Chapter 7.

In keeping with the definitions of mean position and mean velocity we shall define the *average acceleration* vector **a** as the derivative with respect to time of **v**. From (6.3.7) we then obtain

$$\bar{\mathbf{a}}(t) = \frac{d\bar{\mathbf{v}}}{dt} = \frac{\hbar}{2mi} \iiint \frac{\partial}{\partial t} (\varphi^* \nabla\varphi - \varphi \nabla\varphi^*) \, dV. \tag{6.3.11}$$

This leads to the following expression for the x-component $\bar{a}_x$ of **a**:

$$\begin{aligned} \bar{a}_x &= \frac{\hbar}{2mi} \iiint \frac{\partial}{\partial t} \left(\varphi^* \frac{\partial\varphi}{\partial x} - \varphi \frac{\partial\varphi^*}{\partial x}\right) \partial V \\ &= \frac{\hbar}{2mi} \iiint \left[\left(\frac{\partial\varphi^*}{\partial t} \frac{\partial\varphi}{\partial x} - \frac{\partial\varphi}{\partial t} \frac{\partial\varphi^*}{\partial x}\right)\right. \\ &\quad \left. + \left(\varphi^* \frac{\partial^2\varphi}{\partial x\, \partial t} - \varphi \frac{\partial^2\varphi^*}{\partial x\, \partial t}\right)\right] dV, \end{aligned} \tag{6.3.12}$$

and similar expressions for the other two components $\bar{a}_y$, $\bar{a}_z$.

Now multiplying equation (6.1.6) by $-\dfrac{\hbar^2}{4m^2} \dfrac{\partial\varphi}{\partial x}$ and equation (6.1.4) by $-\dfrac{\hbar^2}{4m^2} \dfrac{\partial\varphi^*}{\partial x}$ and adding gives

$$\begin{aligned} \frac{\hbar}{2mi} \left(\frac{\partial\varphi^*}{\partial t} \frac{\partial\varphi}{\partial x} - \frac{\partial\varphi}{\partial t} \frac{\partial\varphi^*}{\partial x}\right) &= \frac{U}{2m} \left(\varphi^* \frac{\partial\varphi}{\partial x} + \varphi \frac{\partial\varphi^*}{\partial x}\right) \\ &\quad - \frac{\hbar^2}{4m^2} \left(\nabla^2\varphi^* \frac{\partial\varphi}{\partial x} + \nabla^2\varphi \frac{\partial\varphi^*}{\partial x}\right). \end{aligned}$$

Further, differentiating equation (6.1.4) with respect to x and multiplying by $\frac{\hbar^2}{4m^2}\varphi^*$, differentiating equation (6.1.6) with respect to x and multiplying by $\frac{\hbar^2}{4m^2}\varphi$ and adding gives

$$\frac{\hbar}{2mi}\left(\varphi^*\frac{\partial^2\varphi}{\partial x\,\partial t}-\varphi\frac{\partial^2\varphi^*}{\partial x\,\partial t}\right)=-\frac{1}{2m}\left(\varphi^*\frac{\partial\,(U\varphi)}{\partial x}+\varphi\frac{\partial\,(U\varphi^*)}{\partial x}\right)$$
$$+\frac{\hbar^2}{4m^2}\left(\varphi^*\frac{\partial\,(\nabla^2\varphi)}{\partial x}+\varphi\frac{\partial\,(\nabla^2\varphi^*)}{\partial x}\right).$$

If these two expressions are substituted into the integrand of (6.3.12) one further obtains

$$\bar{a}_x=-\frac{1}{m}\iiint\frac{\partial U}{\partial x}\varphi\varphi^*\,dV+\frac{\hbar^2}{4m^2}\iiint\left[\varphi^*\,\nabla^2\left(\frac{\partial\varphi}{\partial x}\right)-\frac{\partial\varphi}{\partial x}\nabla^2\varphi^*\right]dV$$
$$+\frac{\hbar^2}{4m^2}\iiint\left[\varphi\,\nabla^2\left(\frac{\partial\varphi^*}{\partial x}\right)-\frac{\partial\varphi^*}{\partial x}\nabla^2\varphi\right]dV. \tag{6.3.13}$$

Finally applying Green's theorem to the last two integrals here shows, as before, that they are zero if extended over all space, so that we are led to the following simple expression for $\bar{a}_x$:

$$\bar{a}_x=-\frac{1}{m}\iiint\frac{\partial U}{\partial x}\varphi\varphi^*\,dV, \tag{6.3.14}$$

and analogous expressions for $\bar{a}_y$ and $\bar{a}_z$. They can be combined in the vector equation

$$\bar{\mathbf{a}}(t)=-\frac{1}{m}\iiint\nabla U\,\varphi\varphi^*\,dV. \tag{6.3.15}$$

The meaning of the relationship (6.3.15) becomes clear if one recalls that, according to (1.1.5), $-\nabla U$ is the force F acting on the particle and thus, by the definition (6.1.11) the integral on the right-hand side of (6.3.15) represents the *mean force* on the particle. Thus (6.3.15) can be written in the form

$$m\bar{\mathbf{a}}=\bar{\mathbf{F}}, \tag{6.3.16}$$

which is the Newtonian equation of motion for the average movement of the particle. This is *Ehrenfest's theorem*.

We may interpret this theorem also in the following alternative way. Assume that a particle starts its movement from a definite position in space. It will then, under the influence of the external field, move along a certain path and arrive in a certain end position after a time t. If this experiment is now repeated several times, observations of the particle will reveal that its path will in general be different each time and so will its end position. However, if averages are taken of the measurements in a great number of such experiments the average path will turn out to be the one predicted by the Newtonian equations and the average end position can be predicted in the same way as in classical dynamics.

This situation is somewhat similar to that obtaining for a "classical" small particle suspended in a fluid, which performs a *Brownian movement* owing to the irregular impacts of the molecules of the fluid. Its path under the influence of an external field of force will therefore again show random deviations from the classical path. But if repeated observations are made on a particle, always starting from the same initial position, the average movement will still follow the Newtonian law.

We may imagine a "quantum particle" to perform a kind of irregular "zigzag" or "wobbling" motion superposed over its classical path. But whereas the Brownian motion of a classical particle depends on temperature, the kinetic energy of this movement being proportional to the absolute temperature T, the quantum mechanical wobbling motion is independent of temperature and persists even at $T = 0$. Its kinetic energy is in fact identical with the "zero-point energy" mentioned in section 5.1 in connection with the harmonic oscillator. We may think of the Heisenberg uncertainty relations as being the result of this random component of the movement of a particle.

It is well known that the phenomenon of *diffusion* of a suspension of equal particles in a fluid is due to the Brownian movement of these particles. If they are at $t = 0$ concentrated in a small volume element in space then they will, after an interval t, have diffused through the fluid owing to the random nature of the Brownian motion so that their *concentration* c at time t will be a definite function of space. This function can be determined by solving the classical differential equation of diffusion

$$\frac{\partial c}{\partial t} = D\,\nabla^2 c, \qquad (6.3.17)$$

where D is the "*diffusion coefficient*".

Similarly, if a large number of quantum particles are let lose at $t = 0$ from a definite initial state, there exists, as was pointed out in section 6.1, a probability $|\varphi(\mathbf{r}, t)|^2\, dV$ for finding any one of the particles within the volume element dV at $\mathbf{r}$ at time t. This may be thought of as a kind of diffusion process, superposed over the average movement according to the equation (6.3.16), $|\varphi|^2$ playing the part of the concentration.

The time-dependent Schrödinger equation (6.1.4) which governs this diffusion process has, indeed, a formal analogy to the classical diffusion equation (6.3.17) in so far as both equations are of the second order in the space coordinates but only of the first order in the time. This analogy is particularly striking in the case where there is no external force acting and thus U in (6.1.4) may be taken to be zero. The equation then reduces to

$$\frac{\partial \varphi}{\partial t} = \frac{i\hbar}{2m} \nabla^2 \varphi, \tag{6.3.18}$$

which has precisely the form of the classical diffusion equation (6.3.17) except for the fact that the "diffusion coefficient" $-\frac{i\hbar}{2m}$ is purely imaginary. This is an essential feature of the quantum mechanical diffusion phenomenon whose implications we shall consider later on.

The formal analogy between the differential equations (6.3.17) and (6.3.18) makes it possible to calculate in mathematical terms the "spreading out" of the path of a quantum mechanical system around the classical path (or the average path) from known solutions of classical diffusion problems. As an example let us assume that at $t = 0$ a large number of identical particles is concentrated in the immediate neighbourhood of the yz-plane with a Gaussian distribution

$$c_0(x) = C \exp\left[-\frac{x^2}{2a^2}\right]. \tag{6.3.19}$$

It can be shown that the solution of the differential equation (6.3.17) for an arbitrary initial condition $c\,(x, 0) = c_0(x)$ and no boundary conditions imposed is

$$c\,(x, t) = \frac{1}{2\sqrt{\pi D t}} \int_{-\infty}^{+\infty} \exp\left[-\frac{(x-\xi)^2}{4Dt}\right] c_0(\xi)\, d\xi\,. \tag{6.3.20}$$

If (6.3.19) is substituted in (6.3.20) one obtains from a simple calculation

$$c\,(x, t) = \frac{C}{\sqrt{\alpha}} \exp\left[-\frac{x^2}{2a^2\alpha}\right], \tag{6.3.21}$$

with

$$\alpha = 1 + 2Dt/a^2. \tag{6.3.22}$$

Hence the solution of (6.3.18) under the initial condition (6.3.19) for φ is obtained by replacing D in (6.3.22) by $i\hbar/2m$, that is by setting

$$\alpha = 1 + \frac{i\hbar}{ma^2}\, t. \tag{6.3.23}$$

The mass density distribution of the particles $\varrho = m\varphi\varphi^*$ is now from (6.3.21) and (6.3.23):

$$\varrho\,(x, t) = \frac{CC^*}{\sqrt{\alpha\alpha^*}} \exp\left[-\frac{x^2}{2a^2}\left(\frac{1}{\alpha} + \frac{1}{\alpha^*}\right)\right] = \frac{m}{a\beta\sqrt{\pi}} \exp\left[-\frac{x^2}{a^2\beta^2}\right], \tag{6.3.24}$$

where we have put

$$\beta = \left(1 + \frac{\hbar^2}{m^2 a^4} t^2\right)^{1/2}, \quad CC^* = \frac{m}{a\sqrt{\pi}}, \tag{6.3.25}$$

the latter in order to make ϱ satisfy the necessary condition

$$\int_{-\infty}^{+\infty} \varrho \, dx = m. \tag{6.3.26}$$

For $t = 0$ (6.3.25) reduces to the initial distribution

$$\varrho_0 = \frac{m}{a\sqrt{\pi}} \exp\left[-\frac{x^2}{a^2}\right], \tag{6.3.27}$$

and it is further seen that with increasing time the "width" $a\beta$ of the distribution increases indefinitely. This illustrates the fact that, as time goes on, the probability of the deviation of the position of a particle from the classical or mean position becomes greater and greater.

We now return to the question of the significance of an imaginary diffusion coefficient in (6.3.18). It is well known that the ordinary diffusion process is *irreversible*. This is also mathematically evident from an inspection of the diffusion equation (6.3.17); for if we reverse the sign of t the right-hand member of this equation remains unchanged while the left-hand member changes sign so that the equation is not invariant with respect to a reversal of time, which means that the diffusion process cannot be made to run backwards.

The "quantum-diffusion" processes, however, consist actually in a spreading out of *De Broglie wave packets* (see section 3.2). Now the propagation of De Broglie waves is governed by the classical wave equation (1.2.2), which is immediately seen to be invariant with respect to a reversal of the sign of t. Thus the De Broglie waves can run forwards and backwards, and hence the propagation of De Broglie wave packets is a strictly *reversible* process.

This can also be directly deduced quite generally from the time-dependent wave equation (6.1.1) of which the equation (6.3.18) is a special case. Reversing the direction of the velocity or the momentum of the system is equivalent to a change of sign of the momentum operators $\frac{\hbar}{i}\frac{\partial}{\partial q_k}$; this results in the transformation of the Hamiltonian operator $\mathscr{H}$ into its conjugate complex $\mathscr{H}^*$. If in addition the sign of t is also reversed equation (6.1.1) goes over into

$$\mathscr{H}^*\varphi = -i\hbar \frac{\partial \varphi}{\partial t}. \tag{6.3.28}$$

If we compare this equation with the conjugate complex of (6.1.1)

$$\mathscr{H}^*\varphi^* = -i\hbar \frac{\partial \varphi^*}{\partial t}, \tag{6.3.29}$$

we see that the two equations only differ in so far as the wave function φ is replaced by its conjugate complex φ^*. But since φ is not directly observable, but only the probability density $|\varphi|^2$ this makes no actual difference as far as the physics of the quantum mechanical processes is concerned. It therefore follows that they are indeed strictly reversible.

We now see that the significance of the imaginary quantum mechanical diffusion coefficient lies in the fact that this "diffusion" process is actually a spreading of De Broglie wave packets, as already stated before. This fact is further illustrated by the results obtained in the solution of the special example studied above. For as the quantity β (6.3.25) depends on t *quadratically* a reversal of the time direction will make no difference to the result (6.3.24). This is in marked contrast to the classical diffusion process where the quantity α (6.3.22) depends *linearly* on t and therefore the process cannot run in the opposite direction.

Problems and exercises

P. 6.1. According to classical electrodynamics the wave field of a variable electric dipole of moment Π at a large distance r from the dipole can be regarded as consisting of plane electromagnetic waves in the direction of $\mathbf{r}$ whose electric and magnetic field strengths E and H are given by

$$E = H = \frac{1}{c^2}\left(\frac{\ddot{\Pi}}{r}\sin\theta\right),$$

where θ is the angle between $\mathbf{r}$ and the direction of Π. Use this formula for deriving the expression (6.2.8) for the intensity I of the dipole radiation.

P. 6.2. Derive expressions for the frequencies of the spectral lines emitted by "electrons in a box" and for their intensities, making use of the formulae (4.1.28) and (6.2.9) and the results of P. 4.9. State in what direction the emitted light is polarized.

P. 6.3. What kind of spectrum is emitted by an assembly of linear harmonic oscillators of oscillation frequency ν in a "mixed" state with wave function (6.1.16)? Use the results of P. 5.2 and formula (6.2.9) for calculating the intensity of the emitted radiation.

P. 6.4. Derive the diffusion equation (6.3.17) from Fick's law of diffusion, according to which the current density $\mathbf{Q}$ of the diffusion flow is proportional to the concentration gradient.

$$\mathbf{Q} = -D\nabla c.$$

P. 6.5. Verify that the function $c = \frac{1}{2\sqrt{\pi D t}}\exp[-x^2/4Dt]$ is a solution of the one-dimensional diffusion equation $\frac{\partial c}{\partial t} = D\frac{\partial^2 c}{\partial x^2}$, which satisfies the condition $\int_{-\infty}^{+\infty} c(x, t)\,dx = 1$ for all time and is zero everywhere at $t = 0$ except at $x = 0$. Hence prove that the expression (6.3.20) is the unique solution of (6.3.17) for the initial condition $c(x, 0) = c_0(x)$.

P. 6.6. Verify that the function

$$c(x, t) = e^{-\lambda^2 D t}(A\cos\lambda x + B\sin\lambda x)$$

is a solution of the linear diffusion equation $\frac{\partial c}{\partial t} = D\frac{\partial^2 c}{\partial x^2}$. Hence prove that the solutions of the wave equation (6.3.18) under the boundary conditions of the "particle in the box" problem have the form of standing De Broglie waves with frequencies $\nu_n = E_n/h$, where E_n is given by (4.1.28).

CHAPTER 7

Operator and Matrix Mechanics

IN THE preceding chapters we have repeatedly come across the notion of *operators* which are symbols for mathematical operations to be performed on functions of the coordinates. We have shown in particular how the quantum mechanical wave equation can be derived from the classical Hamilton–Jacobi equation by the replacement of the quantities momentum and energy by certain differential operators. One may therefore appropriately refer to this mathematical formalism as "*operator mechanics*". In the present chapter it will be shown that the operator aspect can be extended to all "observable" physical quantities by associating them with certain operators.

It was further demonstrated (in section 4.3) how one can construct a *matrix* to any quantity defined as a function of the coordinates from the eigenfunctions of the wave equation for any particular quantum mechanical problem, and subsequently it was shown that the elements of these matrices were closely related to the expectation values of the associated quantities in the various quantum states and the transition probabilities from one such state to another. In this chapter we shall extend this procedure to quantities which are defined as operators, and we shall show how in this way operators may always be "represented" by matrices. It is thus possible to replace the formalism of "operator mechanics" by one of "*matrix mechanics*", and to formulate the laws of quantum mechanics in matrix form as was indicated at the end of section 2.2.

7.1. The notion of operators and their representation by matrices

We begin this section by giving a brief outline of the most important properties of *linear operators*. Generally, an operator $\mathscr{A}$ is a mathematical symbol for an operation performed on a function $\psi(q_1, \ldots, q_N)$ of the coordinates which transforms this function into another function $\chi(q_1, \ldots, q_N)$ according to the symbolic equation

$$\chi = \mathscr{A}\psi. \tag{7.1.1}$$

$\mathscr{A}$ may, for example, represent an algebraic operation, say a multiplication by a factor, in which case $\mathscr{A}$ is simply an ordinary function f of the coordinates; or it may represent a differentiation with respect to the coordinates when it is called a "*differential operator*". If, in particular, $\chi = \psi$, that is to say if the operation transforms a function into itself, the operator is called a "*unit operator*" and shall be denoted by $\mathscr{I}$. Thus

$$\psi = \mathscr{I}\psi \tag{7.1.2}$$

is the expression for a unit or "identity" operation.

An operator is called "*linear*" if for any set of real or complex numbers $c_1, c_2, \ldots,$

$$\mathscr{A}\,(c_1\psi_1 + c_2\psi_2 + \cdots) = c_1\mathscr{A}\psi_1 + c_2\mathscr{A}\psi_2 + \cdots. \tag{7.1.3}$$

In the following we shall always assume that this condition is satisfied.

The *sum* and the *difference* respectively of two operators $\mathscr{A}$ and $\mathscr{B}$ is defined by the relation

$$(\mathscr{A} \pm \mathscr{B})\,\psi \equiv \mathscr{A}\psi \pm \mathscr{B}\psi, \tag{7.1.4}$$

which evidently satisfies the commutation law of ordinary algebra.

The *product* of two operators $\mathscr{A}$ and $\mathscr{B}$ is defined as the succession of the operations $\mathscr{B}$ and $\mathscr{A}$ on the same function ψ *in this order*; thus

$$(\mathscr{A}\mathscr{B})\,\psi \equiv \mathscr{A}\,(\mathscr{B}\psi). \tag{7.1.5}$$

It appears from this definition that, in general, $\mathscr{A}\mathscr{B}$ will *not* be equal to $\mathscr{B}\mathscr{A}$, which means that operator algebra does not obey the commutation law of ordinary algebra. The quantity

$$[\mathscr{A}\mathscr{B}] \equiv (\mathscr{A}\mathscr{B}) - (\mathscr{B}\mathscr{A}), \tag{7.1.6}$$

which is called the "*commutator*" of $\mathscr{A}$ and $\mathscr{B}$, does therefore in general not vanish for two arbitrary operators. If it does vanish the operators are said to "*commute*" with each other. If, on the other hand $(\mathscr{A}\mathscr{B}) = -(\mathscr{B}\mathscr{A})$, the operators are called "*anticommutative*". Any two ordinary functions, of course, always commute.

The operator which symbolizes the inverse transformation to (7.1.1) is called the "*inverse*" or "*reciprocal*" operator to $\mathscr{A}$ and is denoted by $\mathscr{A}^{-1}$. Thus

$$\psi = \mathscr{A}^{-1}\chi. \tag{7.1.7}$$

Combining (7.1.7) with (7.1.1) we obtain

$$\psi = (\mathscr{A}^{-1}\mathscr{A})\,\psi \quad \text{or} \quad \chi = (\mathscr{A}\mathscr{A}^{-1})\,\chi,$$

from which follows, with the help of (7.1.2), that

$$(\mathscr{A}\mathscr{A}^{-1}) = (\mathscr{A}^{-1}\mathscr{A}) = \mathscr{I}. \tag{7.1.8}$$

This indicates that the product of two reciprocal operators is equal to the unit operator and justifies the nomenclature.

As already mentioned, any function of the coordinates and, of course, the coordinates themselves, can be regarded as operators. In section 3.1 we further introduced the notion of the momentum operator [(3.1.7) and (3.1.18)] which replace the classical quantity momentum in quantum mechanics and which is a differential operator. By means of the basic formulae (7.1.4), (7.1.5), and (7.1.7) of operator algebra it is now possible to construct operators corresponding to any physical quantity which is classically defined by a certain function of the coordinates and momenta of a mechanical system.

We have already made extensive use of one important example of this procedure in the "Hamiltonian operator" which replaces the Hamiltonian function in quantum mechanics (section 3.1). It is a fundamental assumption of quantum physics that any *observable* quantity is to be represented by an appropriate operator. Thus the basic equations of quantum mechanics have the form of operator equations of which the most important example so far encountered is the quantum mechanical wave equation.

A differential equation of the form

$$\mathscr{A}\chi = a\chi \tag{7.1.9}$$

in which $\mathscr{A}$ is a differential operator in the coordinates, χ a function of the coordinates, and a a constant will in general, as explained in section 1.2, have solutions only for certain discrete values a_n of a under given boundary conditions. These are called the "*eigenvalues*" of the operator and the corresponding functions χ_n which are solutions of (7.1.9) are called the "*eigenfunctions*" of the operator.

The quantum-mechanical wave equation (3.1.19) is seen to be a special case of (7.1.9) with $\mathscr{A} = \mathscr{H}$ and the eigenfunctions and eigenvalues of the wave equation may thus also be described as the eigenfunctions and eigenvalues of the Hamiltonian operator. This operator represents the energy function of the system and its eigenvalues E_n are, as we have seen, the allowed values of the energy of the system, or, as we may also say, the *observable* values of the energy. More generally it is maintained that the observable values of a physical quantity A, which is quantum mechanically represented by the operator $\mathscr{A}$, are identical with the eigenvalues a_n of the eigenvalue equation (7.1.9). This hypothesis forms the basis of the so-called "*theory of observables*" in quantum physics.

We now turn to another very important feature of quantum mechanics, namely the close relationship between operators and matrices. The general notion of matrices was introduced in section 4.3 and it was shown how one can associate with any function $f(\mathbf{r})$ of the coordinates of a particle a matrix by means of the eigenfunctions ψ_k (4.3.8). Extensive use was made in the preceding chapters of the coordinate matrix (4.3.15) which is a special case of this type of matrices. This concept can easily be extended to general mechanical systems with generalized coordinates $q_1, \ldots, q_N$ by defining matrices $\mathfrak{F}$ to functions $f(q_1, \ldots, q_N)$ whose elements are given by the relations

$$f_{kj} = \int \cdots \int \psi_k^* f \psi_j \, dq_1 \cdots dq_N. \tag{7.1.10}$$

Setting

$$f\psi_j = \chi_j \tag{7.1.11}$$

and using the notion of the "inner product" of two functions, defined by (4.1.31) we may also write (7.1.10) in the symbolic form

$$f_{kj} = (\psi_k \cdot \chi_j). \tag{7.1.12}$$

On account of the preceding considerations concerning operators it suggests itself to extend the definition (7.1.12) to functions χ_j defined by equation (7.1.1), and thus to associate a matrix $\mathfrak{A}$ with an operator $\mathscr{A}$ by the relations

$$a_{kj} = (\psi_k \chi_j) = (\psi_k \cdot \mathscr{A}\psi_j) \tag{7.1.13}$$

for the elements of $\mathfrak{A}$ or, written *in extenso*,

$$a_{kj} = \int \cdots \int \psi_k^* \mathscr{A}\psi_j \, dq_1 \cdots dq_N. \tag{7.1.14}$$

The matrix $\mathfrak{A}$ is said to "represent" the operator $\mathscr{A}$. What this representation means from the physical point of view will be discussed later on.

It is evident from the definition of the sums and differences of operators and of matrices and from (7.1.14) that to a linear combination of operators $c_1\mathscr{A}_1 + c_2\mathscr{A}_2 + \cdots$ belongs the matrix $c_1\mathfrak{A}_1 + c_2\mathfrak{A}_2 + \cdots$. We shall now show that the product $\mathscr{C} = (\mathscr{A}\mathscr{B})$ of two operators $\mathscr{A}$ and $\mathscr{B}$ is associated with the product $\mathfrak{C} = (\mathfrak{A}\mathfrak{B})$ of the representing matrices $\mathfrak{A}$ and $\mathfrak{B}$, as defined by the formula (4.3.2).

According to (7.1.13) the matrix elements c_{kj} are defined by the inner products

$$c_{kj} = (\psi_k \cdot \mathscr{A}\mathscr{B}\psi_j). \tag{7.1.15}$$

If we now expand the function $\mathscr{B}\psi_j$ in a series of eigenfunctions according to the theorem (4.2.8) we obtain, with the help of (4.2.10)

$$\mathscr{B}\psi_j = \sum_i a_i \psi_i = \sum_i (\psi_i \cdot \mathscr{B}\psi_j)\, \psi_i = \sum_i b_{ij}\psi_i. \tag{7.1.16}$$

Substituting from (7.1.16) into (7.1.15) and exchanging the sequence of integration and summation we obtain at once

$$c_{kj} = \sum_i (\psi_k \cdot \mathscr{A}\psi_i)\, b_{ij} = \sum_i a_{ki} b_{ij}, \tag{7.1.17}$$

which is indeed identical with the rule (4.3.2) for the multiplication of two matrices, except for the fact that the present matrices are infinite.

Thus we see that any algebraic manipulation of operators is completely reflected by the corresponding manipulations of the representing matrices which makes it possible to translate relationships between operators into the language of matrices and vice versa. In particular if two operators $\mathscr{A}$ and $\mathscr{B}$ "commute" the product of the associated matrices is also commutative, that is $\mathfrak{A}\mathfrak{B} = \mathfrak{B}\mathfrak{A}$, and if $\mathscr{A}$ and $\mathscr{B}$ are anticommutative it follows that $\mathfrak{A}\mathfrak{B} = -\mathfrak{B}\mathfrak{A}$.

In section 4.3 we introduced the concept of "self-adjoined" or "Hermitian" matrices and we proved that the matrices of the form (7.1.10) were always Hermitian. If a matrix of the more general type (7.1.14) is Hermitian, we call the corresponding *operator* also *Hermitian*. The condition for this to be the case is therefore according to (4.3.7) and (7.1.13)

$$(\psi_k \cdot \mathscr{A}\psi_j) = (\psi_j \cdot \mathscr{A}\psi_k)^* = (\mathscr{A}\psi_k \cdot \psi_j),$$

and an operator $\mathscr{A}$ will thus be Hermitian if for any two functions f and g of the coordinates the equation

$$(f \cdot \mathscr{A}g) = (\mathscr{A}f \cdot g) \tag{7.1.18}$$

is fulfilled.

On account of the Hermitian character of the matrices (7.1.10) it was possible to put forward in section 4.3 the statistical interpretation of these matrices which consists in identifying the diagonal elements f_{kk} with the average or expectation values of the quantity f in the quantum states k and the arithmetic means of symmetrical off-diagonal elements $(f_{kj} + f_{jk})/2$ with the averages of f in the transitional process between the quantum states k and j.

It seems logical to extend this statistical interpretation to all matrices of the form (7.1.14) provided they are self-adjoined, that is provided that the corresponding operators are Hermitian and satisfy the condition (7.1.18). If this interpretation is accepted as valid it follows that a necessary condition for an operator to represent a physical observable quantity is that it must be Hermitian. It will be proved in the following section 7.2 that the linear momentum and the energy operators are indeed Hermitian.

We shall further assume that, in generalization of the relation (6.1.11), the time-dependent expectation value of a physical quantity A in a non-stationary system which is represented by the operator $\mathscr{A}$ is given by the formula

$$\bar{A}(t) = \int \cdots \int \varphi^*(q_1, \ldots, q_N)\, \mathscr{A}\varphi\, (q_1, \ldots, q_N)\, dq_1 \cdots dq_N, \tag{7.1.19}$$

where φ is the time-dependent wave function of the system.

We shall now prove that the eigenvalues of Hermitian operators are always real and its eigenfunctions always orthogonal to each other, in generalization of the theorem concerning the eigenvalues and eigenfunctions of the Hamiltonian operator which was proved in section 4.2.

From (7.1.9) it follows that for any eigenvalue a_k

$$\mathscr{A}\chi_k = a_k\chi_k. \tag{7.1.20}$$

If further we put $k = j$ in equation (7.1.18) and substitute χ_k for f and g, we obtain

$$(\chi_k \cdot \mathscr{A}\chi_k) = (\mathscr{A}\chi_k \cdot \chi_k). \tag{7.1.21}$$

Combination of these two equations results in

$$(a_k - a_k^*) \int \cdots \int \chi_k^* \chi_k \, dq_1 \cdots dq_N = 0,$$

and as the integral in this equation is necessarily finite we conclude that

$$a_k = a_k^*, \tag{7.1.22}$$

which means that the eigenvalues are indeed real.

We now put $k \gtrless j$ and assume that the quantum states k and j are not degenerate so that $a_k \gtrless a_j$. We now have instead of (7.1.21)

$$(\chi_k \cdot \mathscr{A}\chi_j) = (\mathscr{A}\chi_k \cdot \chi_j), \tag{7.1.23}$$

from which follows with the help of (7.1.20)

$$(a_k - a_j) \int \cdots \int \chi_k^* \chi_j \, dq_1 \cdots dq_N = 0,$$

and hence

$$(\chi_k \cdot \chi_j) = \int \cdots \int \chi_k^* \chi_j \, dq_1 \cdots dq_N = 0. \tag{7.1.24}$$

This equation means that the eigenfunctions belonging to different eigenvalues are indeed mutually orthogonal to each other.

It follows from the linearity of the equation (7.1.20) for linear operators that one may normalize the eigenfunctions χ_k by making their "norms" equal to one. If this is done the eigenfunctions of (7.1.9) form an *orthonormal set.*

It is possible to represent an operator in many different ways by a matrix when, instead of the eigenfunctions ψ_k of the Hamiltonian operator, the eigenfunctions χ_k of another Hermitian differential operator, say $\mathscr{B}$, are used in (7.1.14) which are the solutions of the differential equation

$$\mathscr{B}\chi = b\chi.$$

We have just shown that these can be made to form an orthonormal set, and hence the conclusions reached above concerning the complete parallelity between operator algebra and the algebra of the representing matrices remains in force. The particular representation in which $\mathscr{B} = \mathfrak{H}$ is called "*energy representation*".

If one makes $\mathscr{B} = \mathscr{A}$, that is if in (7.1.13) the eigenfunctions of $\mathscr{A}$ itself are used instead of the eigenfunctions of the wave equation, one obtains with the help of (7.1.20)

$$a_{kj} = (\chi_k \cdot \mathscr{A}\chi_j) = a_j (\chi_k \cdot \chi_j), \tag{7.1.25}$$

and owing to the fact that

$$(\chi_k \cdot \chi_j) = 0 \quad \text{for} \quad k \gtrless j, \quad (\chi_k \cdot \chi_k) = 1$$

it is seen that the matrix $\mathfrak{A}$ is *diagonal* and that its elements are identical with the eigenvalues of the operator $\mathscr{A}$.

If it is thus possible to find a suitable representation for an operator in which its matrix turns out to be diagonal the elements of that matrix are then identical with the quantized and thus the observable values of the physical quantity in question. This is in essence the method used for solving quantum mechanical problems in the theory developed by Heisenberg, Born, and Jordan (1925–6), originally termed "*quantum mechanics*" but better denoted as "*matrix mechanics*".

We finish this section with a further comment on the relation (7.1.19) which is supposed to represent the average of an observable quantity A in a non-stationary system. According to the theorem (4.2.8) it is possible to expand the wave function φ at any particular time t in a series of eigenfunctions of the differential equation (7.1.9) which, as we have proved, form an orthogonal system. Thus we can write

$$\varphi(t) = \sum_i c_i(t)\,\chi_i, \tag{7.1.26}$$

where the coefficients depend on time. If the expressions (7.1.26) and its conjugate complex are substituted in (7.1.19) one obtains with the help of (7.1.20)

$$\begin{aligned} \bar{A} &= \int \cdots \int \Big(\sum_i c_i^* \chi_i^*\Big)\ \mathscr{A}\ \Big(\sum_k c_k \chi_k\Big)\, dq_1 \cdots dq_N \\ &= \int \cdots \int \Big(\sum_i c_i^* \chi_i^*\Big)\Big(\sum_k c_k a_k \chi_k\Big)\, dq_1 \cdots dq_N \\ &= \sum_{ik} c_i^* c_k a_\kappa (\chi_i \cdot \chi_k) = \sum_k a_\kappa |c_\kappa|^2. \end{aligned} \tag{7.1.27}$$

The meaning of this result can best be explained by again assuming, as we have done before, that observations are made on a very large number of identical, non-stationary systems. Each of these will, at any time, be in a state where the observable quantity A has one of the values a_k. The *average* value $\bar{A}$ of A, however, will in general *not* coincide with any of these values and will be equal to the "weighted" average of all the a_k's

$$\bar{A}(t) = \sum_k a_k w_k(t), \tag{7.1.28}$$

the weighting factor w_k being equal to the relative number of systems in the quantum state k or the probability of finding any particular system in that state. Comparing (7.1.27) and (7.1.28) we see that $w_k = |c_k|^2$, which is a generalization of the theorem discussed in section 6.1.

7.2. The linear momentum and the energy operators

In this section we shall first examine in some detail the properties of the linear momentum operator which represents the classical linear momentum in quantum mechanics. For a single particle the momentum vector $\mathbf{p}$ is replaced by the vector operator (3.1.7)

$$\not{p} = \frac{\hbar}{i}\,\nabla, \tag{7.2.1}$$

and for a general system the generalized momenta p^s are represented by the operators (3.1.18)

$$\not{p}_s = \frac{\hbar}{i}\,\frac{\partial}{\partial q_s} \quad (s = 1, 2, \ldots, N). \tag{7.2.2}$$

When the expression (7.2.1) for $\not{p}$ is substituted in (7.1.14) one obtains the expressions for the elements of the "*momentum matrix*" $\mathfrak{p}$ of a particle:

$$p_{kj} = \frac{\hbar}{i}\iiint \psi_k^*\,\nabla\psi_j\,dV, \tag{7.2.3}$$

and, similarly, one can represent the components of the operator (7.2.2) by the matrices

$$\overset{s}{p}_{kj} = \frac{\hbar}{i}\int\cdots\int \psi_k^*\,\frac{\partial\psi_j}{\partial q_s}\,dq_1\cdots dq_N. \tag{7.2.4}$$

In the preceding section we postulated that an operator representing an observable quantity must be self-adjoined or Hermitian. We shall now prove that the matrices (7.2.4) are Hermitian, which implies that the operators (7.2.2) must also be Hermitian.

The expressions (7.2.4) can readily be transformed as follows:

$$\overset{s}{p}_{kj} = \frac{\hbar}{i}\int\cdots\int \frac{\partial\,(\psi_k^*\psi_j)}{\partial q_s}\,dq_1\cdots dq_N - \frac{\hbar}{i}\int\cdots\int \psi_j\,\frac{\partial\psi_k^*}{\partial q_s}\,dq_1\cdots dq_N.$$

All the integrals

$$\int_{-\infty}^{+\infty}\frac{\partial\,(\psi_k^*\psi_j)}{\partial q_s}\,dq_s$$

are zero for any values of $q_1 \cdots q_{s-1}, q_{s+1} \cdots q_N$, because ψ_k and ψ_j vanish at infinity. Hence the first integral on the right-hand side in the above equation is also zero and we obtain an alternative expression for $\overset{s}{p}_{kj}$:

$$\overset{s}{p}_{kj} = -\frac{\hbar}{i}\int\cdots\int \psi_j\,\frac{\partial\psi_k^*}{\partial q_s}\,dq_1\cdots dq_N. \tag{7.2.5}$$

On the other hand, interchanging the subscripts in (7.2.4) we have

$$\overset{s}{p}_{jk} = \frac{\hbar}{i} \int \cdots \int \psi_j^* \frac{\partial \psi_k}{\partial q_s}\, dq_1 \cdots dq_N, \tag{7.2.6}$$

where the right-hand member is the conjugate complex of that of (7.2.5). Thus we have

$$\overset{s}{p}_{kj} = (\overset{s}{p}_{jk})^*, \tag{7.2.7}$$

which is the condition for the matrices $\mathfrak{p}^s$ to be Hermitian according to the definition (4.3.7).

In keeping with the general ideas developed in section 7.1, we shall identify the diagonal elements of the momentum matrix (7.2.3) with the mean or expectation values of the momentum vector of a particle in a stationary state, characterized by the quantum number k:

$$\bar{\mathbf{p}}_k = \frac{\hbar}{i} \iiint \psi_k^* \, \nabla \psi_k \, dV. \tag{7.2.8}$$

Starting from a different consideration we had in section 6.3 obtained the expression (6.3.10) for the average of the momentum vector in a non-stationary state. This formula holds, of course, also for a stationary state when the expression (6.1.5) is to be used for φ which transforms it indeed into the present formula (7.2.8). This result, once again, is a proof for the logical consistency of the theory.

Let us now consider the differential equation (7.1.9) for the "*momentum eigenfunctions*" $\chi(\mathbf{r})$ of a particle; the eigenvalues a are here to be replaced by $\mathbf{p}$. Substituting the operator (7.2.1) for $\mathscr{A}$ we obtain the partial differential equation

$$\nabla \chi = \frac{i\mathbf{p}}{\hbar} \chi, \tag{7.2.9}$$

whose general solution is readily seen to be

$$\chi(\mathbf{r}) = C\, e^{\frac{i}{\hbar}(\mathbf{p}\cdot\mathbf{r})}, \tag{7.2.10}$$

where C is an arbitrary constant. This function has the form (1.2.8) and therefore represents a plane wave in the direction of $\mathbf{p}$ with wave vector $\mathbf{k}$ which is related to $\mathbf{p}$ by the De Broglie relation (2.2.18)

$$\mathbf{p} = 2\pi\hbar\mathbf{k} = h\mathbf{k}.$$

If no boundary conditions are imposed on the function, equation (7.2.9) has a "continuous eigenvalue spectrum" which means that the measurement of the momentum of a *free particle* can give *any* value of $\mathbf{p}$. But as χ must necessarily be finite everywhere, the exponent in (7.2.10) must be imaginary and hence p must be real. If χ has to satisfy certain boundary conditions this may result in a *discrete* eigenvalue spectrum for the momentum. If, for example, the particle is confined to the space between two parallel planes a distance l apart and normal to the direction of **k** only those De Broglie waves can establish themselves as normal modes of vibration in a stationary state whose wave numbers are integer multiples of $1/2l$, as we have found in the treatment of the problem of the "particle in the box" [eqn. (4.1.26)]. The eigenvalues p_n of the magnitude of the momentum therefore form in this case a discrete series

$$p_n = hn/2l \quad (n = 1, 2, \ldots). \tag{7.2.11}$$

We have already mentioned in section 7.1 that the generalized coordinates q_s of a system can also be regarded as operators, and since they evidently satisfy the condition (7.1.18) they are Hermitian operators which commute with each other because $q_s \cdot q_r = q_r \cdot q_s$. It also follows immediately from the definition of the linear momentum operator (7.2.2) that these operators commute with each other. We therefore have from (7.1.6)

$$[q_s \cdot q_r] = 0, \quad [\not{p}_s \cdot \not{p}_r] = 0, \tag{7.2.12}$$

for any pair of subscripts *s*, *r*. The question arises whether coordinate and momentum operators commute amongst each other.

Let us first consider the commutator belonging to the coordinate q_s and its associated momentum $\not{p}_s$. It follows from (7.2.2) that

$$\not{p}_s q_s \psi = \frac{\hbar}{i} \frac{\partial}{\partial q_s} (q_s \psi) = \frac{\hbar}{i} \left(\psi + q_s \frac{\partial \psi}{\partial q_s} \right)$$

and

$$q_s \not{p}_s \psi = q_s \frac{\hbar}{i} \frac{\partial \psi}{\partial q_s}.$$

Hence

$$(\not{p}_s q_s - q_s \not{p}_s)\, \psi = \frac{\hbar}{i} \psi \quad (s = 1, 2, \ldots, N), \tag{7.2.13}$$

or in operator form

$$[\not{p}_s q_s] = \frac{\hbar}{i} \mathscr{I} \quad (s = 1, 2, \ldots, N). \tag{7.2.14}$$

Thus the two operators $\not{p}_s$ and q_s do *not* commute. By the same argument one finds at once that

$$[\not{p}_s q_r] = 0 \quad \text{for} \quad s \gtrless r. \tag{7.2.15}$$

According to what has been said in section 7.1 the operator equations (7.2.12), (7.2.14), and (7.2.15) can also be written in matrix form, and we see that the product of the matrices $\mathfrak{p}_s$ and $\mathfrak{q}_s$ is non-commutative, all the others are commutative.

In classical physics, of course, all observables are mathematically represented by ordinary functions which obey the commutative law of ordinary algebra. There is, further, at least in principle, no limitation to the accuracy of the simultaneous measurement of any two or more of such observable quantities. On the other hand, we have shown in section 3.2 that as a result of the Heisenberg uncertainty principle (3.2.12) the simultaneous measurement of the position and the momentum of a material particle is only possible with a limited degree of accuracy, the uncertainty being of the order of magnitude of Planck's constant h. This suggests that the "*commutation rule*" (7.2.14) is the precise quantum mechanical formulation of the Heisenberg uncertainty relations for position and momentum, extended to general mechanical systems. If this is taken to be the proper interpretation of this commutation rule the generalization of the Heisenberg principle expressed by the second of the equations (3.2.13) is indeed justified.

We shall now prove quite generally that the simultaneous measurement of any two observables A and B, whose operators obey the commutation rule

$$[\mathscr{A}\mathscr{B}] = \frac{\hbar}{i}\mathscr{I} \tag{7.2.16}$$

is only possible with a limited accuracy. Let $\bar{A}$ be the average of A and $\bar{B}$ the average of B and introduce the operators

$$\Delta\mathscr{A} = \mathscr{A} - \bar{A}, \quad \Delta\mathscr{B} = \mathscr{B} - \bar{B}, \tag{7.2.17}$$

representing the deviations of A and B from their averages. As $\bar{A}$ and $\bar{B}$ are ordinary numbers it follows from (7.2.16) that

$$[\Delta\mathscr{A}\,\Delta\mathscr{B}] = (\mathscr{A} - \bar{A})\,(\mathscr{B} - \bar{B}) - (\mathscr{B} - \bar{B})\,(\mathscr{A} - \bar{A}) = \frac{\hbar}{i}\mathscr{I}. \tag{7.2.18}$$

Now consider the imaginary part of the inner product of the two functions $\Delta\mathscr{A}\psi$ and $\Delta\mathscr{B}\psi$, where ψ is an arbitrary function with norm equal to unity. From (4.1.31) we have

$$\begin{aligned} 2\,\mathrm{Im}\,(\Delta\mathscr{A}\psi \cdot \Delta\mathscr{B}\psi) &= (\Delta\mathscr{A}\psi \cdot \Delta\mathscr{B}\psi) - (\Delta\mathscr{A}\psi \cdot \Delta\mathscr{B}\psi)^* \\ &= (\Delta\mathscr{A}\psi \cdot \Delta\mathscr{B}\psi) - \Delta\mathscr{B}\psi \cdot \Delta\mathscr{A}\psi). \end{aligned} \tag{7.2.19}$$

Since the operators $\mathscr{A}$ and $\mathscr{B}$ are Hermitian we have from (7.1.18)

$$\left.\begin{aligned} (\Delta\mathscr{A}\psi \cdot \Delta\mathscr{B}\psi) &= (\psi \cdot \Delta\mathscr{A}\,\Delta\mathscr{B}\psi), \quad (\Delta\mathscr{B}\psi \cdot \Delta\mathscr{A}\psi) = (\psi \cdot \Delta\mathscr{B}\,\Delta\mathscr{A}\psi) \\ (\Delta\mathscr{A}\psi \cdot \Delta\mathscr{A}\psi) &= (\psi \cdot (\Delta\mathscr{A})^2\psi), \quad (\Delta\mathscr{B}\psi \cdot \Delta\mathscr{B}\psi) = (\psi \cdot (\Delta\mathscr{B})^2\psi), \end{aligned}\right\} \tag{7.2.20}$$

and thus (7.2.19) can, with the help of (7.2.18), be transformed into

$$\mathrm{Im}\,(\Delta\mathscr{A}\psi \cdot \Delta\mathscr{B}\psi) = \frac{1}{2}\,(\psi \cdot [\Delta\mathscr{A} \cdot \Delta\mathscr{B}]\,\psi) = \frac{1}{2}\left(\psi \cdot \frac{\hbar}{i}\,\psi\right) = -\frac{\hbar i}{2}.$$

Now the modulus of $(\Delta\mathscr{A}\psi \cdot \Delta\mathscr{B}\psi)$ can certainly not be smaller than the modulus of its imaginary part; hence the inequality

$$|(\Delta\mathscr{A}\psi \cdot \Delta\mathscr{B}\psi)| \geqslant \frac{\hbar}{2} \tag{7.2.21}$$

will hold. Further, according to the so-called "*Schwartz inequality*", the modulus of the inner product of any two functions is never larger than the product of the square roots of their norms. The norm of $\Delta\mathscr{A}\psi$ is, by its definition, the inner product of $\Delta\mathscr{A}\psi$ with itself, and hence, by virtue of the third of the equations (7.2.20), and the relation (7.1.19), one has

$$\left.\begin{aligned} (\Delta\mathscr{A}\psi \cdot \Delta\mathscr{A}\psi) &= (\psi \cdot (\Delta\mathscr{A})^2\psi) \\ &= \int \cdots \int \psi^*(\Delta\mathscr{A})^2\psi \, dq_1 \cdots dq_N = \overline{(\Delta \underline{A})^2}\,, \end{aligned}\right\} \tag{7.2.22}$$

similarly

$$(\Delta\mathscr{B}\psi \cdot \Delta\mathscr{B}\psi) = \overline{(\Delta \underline{B})^2}. \tag{7.2.23}$$

Thus finally one derives from (7.2.21), (7.2.22), (7.2.23), and Schwartz's theorem the inequality

$$\sqrt{\overline{(\Delta A)^2}} \cdot \sqrt{\overline{(\Delta B)^2}} \geqslant \frac{\hbar}{2} = \frac{h}{4\pi}, \tag{7.2.24}$$

which expresses the fact that the combined error of the simultaneous measurement of the observable quantities A and B cannot be made smaller than $h/4\pi$. This theorem holds in particular for the measurement of any of the generalized coordinates q_s and its associated momentum p_s, and thus constitutes a more rigorous formulation of the Heisenberg uncertainty principle (3.2.13).

Next in this section we wish to investigate the properties of the energy operator. Let us first consider the Hamiltonian operator which belongs to a time-independent Hamiltonian function. We know that it satisfies the wave equation (3.1.19)

$$\mathscr{H}\psi = E\psi \tag{7.2.25}$$

which, as already remarked in section 7.1, is a special case of the eigenvalue equation (7.1.9). The eigenvalues and eigenfunctions of $\mathscr{H}$ are thus identical with the eigenvalues E_n of the energy of the system, and the corresponding eigenfunctions ψ_n are the eigenfunctions of the wave equation.

Again we may represent the operator $\mathscr{H}$ by a matrix which we call the "*energy matrix*". In the "energy representation" this matrix is given by

$$H_{kj} = (\psi_k \cdot \mathscr{H}\psi_j) = \int \cdots \int \psi_k^* \mathscr{H}\psi_j \, dq_1 \cdots dq_N, \tag{7.2.26}$$

and it follows from (7.2.25) that this matrix is *diagonal*, the diagonal elements being identical with the energy eigenvalues E_k. This result follows, of course, also directly from the general theorem (7.1.25).

A diagonal matrix with real elements is evidently always self-adjoined. It follows that the Hamiltonian time-independent operator is Hermitian and thus indeed properly represents an observable quantity, the energy of the system concerned. The expectation values of the energy in the various possible quantum states are further seen to be identical with the energy eigenvalues E_k; this is, of course, trivial, since E_k is the only value of energy the system is capable of in the quantum state k.

We turn now to the time-dependent Hamiltonian operator which represents the Hamiltonian function for a non-stationary system. It can readily be shown that this operator is also Hermitian. Let ψ_k be the eigenfunctions of $\mathscr{H}$ at some arbitrary time. Let further f and g be two arbitrary functions of the coordinates which we may expand in series with respect to these eigenfunctions as follows:

$$f = \sum_k a_k\psi_k, \quad g = \sum_j b_j\psi_j, \tag{7.2.27}$$

where the a_k and b_j are (in general complex) constants. It follows from (7.2.25) and (7.2.27), in view of the fact that the functions ψ_k are orthogonal to each other, that

$$(f \cdot \mathscr{H}g) = \sum_{kj} a_k^* b_j (\psi_k \cdot \mathscr{H}\psi_j) = \sum_k a_k^* b_k E_k$$

and

$$(\mathscr{H}f \cdot g) = \sum_{kj} a_k^* b_j (\mathscr{H}\psi_k \cdot \psi_j) = \sum_k a_k^* b_k E_k$$

so that

$$(f \cdot \mathscr{H}g) = (\mathscr{H}f \cdot g), \tag{7.2.28}$$

which is, indeed, identical with the condition (7.1.18) for self-adjointness of $\mathscr{H}$.

Equation (7.2.28) must, of course, also hold if we put $f = g = \varphi$, where φ is the time-dependent wave function which satisfies the time-dependent wave equation (6.1.1)

$$\mathscr{H}\varphi = i\hbar \frac{\partial\varphi}{\partial t}; \tag{7.2.29}$$

thus

$$(\varphi \cdot \mathscr{H}\varphi) = (\mathscr{H}\varphi \cdot \varphi). \tag{7.2.30}$$

Now consider the derivative with respect to time of the "norm" of φ, that is the inner product of φ with itself. With the help of (7.2.29) we obtain

$$\frac{d\,(\varphi \cdot \varphi)}{dt} = \left(\frac{\partial\varphi}{\partial t} \cdot \varphi\right) + \left(\varphi \cdot \frac{\partial\varphi}{\partial t}\right) = \frac{i}{\hbar}(\mathscr{H}\varphi \cdot \varphi) - \frac{i}{\hbar}(\varphi \cdot \mathscr{H}\varphi),$$

and we see at once that in virtue of (7.2.30) the right-hand side of this equation is zero, so that

$$\frac{d\,(\varphi \cdot \varphi)}{dt} = \frac{d}{dt}\int \cdots \int \varphi^*\varphi \, dq_1 \cdots dq_N = 0. \tag{7.2.31}$$

Thus we may normalize the function φ in the usual way at any moment and it follows from (7.2.31) that the normalization will be preserved in the course of time. This justifies *a posteriori* the interpretation of $|\varphi|^2$ as a probability density and is the general proof of the theorem which was shown to hold for a special case in section 6.1.

The expectation value $\bar{E}$ of the energy of a non-stationary system follows direct from the relation (7.1.27) if one substitutes E_k for a_k:

$$\bar{E} = \sum_k E_k \, |c_k|^2. \tag{7.2.32}$$

Here the quantities c_k are the coefficients in the expansion

$$\varphi = \sum_k c_k \varphi_k, \tag{7.2.33}$$

which are given by the generalization of the expressions (6.1.17)

$$c_k = (\varphi_k \cdot \varphi). \tag{7.2.34}$$

This clearly demonstrates that our interpretation of the quantities $|c_k|^2$ as the probabilities for finding a member of an ensemble of non-stationary systems in the quantum state k is correct.

In section 6.1 we introduced the "*energy operator*" (6.1.2)

$$\mathscr{E} = i\hbar \frac{\partial}{\partial t}$$

with the help of which the wave equation (7.2.29) can be written in the form of the very simple operator equation

$$\mathscr{H} = \mathscr{E}. \tag{7.2.35}$$

Since $\mathscr{H}$ was shown to be a Hermitian operator it follows that the operator $\mathscr{E}$ must also be Hermitian and therefore correctly represents the observable quantity energy, similarly as the momentum operators (7.2.1) and (7.2.2) represent the linear momentum.

Since t and the coordinates q_k are independent variables $\mathscr{E}$ evidently commutes with the coordinates and the momentum operators. t itself may, of course, also be regarded as a number operator, like the coordinates, and in strict analogy to the argument concerning the commutation properties of a coordinate and the associated momentum, it follows that t and $\mathscr{E}$ are non-commuting and satisfy the commutation rule

$$[t \cdot \mathscr{E}] = \frac{\hbar}{i} \mathscr{I}. \tag{7.2.36}$$

This relation constitutes the precise quantum mechanical formulation of the Heisenberg uncertainty principle for the measurement of the energy of a system at a certain time as expressed by the first of the relations (3.2.12) and (3.2.13). Furthermore, since the commutation rule (7.2.36) has the form of the relation (7.2.16), it follows from (7.2.24) that this measurement cannot be carried out with a smaller combined error in t and E than $h/4\pi$.

The expectation value of the energy of a general, non-stationary system is, according to (7.1.19), given by

$$\bar{E}(t) = \int \cdots \int \varphi^* \mathscr{H} \varphi \, dq_1 \cdots dq_N . \tag{7.2.37}$$

If the system is closed, say, if it consists of an isolated atom, it will settle in a stationary state with time-independent wave function ψ_k and energy E_k^0. Let the corresponding Hamiltonian operator be $\mathscr{H}^0$. If now the system is subjected to an external disturbance the expectation value of the energy will be changed to $\bar{E}_k = E_k^0 + E_k'$. Let the Hamiltonian of the "perturbed" system be $H^0 + H'$ and assume that H' is small compared with H^0, that is, that the perturbation is small. Under these circumstances we may, for a first approximation, assume the wave function still to be essentially unchanged and calculate $\bar{E}_k$ from (7.2.37)

$$\begin{aligned} \bar{E}_k &= \int \cdots \int \psi_k^* \, (\mathscr{H}^0 + \mathscr{H}') \, \psi_k \, dq_1 \cdots dq_N \\ &= E_k^0 + \int \cdots \int \psi_k^* \mathscr{H}' \psi_k \, dq_1 \cdots dq_N . \end{aligned} \tag{7.2.38}$$

Thus the energy difference between the perturbed and unperturbed state of the system is, in first approximation, obtained from the relation

$$E_k' = \int \cdots \int \psi_k^* \mathscr{H}' \psi_k \, dq_1 \cdots dq_N . \tag{7.2.39}$$

Formula (7.2.39) is the simplest example of the mathematical method used in the so-called "*perturbation theory*" which finds very extensive use in many problems of atomic physics. We shall make application of it in the sections 7.3 and 8.2.

7.3. The angular momentum operator

The angular momentum P of a particle, performing an orbital movement along a circle of radius a about the centre of this circle, is classically defined by

$$P = ma^2 v = ap , \tag{7.3.1}$$

where p is its linear momentum along the path. The operator $\mathscr{P}$ representing the observable P is, as usual, obtained by the replacement of p by the operator $\frac{\hbar}{i} \frac{d}{ds}$, where ds is the line element on the circle. The corresponding azimuth angle $d\varphi$ is connected to ds by $ds = a d\varphi$; hence the expression for the "*angular momentum operator*" is

$$\mathscr{P} = \frac{\hbar}{i} \frac{\partial}{\partial \varphi}. \tag{7.3.2}$$

As this formula contains only the derivative with respect to φ it also holds more generally for the angular momentum of any rigid body rotating about an axis, fixed in space, with respect to that axis.

Formula (7.3.2) could, of course, have been directly derived from the general formula (7.2.2) by using φ as (the only) generalized coordinate with which the angular momentum P is classically associated. Hence no further proof for the Hermitian character of $\mathscr{P}$ is needed and for the validity of the commutation rule

$$[\mathscr{P}\varphi] = \frac{\hbar}{i} \mathscr{I}. \tag{7.3.3}$$

The differential equation (7.1.9) for the eigenfunctions $\chi(\varphi)$ takes the form

$$\frac{d\chi}{d\varphi} = \frac{iP}{\hbar} \chi, \tag{7.3.4}$$

whose general solution is

$$\chi = A\, e^{\frac{i}{\hbar} P\varphi}. \tag{7.3.5}$$

As χ has to be a univalued function of φ, $P/\hbar$ must be an integer. Thus the eigenvalues of $\mathscr{P}$ are

$$P_u = \hbar u \quad (u = 0, \pm 1, \pm 2, \ldots). \tag{7.3.6}$$

Equation (7.3.6) is formally identical with Bohr's quantum condition (2.2.10), and also coincides with the equation (5.2.8). But whereas the latter formula was, in section 5.2, derived for a *single* particle only, the present derivation shows that it also holds quite generally for any *one-dimensional rotator* consisting of an arbitrarily shaped rigid body.

The eigenfunctions χ_u of $\mathscr{P}$ are from (7.3.5) and (7.3.6)

$$\chi_u = A\, e^{iu\varphi}, \quad AA^* = 1/2\pi, \tag{7.3.7}$$

and therefore, according to (5.2.10), identical with the eigenfunctions of the Schrödinger equation for the one-dimensional rotator. It follows that the angular momentum matrix $\mathfrak{P}$ belonging to $\mathscr{P}$ in the "energy representation" must be identical with the matrix in the "angular momentum representation", and thus, according to the theorem (7.1.25) must be diagonal with the eigenvalues (7.3.6) as elements.

The angular momentum of a particle with position vector $\mathbf{r}$ which performs an arbitrary movement with velocity vector $\mathbf{v}$ in space, referred to the origin of coordinates, is a so-called "axial vector" (actually an "antisymmetric tensor") defined by

$$\mathbf{P} = m(\mathbf{r} \wedge \mathbf{v}). \tag{7.3.8}$$

This may also be written in terms of the linear momentum vector $\mathbf{p}$,

$$\mathbf{P} = \mathbf{r} \wedge \mathbf{p}, \tag{7.3.9}$$

or in component form:

$$\left.\begin{aligned} P_x &= yp_z - zp_y, \\ P_y &= zp_x - xp_z, \\ P_z &= xp_y - yp_x. \end{aligned}\right\} \tag{7.3.10}$$

Replacing $\mathbf{p}$ by the linear momentum vector operator (7.2.1) we obtain now the expression for the angular momentum operator of a particle with respect to the origin of coordinates

$$\mathscr{P} = \frac{\hbar}{i}(\mathbf{r} \wedge \nabla) \tag{7.3.11}$$

with components

$$\left.\begin{aligned} \mathscr{P}_x &= \frac{\hbar}{i}\left(y\frac{\partial}{\partial z} - z\frac{\partial}{\partial y}\right), \\ \mathscr{P}_y &= \frac{\hbar}{i}\left(z\frac{\partial}{\partial x} - x\frac{\partial}{\partial z}\right), \\ \mathscr{P}_z &= \frac{\hbar}{i}\left(x\frac{\partial}{\partial y} - y\frac{\partial}{\partial x}\right). \end{aligned}\right\} \tag{7.3.12}$$

As $\mathscr{P}_x = y\mathscr{p}_z - z\mathscr{p}_y$ and y commutes with $\mathscr{p}_z$ and z with $\mathscr{p}_y$ one sees at once, owing to the fact that the linear momentum operators are self-adjoined, that $\mathscr{P}_x$ must also be self-adjoined and the same holds for $\mathscr{P}_y$ and $\mathscr{P}_z$. Thus the operator (7.3.11) or (7.3.12) respectively fulfils the general requirements for being the representative of the observable quantity $\mathbf{P}$.

From (7.3.12) one obtains for the commutator of $\mathscr{P}_x$ and $\mathscr{P}_y$

$$\begin{aligned} [\mathscr{P}_x\mathscr{P}_y]\,\psi &= (\mathscr{P}_x \cdot \mathscr{P}_y\psi) - (\mathscr{P}_y \cdot \mathscr{P}_x\psi) \\ &= -\hbar^2\left\{\left(y\frac{\partial}{\partial z} - z\frac{\partial}{\partial y}\right)\left(z\frac{\partial\psi}{\partial x} - x\frac{\partial\psi}{\partial z}\right)\right. \\ &\quad \left. - \left(z\frac{\partial}{\partial x} - x\frac{\partial}{\partial z}\right)\left(y\frac{\partial\psi}{\partial z} - z\frac{\partial\psi}{\partial y}\right)\right\} \\ &= -\hbar^2\left(y\frac{\partial\psi}{\partial x} - x\frac{\partial\psi}{\partial y}\right) = i\hbar\mathscr{P}_z\psi, \end{aligned}$$

and analogous equations for the commutator of $\mathscr{P}_y$ and $\mathscr{P}_z$, and that of $\mathscr{P}_z$ and $\mathscr{P}_x$. Thus $\mathscr{P}_x, \mathscr{P}_y, \mathscr{P}_z$ satisfy the commutation rules

$$[\mathscr{P}_x\mathscr{P}_y] = i\hbar\mathscr{P}_z, \quad [\mathscr{P}_y\mathscr{P}_z] = i\hbar\mathscr{P}_x, \quad [\mathscr{P}_z\mathscr{P}_x] = i\hbar\mathscr{P}_y. \tag{7.3.13}$$

The meaning of these relations is that not all three components of the angular momentum can be measured simultaneously with unlimited accuracy, in contrast to the linear momentum, whose components according to (7.2.12) all commute with each other and thus can be accurately measured at one and the same time.

We now proceed to derive an expression for the operator $\mathscr{P}^2$ which represents the square of the magnitude P of the angular momentum vector. The operator $\mathscr{P}_x^2$ belonging to P_x^2 is obtained from the first of the equations (7.3.12)

$$\begin{aligned}\mathscr{P}_x^2 &= \frac{\hbar}{i}\left(y\frac{\partial}{\partial z} - z\frac{\partial}{\partial y}\right)\frac{\hbar}{i}\left(y\frac{\partial}{\partial z} - z\frac{\partial}{\partial y}\right)\\ &= -\hbar^2\left(y^2\frac{\partial^2}{\partial z^2} + z^2\frac{\partial^2}{\partial y^2} - 2yz\frac{\partial^2}{\partial y\,\partial z} - y\frac{\partial}{\partial y} - z\frac{\partial}{\partial z}\right),\end{aligned} \tag{7.3.14}$$

and the corresponding expressions for $\mathscr{P}_y^2$ and $\mathscr{P}_z^2$ follow from (7.3.14) by means of cyclic permutation of the subscripts.

As $P^2 = P_x^2 + P_y^2 + P_z^2$ we define $\mathscr{P}^2$ as

$$\mathscr{P}^2 = \mathscr{P}_x^2 + \mathscr{P}_y^2 + \mathscr{P}_z^2. \tag{7.3.15}$$

Using the formula (7.3.14) for $\mathscr{P}_x^2$ and the analogous ones for $\mathscr{P}_y^2$ and $\mathscr{P}_z^2$ we find, after some simple manipulations, from (7.3.15) the following expression for $\mathscr{P}^2$:

$$\begin{aligned}\mathscr{P}^2 &= -\hbar^2\left\{r^2\nabla^2 - \left(x^2\frac{\partial^2}{\partial x^2} + y^2\frac{\partial^2}{\partial y^2} + z^2\frac{\partial^2}{\partial z^2} + 2yz\frac{\partial^2}{\partial y\,\partial z}\right.\right.\\ &\qquad\left.\left. + 2zx\frac{\partial^2}{\partial z\,\partial x} + 2xy\frac{\partial^2}{\partial x\,\partial y}\right) - 2(\mathbf{r}\nabla)\right\}\\ &= -\hbar^2\{r^2\nabla^2 - (\mathbf{r}\nabla)^2 - (\mathbf{r}\nabla)\} = -\hbar^2\left\{r^2\nabla^2 - r\frac{\partial}{\partial r}\left(r\frac{\partial}{\partial r}\right) - r\frac{\partial}{\partial r}\right\}\\ &= -\hbar^2\left\{r^2\nabla^2 - \frac{\partial}{\partial r}\left(r^2\frac{\partial}{\partial r}\right)\right\}.\end{aligned} \tag{7.3.16}$$

For what is to follow it is convenient to introduce spherical polar coordinates r, θ, φ with the polar axis coinciding with the z-axis.

The last of the equations (7.3.12) can easily be transformed to the new coordinate system and then takes the form

$$\mathscr{P}_z = \frac{\hbar}{i}\frac{\partial}{\partial\varphi} \tag{7.3.17}$$

which contains the derivative with respect to φ only and is identical with (7.3.2). Thus the z-component P_z of the angular momentum is quantized, the quantum condition being

$$P_{zu} = \hbar u \quad (u = 0, \pm 1, \pm 2, \ldots), \tag{7.3.18}$$

and the eigenfunctions being given by (7.3.7).

In order to deal with the eigenvalue problem of $\mathscr{P}^2$ we make use of the expression (5.2.2) for the Laplace operator and substitute it in (7.3.16). This results in the following equation for $\mathscr{P}^2$ in terms of polar coordinates

$$\mathscr{P}^2 = -\hbar^2 \left\{ \frac{1}{\sin\theta} \frac{\partial}{\partial\theta} \left(\sin\theta \frac{\partial}{\partial\theta} \right) + \frac{1}{\sin^2\theta} \frac{\partial^2}{\partial\varphi^2} \right\} \tag{7.3.19}$$

which is *independent* of r.

The eigenvalues P^2 and eigenfunctions χ of $\mathscr{P}^2$ are, as usual, obtained from the differential equation

$$\mathscr{P}^2\chi = P^2\chi \tag{7.3.20}$$

which, because of

$$P^2 = 2mr^2K,$$

turns out to be identical with the Schrödinger equation for the two-dimensional rotator (5.2.12) when the wave function ψ there is replaced by χ. Thus P^2 is also quantized and the quantum levels are given by the expression (5.2.25), namely

$$P_l^2 = \hbar^2 l(l+1) \quad (l = 0, 1, 2, \ldots; |u| \leqslant l). \tag{7.3.21}$$

The corresponding eigenfunctions of $\mathscr{P}^2$ are given by (5.2.26).

The calculations in section 5.2 were restricted to a particle moving at a *fixed* distance from a centre. The present derivation of the quantum conditions (7.3.18) and (7.3.21), however, are valid quite *generally*. In fact it can be shown that they are not only valid for the angular momentum of a *single* particle but also for the resulting angular momentum of a *closed system* of such particles, in particular for the angular momentum of an arbitrarily shaped rigid body which can rotate about a fixed centre.

The fact that the magnitude of the angular momentum vector and also its component in the z-direction (which is, of course, arbitrary in the absence of external forces) are both simultaneously quantized may be interpreted as meaning that the direction of this vector is restricted to a finite number of angles θ. This "direction quantization" was already a feature of the older quantum theory as applied to atoms, following from the Sommerfeld quantum conditions.

Because, as mentioned above, the eigenfunctions of $\mathscr{P}^2$ are the same as the energy eigenfunctions of the two-dimensional rotator, the matrix $\mathfrak{P}^2$ belonging to the operator $\mathscr{P}^2$ in the energy representation must be diagonal with elements equal to the eigenvalues P^2 (7.3.21). The matrix $\mathfrak{P}_z$ belonging to $\mathscr{P}_z$ is, for a given quantum number l, *finite* as a result of the condition $|u| \leqslant l$, having $2l + 1$ rows and columns, while the matrix $\mathfrak{P}^2$ is infinite, just as all the other matrices representing observables which we have come across so far.

It was already mentioned in section 2.2 that, apart from the angular momentum due to its orbital movement, an electron also possesses an "intrinsic" angular spin momentum S about an internal axis. There is overwhelming experimental evidence for the assumption that *all* elementary particles have a spin which in fact can be measured. The spin thus constitutes an observable parameter in addition to the position coordinates. We therefore associate with it an operator $\mathscr{s}$, called the "*spin operator*", and assume that $\mathscr{s}$ satisfies the commutation rules (7.3.13)

$$[\mathscr{s}_x \mathscr{s}_y] = i\hbar \mathscr{s}_z, \quad [\mathscr{s}_y \mathscr{s}_z] = i\hbar \mathscr{s}_x, \quad [\mathscr{s}_z \mathscr{s}_x] = i\hbar \mathscr{s}_y. \tag{7.3.22}$$

However, relations of the type (7.3.12), which express the angular momentum operator in terms of derivatives with respect to coordinates, are not valid here because an elementary particle, by its very nature, cannot be considered to be made up of smaller parts whose movements could be observed. We therefore cannot write $\mathscr{s}_z$ in the form (7.3.17) and hence cannot conclude that the eigenvalues of $\mathscr{s}_z$ must be integer multiples of $\hbar$. But it will be shown now that, nevertheless, the eigenvalues of $\mathscr{s}_z$ form a sequence of spin momenta with *differences* equal to $\hbar$.

It follows from the commutation rules (7.3.22) that

$$\begin{aligned} \mathscr{s}_z (\mathscr{s}_x \pm \mathscr{s}_y) &= [\mathscr{s}_z \mathscr{s}_x] \pm i [\mathscr{s}_z \mathscr{s}_y] + (\mathscr{s}_x \mathscr{s}_z) \pm i (\mathscr{s}_y \mathscr{s}_z) \\ &= ((\mathscr{s}_x \pm i\mathscr{s}_y)\, \mathscr{s}_z) \pm \hbar\, (\mathscr{s}_x + i\mathscr{s}_y) = (\mathscr{s}_x \pm i\mathscr{s}_y)\, (\mathscr{s}_z \pm \hbar). \end{aligned} \tag{7.3.23}$$

Now assume χ to be an eigenfunction of s_z with eigenvalue σ, that is

$$\mathscr{s}_z \chi = \sigma \chi, \tag{7.3.24}$$

and put

$$\psi = (\mathscr{s}_x + i\mathscr{s}_y)\, \chi. \tag{7.3.25}$$

From (7.3.23), (7.3.24), and (7.3.25) we then obtain the equation

$$\mathscr{s}_z \psi = (\mathscr{s}_x \pm i\mathscr{s}_y)\, (\sigma \pm \hbar)\, \chi = (\sigma \pm \hbar)\, \psi, \tag{7.3.26}$$

which expresses the fact that $\sigma + \hbar$ and $\sigma - \hbar$ are also eigenvalues of $\mathscr{s}_z$. Thus our contention is indeed proved.

We can still write the eigenvalues of $\mathscr{s}_z$ in the form

$$S_z = s\hbar, \tag{7.3.27}$$

but the "*spin quantum number*" s will in general not be integer, only the differences between successive values of s will be equal to one. Let s_m be the largest value of s. Then s can altogether assume $2s_m + 1$ values which are distributed symmetrically about zero. It follows that the quantum numbers s can have only either *integer* or *half-integer* values. In the first case s can be equal to zero, that is the particle can exist in a spin-free state, in the second case this is not possible. As already mentioned in section 2.2, it follows from experiment that for the electron $s_m = \frac{1}{2}$. Thus the spin of an electron can have only two values in a definite direction in space, either $\hbar/2$ or $-\hbar/2$. The same apparently holds for all light elementary particles and the proton and the neutron, whereas for photons and "π-mesons" $s_m = 1$.

It can be shown that the formula (7.3.21) for the eigenvalues of the square of the angular momentum remains valid for the spin as well if l is replaced by s_m, that is

$$S^2 = \hbar^2 s_m (s_m + 1). \tag{7.3.28}$$

For $s_m = \frac{1}{2}$ in particular the relation yields the value $S^2 = \frac{3}{4}\hbar^2$. This looks peculiar at first sight as one would expect S^2 to be equal to the square of the magnitude of the spin, namely $\hbar^2/4$. The explanation of this apparent contradiction is that the spin axis may be supposed to perform a random rotation in space in the absence of external forces which is the analogue of the random translational "wobbling" motion of the particle mentioned in section 6.3. In order to measure S^2 one has to make separate measurements of S in three mutually orthogonal directions. The expectation values for these measurements are all equal to $\hbar/2$. Hence, since these three observations are independent of each other the expectation value of S^2 will indeed be equal to $3 \times \hbar^2/4$.

The spin operators $\mathfrak{s}_x, \mathfrak{s}_y, \mathfrak{s}_z$ as well as $\mathfrak{s}^2$ may also be represented by finite matrices. In particular for $s_m = \frac{1}{2}$. Pauli proposed the following representation

$$\mathfrak{S}_x = \frac{\hbar}{2}\begin{pmatrix} 0 & 1 \\ 1 & 0 \end{pmatrix}, \quad \mathfrak{S}_y = \frac{\hbar}{2}\begin{pmatrix} 0 & -i \\ i & 0 \end{pmatrix}, \quad \mathfrak{S}_z = \frac{\hbar}{2}\begin{pmatrix} 1 & 0 \\ 0 & -1 \end{pmatrix}. \tag{7.3.29}$$

The matrices (7.3.29), which are called "*Pauli matrices*", can easily be shown to obey the commutation rule (7.3.22), and are mutually anticommutative (see section 7.1). For example,

$$(\mathfrak{S}_x\mathfrak{S}_y) = \frac{\hbar^2}{4}\begin{pmatrix} i & 0 \\ 0 & -i \end{pmatrix}, \quad (\mathfrak{S}_y\mathfrak{S}_x) = \frac{\hbar^2}{4}\begin{pmatrix} -i & 0 \\ 0 & i \end{pmatrix}$$

thus

$$(\mathfrak{S}_x\mathfrak{S}_y) = -(\mathfrak{S}_y\mathfrak{S}_x), \quad (\mathfrak{S}_x\mathfrak{S}_y - \mathfrak{S}_y\mathfrak{S}_x) = \frac{\hbar^2 i}{2}\begin{pmatrix} 1 & 0 \\ 0 & 1 \end{pmatrix} = i\hbar\mathfrak{S}_z.$$

It also follows that

$$\mathfrak{S}^2 \equiv \mathfrak{S}_x^2 + \mathfrak{S}_y^2 + \mathfrak{S}_z^2 = \frac{3\hbar^2}{4}\mathfrak{I}, \tag{7.3.30}$$

in agreement with the result obtained above.

The spin quantum number s is to be added to the quantum numbers n, l, and u of an atomic particle in order to define its quantum states completely, according to its four degrees of freedom of which three are translational and one rotational. Consequently the total number of quantum states belonging to a given principal quantum number n will be twice as large as stated in section 5.3 namely $2n^2$. Thus the ground state $n = 1$ consists actually of two states with $s = \frac{1}{2}$ and $s = -\frac{1}{2}$, to $n = 2$ belong 8 states, to $n = 3$ belong 18 states, etc. We shall discuss the significance of this result later in section 8.1.

We conclude this section by briefly discussing the influence of an external magnetic field on the energy levels of an atomic electron, a problem that has a direct bearing on its angular momentum. In section 3.1 we derived the expression (3.1.26) for the Hamiltonian operator of an electron in the presence of a magnetic field with vector potential **A**. If this field is sufficiently weak we can neglect the third term in this expression in comparison with the second and thus write for the Hamiltonian operator $\mathscr{H}'$ of the "perturbation" (see section 7.2)

$$\mathscr{H}' = \frac{ie\hbar}{2mc}(\nabla \cdot \mathbf{A} + \mathbf{A} \cdot \nabla). \tag{7.3.31}$$

If the field is homogeneous in the direction of the z-axis with magnitude H we may put

$$A_x = -\mathsf{H}y/2, \quad A_y = \mathsf{H}x/2, \quad A_z = 0, \tag{7.3.32}$$

and thus obtain from (7.3.31) and (7.3.12)

$$\mathscr{H}'\psi = \frac{ie\hbar}{2mc}\left(-y\frac{\partial\psi}{\partial x} + x\frac{\partial\psi}{\partial y}\right)\mathsf{H} = -\frac{e\mathsf{H}\mathscr{P}_z}{2mc}\psi. \tag{7.3.33}$$

The expectation value for E_k', the energy increase of the kth quantum state of the electron by the application of the magnetic field follows from (7.2.39) by substitution from (7.3.33)

$$E_k' = \iiint \psi_k^* \mathscr{H}'\psi_k \, dV = -\frac{e\mathsf{H}}{2mc}\iiint \psi_k^* \mathscr{P}_z \psi_k \, dV = -\frac{e}{2mc}P_{zk}\mathsf{H}. \tag{7.3.34}$$

The energy levels E_k are thus "split up" into two levels one of which is lower than E_k (when P_z is in the direction of H) and the other higher (when P_z is in the opposite direction to H).

This result may also be derived from a different consideration. We have shown in section 6.2 that an electron performing an orbital movement within an atom produces a magnetic moment M_u in the z-direction of magnitude (6.2.23)

$$M_u = \frac{\hbar e}{2mc}u = \frac{he}{4\pi mc}u, \tag{7.3.35}$$

which, combined with (7.3.6), can be written in the form

$$M_u = \frac{e}{2mc}P_u. \tag{7.3.36}$$

The potential energy E_u of the magnetic dipole in the magnetic field H is

$$E_u = -M_u\mathsf{H} = -\frac{heu}{4\pi mc}\mathsf{H} = -\frac{e}{2mc}P_u\mathsf{H}, \tag{7.3.37}$$

a relation which is indeed identical with (7.3.34).

Two applications of these results may be shortly mentioned. Firstly, if the atom undergoes a transition simply consisting in a reversal of P_z, this will give rise either to the emission of radiation with a frequency

$$\nu = \frac{E_u}{h} = \frac{eu}{4\pi mc}\mathsf{H}, \tag{7.3.38}$$

or an absorption of radiation of this frequency which, for magnetic field strengths available in the laboratory, is in the "microwave" region of the spectrum. Secondly, any spectral line, due to an allowed transition, will be split up into a pair of spectral lines with a frequency separation equal to twice the value (7.3.38) which is proportional to the "magnetic quantum number" u and the magnetic field strength H. This is the "*normal Zeeman effect*".

Problems and exercises

P.7.1. If $\mathscr{A}$ is a vector operator then the product $\mathscr{A}\mathscr{A}^{-1}$ in equation (7.1.8), defining the reciprocal operator $\mathscr{A}^{-1}$ is to be replaced by the scalar product of $\mathscr{A}$ and $\mathscr{A}^{-1}$. What, according to this definition, is the reciprocal to the gradient operator ∇?

P.7.2. Show that any real function $F(q_1, \ldots, q_N)$ of the coordinates is a Hermitian number operator.

P.7.3. Find the conditions under which the operators $\mathscr{A} = \alpha_0(x) + \alpha_1(x)(d/dx)$ and $\mathscr{B} = \beta_0(x) + \beta_1(x)(d/dx)$ are commutative or anticommutative (a) generally, (b) if α_0 and β_0 are linear functions of x, and α_1 and β_1 are constants.

P.7.4. Derive the conditions under which the operator $\mathscr{A} = \alpha' x + \beta(d/dx)$ (α', β constants) is Hermitian, using the relation (7.1.18). What is the physical meaning of the vector operator $\mathscr{A} = \mathbf{r} + \frac{\hbar\,\Delta t}{im}\nabla$ relating to a particle of mass m?

P.7.5. Solve the eigenvalue equation (7.1.9) of the Hermitian operator

$$\mathscr{A} = \alpha x + i\beta(d/dx),$$

where α and β are real constants, under the condition that $\chi(x)$ has the same value at $x = 0$ and $x = l$.

P.7.6. Calculate the product of the errors $\sqrt{\overline{(\Delta \underline{x})^2}}$ and $\sqrt{\overline{(\Delta \underline{p_x})^2}}$ of the simultaneous measurement of the coordinate x of a particle in its movement along the x-axis and the corresponding momentum p_x for a Gaussian wave function $\psi = Ce^{-x^2/2a^2}$, where a is a real and C a complex constant. Discuss the relationship of this problem to the general formulation (7.2.24) of Heisenberg's uncertainty principle.

P.7.7. Evaluate the momentum matrix $\mathfrak{p}_{ik}$ belonging to the eigenfunctions (4.1.27), (4.1.29) of the "particle in the box" problem, in analogy to the procedure in P.4.9. Show that the matrix is Hermitian.

P.7.8. Derive the precise formulation of the Heisenberg uncertainty principle for the various quantum states of the "particle in the box" in analogy to the procedure in P.7.6.

P.7.9. Transform the expression (7.3.12) for the components of the angular momentum operator to spherical polar coordinates, and verify the relationship (7.3.17).

P.7.10. Derive the expression (7.3.19) for the square of the angular momentum operator $\mathscr{P}^2$ using the results of P.7.9.

P.7.11. Show that the magnitudes of the Pauli spin matrices (7.3.29) are equal to $\hbar/2$ and that the commutation rules (7.3.22) are obeyed for the combinations $\mathscr{P}_y\mathscr{P}_z$ and $\mathscr{P}_z\mathscr{P}_x$ as well as for $\mathscr{P}_x\mathscr{P}_y$, as proved in section 7.3.

CHAPTER 8

Many-particle Problems

THE fundamental quantum mechanical principles developed so far are valid for any type of mechanical system with a finite number of degrees of freedom. We have, in particular, applied them to systems consisting of a *single particle* under the influence of a field of force. In this chapter we shall consider closed systems made up of *many identical particles*. The term "*identical*" is used here in the sense that it shall be impossible by any sort of observation to distinguish between two or more identical particles.

All "elementary" particles of the same type are indistinguishable from each other and therefore identical, and so are complex particles, like nuclei or whole atoms, if they are all in the same quantum state.

According to the principles of classical physics, particles interact with each other when they exert forces upon each other, and in the absence of such forces the particles are *independent*. This is not so in quantum mechanics because there the particles have the properties of De Broglie waves which interfere with each other even in the absence of forces. In the following we shall first deal with systems of identical particles with negligibly weak interaction forces and then explain some of the methods for dealing with systems of interacting particles with a few applications to problems of atomic physics.

8.1. Non-interacting particles

An "*assembly*" of ν identical particles forms a mechanical system of 3ν degrees of freedom for their translational movements and another 3ν degrees of freedom for their rotational movements. We may thus describe the configuration of the system at any moment by the ν triplets of rectangular coordinates x^i, y^i, z^i and the ν triplets of rotational angles α^i, β^i, γ^i about the coordinate axes for the individual particles. With these quantities we associate ν triplets of linear momenta p^i_x, p^i_y, p^i_z and ν triplets of angular momenta P^i_x, P^i_y, P^i_z. The Hamiltonian function H of the system is the sum of the total kinetic energy K of translation and rotation of the particles and the total potential energy U of the system. K is a function of the momenta only and U a function of the coordinates only.

The Hamiltonian operator $\mathscr{H}$ of the system is, as usual, derived from the Hamiltonian function H, and if $\mathscr{H}$ is made to operate on the wave function Ψ of the system, which is a function of the 6ν coordinates, we can write the wave equation of the system in a stationary state in the form (3.1.19)

$$\mathscr{H}\Psi = E\Psi, \tag{8.1.1}$$

where E is the total energy of the system.

The kinetic energy is clearly the sum of the kinetic energies K^i of the individual particles

$$K = \sum_i K^i(\mathbf{p}^i, \mathbf{P}^i), \tag{8.1.2}$$

but the potential energy, which is made up of the mutual potential energies of the particles and the potential energies of the particles with respect to an external field, cannot be split up into a sum of the form (8.1.2). It is therefore not possible to split the wave equation (8.1.1) into ν separate wave equations for the individual particles with separate wave functions ψ^i, and the general many-particle problem can thus only be solved by solving the wave equation of the whole system under appropriate boundary conditions.

However, if the interaction forces between the particles are negligibly small, the total potential energy U of the system can be resolved into the potential energies U^i of the individual particles in the external field (which we shall assume to be time-independent) so that

$$U = \sum_i U^i(x^i, y^i, z^i; \alpha^i, \beta^i, \gamma^i). \tag{8.1.3}$$

Consequently the Hamiltonian H of the system can be expressed as the sum of Hamiltonians H^i for the particles and similarly the Hamiltonian operator $\mathscr{H}$ can be written in the form

$$\mathscr{H} = \sum_i \mathscr{H}^i \tag{8.1.4}$$

where $\mathscr{H}^i$ is an operator in the co-ordinates of the ith particle.

Let us now assume that the wave function Ψ of the system is the product of wave functions ψ^i

$$\Psi = \prod_i \psi^i \tag{8.1.5}$$

such that ψ^i is a function of the parameters x^i, y^i, z^i, α^i, β^i, γ^i only. It then follows from (8.1.4) and (8.1.5) that

$$\frac{1}{\Psi}\mathscr{H}\Psi = \frac{1}{\prod_i \psi^i}\sum_i \mathscr{H}^i \prod_i \psi^i = \sum_i \frac{1}{\psi_i}\mathscr{H}^i\psi^i,$$

and thus the wave equation (8.1.1) can for such a system be split into ν separate wave equations

$$\mathcal{H}^i \psi^i = E^i \psi^i \quad (i = 1, 2, \ldots, \nu) \tag{8.1.6}$$

with

$$E = \sum_i E^i. \tag{8.1.7}$$

The equations (8.1.6) have all the same form and hence the same set of eigenvalues E_k and the same set of eigenfunctions ψ_k^i in the coordinates of the various particles. Thus once (8.1.6) has been solved for the system in question, the wave functions Ψ of the whole system follow from (8.1.5),

$$\Psi = \prod_i \psi_k^i, \tag{8.1.8}$$

in which the functions ψ_k^i have to be chosen in such a way that the sum of the corresponding eigenvalues E_k^i is [according to (8.1.7)] equal to the given total energy E of the system.

A solution of our problem in the form (8.1.8) implies that the quantum states of all the individual constituent particles of the assembly are known. This, however, conflicts with the notion of the indistinguishability of the particles. If this is taken into account it becomes clear that the quantum state of the whole system is in fact completely determined if the *numbers* ν_k of those particles are known which, at a certain moment, are in the various possible eigenstates with quantum numbers k. The set of relative numbers ν_k/ν constitutes what is called a "*distribution*" of the assembly. The wave function Ψ that belongs to a given distribution may be constructed by linearly combining functions of the form (8.1.8)

$$\Psi = \sum_r c_r \Psi_r \tag{8.1.9}$$

with constant coefficients c_r, provided that the products Ψ_r are made up of wave functions ψ_k^i in such a way that the subscripts k appear just ν_k times in them and that the sum extends over all possible products of this type.

The modulus square of Ψ represents, as usual, the 6ν-dimensional configuration of the system in question. Now if the solution (8.1.9) is to comply with the postulate of non-distinguishability of the particles $|\Psi|^2$ must be invariant with respect to any interchange of two or more particles, and a necessary condition for this to be the case is that Ψ itself must either remain unchanged or change its sign when any two particles are interchanged. A function Ψ which satisfies the first of these conditions is called "*symmetric*" in the coordinates of the particles, and if it satisfies the second condition it is called "*antisymmetric*" in the coordinates of the particles.

Let us first consider what conditions the coefficients c_r in (8.1.9) have to fulfil in order to make Ψ symmetric. The various terms Ψ_r differ from each other only in so far as they represent different distributions of the quantum numbers k over the individual particles. Hence any interchange of particles will only result in an interchange of the quantities Ψ_r amongst each other, and if all such interchanges are to leave Ψ invariant the coefficients c_r must all be equal. Thus there is, indeed, only *one* quantum state belonging to a given distribution whose wave function (apart from an arbitrary factor which can be determined by the normalization procedure) is given by

$$\Psi_s = \sum_r \prod_i^r \psi_k^i. \tag{8.1.10}$$

We shall now show that for Ψ to be antisymmetric it must have the form of the following determinant:

$$\Psi_A = \begin{vmatrix} \psi_1^1 & \psi_1^2 & \psi_1^3 & \cdots & \psi_1^\nu \\ \psi_2^1 & \psi_2^2 & \psi_2^3 & \cdots & \psi_2^\nu \\ \cdot & \cdot & \cdot & \cdots & \cdot \\ \psi_\nu^1 & \psi_\nu^2 & \psi_\nu^3 & \cdots & \psi_\nu^\nu \end{vmatrix}, \tag{8.1.11}$$

where the superscripts, as before, refer to the particles and the subscripts to the quantum numbers constituting the given distribution. Clearly, the expansion of the determinant is a sum of the form (8.1.9), where the coefficients have the values $+1$ and -1. An interchange of two particles amounts to the interchange of two columns of the determinant (8.1.11) which causes the sign of Ψ_A to be changed. Thus ψ_A is indeed antisymmetric.

Again (8.1.11) is, if normalized in the usual way, the *only* wave function that belongs to the given distribution. But whereas in the symmetric case a wave function (8.1.10) can be found for any arbitrary distribution, this is not so in the antisymmetric case. For the determinant (8.1.11) vanishes if two or more of its rows are identical, that is if two or more of the particles of the assembly are in the same quantum state. We therefore see that only such distributions are compatible with an antisymmetric wave function of the system in which each of the quantum states k is either unoccupied or occupied by one particle only.

The fact that assemblies of certain types of identical particles have symmetric wave functions and others antisymmetric wave functions has a very important bearing on the statistical behaviour of these systems and hence on their macroscopic physical properties. Vice versa it may be inferred from the observation of these properties whether a particular type of particle is associated with symmetric or antisymmetric wave functions. It has thus been possible to establish the rule that assemblies of particles with half-integer spin (see section 7.3) have always antisymmetric wave functions and assemblies of particles with integer spin have always symmetric wave functions. This will be discussed more fully in sections 14.2 and 14.3.

Of the "elementary particles" the electrons, for instance, have half-integer spins. Hence assemblies of electrons have antisymmetric wave functions. This fact has important consequences on the distribution of atomic electrons which we shall briefly discuss in the following section 8.2 and on the properties of free electrons in metals which will be discussed in section 15.1. On the other hand, photons are elementary particles with integer spin, and thus assemblies of photons have symmetric wave functions. This fact determines the properties of the so-called "black-body radiation" which we shall discuss in section 15.3. Finally, atoms are composed of protons, neutrons, and electrons which all have half-integer spins. The whole atom may therefore have either an integer or a half-integer spin, depending on the number of constituents. Consequiently atomic gases are assemblies which also have eithe symmetric or antsymmetric wave functions. This becomes significant for their mechanical and thermodynamic behaviour at very low temperatures as will be discussed in section 15.1.

8.2. Interacting particles

The methods used in the preceding section for the treatment of quantum mechanical problems concerning systems of identical particles can only be used if the interaction forces between the particles are negligibly small. If this is not the case the wave function Ψ of the system cannot be composed from eigenfunctions ψ_k^i for the single particles. Nevertheless, the postulate of the indistinguishability of the particles still demands that Ψ be invariant with respect to any interchange of particles, and hence Ψ, as before, must be either symmetric or antisymmetric in the coordinates of the particles.

Although in a system of interacting particles it is not possible to assign separate wave functions to the individual particles, one may nevertheless be able to assign to them sets of quantum numbers defining definite quantum states. If this is so, then the conclusions arrived at in the previous section concerning particles with half-integer spin will still hold, namely that any of these quantum states cannot be occupied by more than one particle at a time.

An example of such a system is the outer electronic part of an atom. The analysis of atomic spectra has indeed revealed that all the spectral lines are due to transitions of the electrons between different quantum states which, as in the case of the one-electron problem discussed in section 7.3, are described by the assignment of four definite quantum numbers n, l, u, s to the various electrons. It is very remarkable that, even before the advent of quantum mechanics, and on the basis of the older Bohr–Sommerfeld quantum theory, Pauli in 1925 was able to pronounce a new and general principle, usually called the "*Pauli exclusion principle*", according to which no two of the atomic electrons must have the same set of four quantum numbers. We now see that this exclusion principle is actually the outcome of the fact that the wave function of the whole system of atomic electrons must be antisymmetric in the (position and spin) coordinates of the electrons.

The Pauli exclusion principle also provides the clue for the understanding of the structure of the electronic system surrounding the nucleus of an atom and for the understanding of the periodic table of the elements. The electrons sharing the same principal quantum number n are said to form a "*shell*" and the different shells are denoted by letters. Thus the shell belonging to $n = 1$, having the lowest energy and being nearest to the nucleus, is called the "K-shell", the shell belonging to $n = 2$, having a higher energy and surrounding the K-shell, is called the "L-shell", etc. In section 7.3 we showed that the number of separate quantum states belonging to the principal quantum number n is equal to $2n^2$. Hence a shell cannot accommodate more than $2n^2$ electrons, and if it does contain this maximum number of electrons it is called "*complete*". Thus the completed K-shell contains two electrons, the completed L-shell eight electrons, etc. The spin axes of half the electrons in a complete shell point in one direction and those of the other half in the opposite direction, and it follows from reasons of symmetry that the total angular momentum (orbital plus spin) of a complete shell is always zero. Consequently its magnetic moment is also zero.

In order to calculate the energy levels of the various quantum states of an atom containing more than one electron one can make use of the perturbation method introduced in section 7.2. We shall restrict ourselves here to the application of this method to the helium atom with $Z = 2$ and two electrons in the complete K-shell.

Let $\psi_k^1\,(r, \theta, \varphi)$ be the eigenfunction of electron 1 in the quantum state k in the absence of the other electron, k standing for the set of quantum numbers n, l, u. The explicit form of this function is given by the relations (5.2.26) and (5.3.22). Similarly, let ψ_j^2 be the eigenfunction of electron 2 in the quantum state j in the absence of electron 1. If the interaction between the two electrons is neglected the eigenfunctions ψ_{kj} for the system of both electrons is, from (8.1.10) and (8.1.11),

$$\Psi_{kj} = C\,(\overset{1}{\psi_k}\overset{2}{\psi_j} \pm \overset{2}{\psi_k}\overset{1}{\psi_j}), \tag{8.2.1}$$

where the upper sign holds if ψ is symmetric in the position coordinates of the two electrons, that is if the spins of the two electrons are *antiparallel* ("*parahelium*"), and the lower sign holds if ψ is antisymmetric in these coordinates, that is if their spins are *parallel* ("*orthohelium*"); C is a normalization factor.

The Coulomb interaction between the two electrons constitutes the perturbation and the potential energy of the perturbation is $U' = e^2/|\mathbf{r}_1 - \mathbf{r}_2|$, where $\mathbf{r}_1$ and $\mathbf{r}_2$ are the position vectors of the electrons. Thus the Hamiltonian operator for the perturbation is simply

$$\mathscr{H}' = \frac{e^2}{|\mathbf{r}_1 - \mathbf{r}_2|}, \tag{8.2.2}$$

and the shift E'_{kj} of the energy levels from their values in the absence of interaction between the electrons is, according to (7.2.39), obtained from

$$E'_{kj} = \int \cdots \int \Psi^*_{kj} \mathscr{H}' \Psi_{kj} dV_1 dV_2. \tag{8.2.3}$$

In particular, for the ground state of the helium atom, one has from (5.3.25)

$$\psi_1^1 = \frac{1}{\sqrt{\pi a^3}} e^{-r_1/a}, \quad \psi_1^2 = \frac{1}{\sqrt{\pi a^3}} e^{-r_2/a}, \tag{8.2.4}$$

where a is the Bohr radius for the helium ion. It is evident from equation (8.2.1) that for the ground state Ψ_{11} must be symmetric and the spins of the electrons antiparallel, which complies with Pauli's principle:

$$\Psi_{11} = \frac{1}{\pi a^3} e^{-(r_1+r_2)/a}. \tag{8.2.5}$$

Thus, from (8.2.2) and (8.2.3),

$$E'_{11} = \frac{e^2}{\pi^2 a^6} \int \cdots \int \frac{1}{|\mathbf{r}_1 - \mathbf{r}_2|} e^{-2(r_1+r_2)/a} dV_1 dV_2. \tag{8.2.6}$$

Evaluation of the integral in (8.2.6) leads finally to the value

$$E'_{11} = \frac{5}{8} \frac{e^2}{a}, \tag{8.2.7}$$

which indicates that, as a result of the repulsion between the electrons, the energy level of the ground state is increased by this amount. The energy of the ground state itself, when this interaction is neglected, follows from (5.3.17) and (5.3.18):

$$E^0_{11} = -\frac{2e^2}{a}; \tag{8.2.8}$$

thus the energy shift amounts to about 30 per cent.

Let us now consider a system of two isolated atoms in close proximity. Electrons may then change over from the outermost shell of one of the atoms to that of the other and vice versa. Processes of this kind are responsible for the *chemical* interaction between atoms and the *formation of molecules*. There are two extreme cases of such electron interchanges between two atoms. Firstly, one or more electrons may detach themselves permanently from one of the atoms and occupy quantum states in the outermost shell of the other atom. This results in the formation of *ions* and an attracting electrostatic force between the oppositely charged ions, leading to a "*heteropolar*" chemical compound. Secondly, equal numbers of electrons may be continuously exchanged between two atoms. This process leaves the atoms electrically neutral but, as was first pointed out by Heitler and London in 1927, gives rise to so-called "*exchange forces*" between the atoms which are responsible for the formation of "*homoeopolar*" chemical compounds. In general, however, chemical "*valencies*" or "*bonds*" will be due to a mixture of both types of electron interchange processes. The development of the quantum mechanical theory of chemical processes is mainly due to Pauling and Coulson.

The electronic cloud of an isolated pair of atoms evidently forms a closed system of interacting identical particles of the type considered in the present section and, from what has been said before, it appears that its wave function must be antisymmetric in the positions and spins of the two electrons. In order to calculate the interaction forces between the atoms one has to solve the wave equation (8.1.1) for a given distance R between the nuclei of the two atoms and to determine the eigenvalue $E_1(R)$ of the energy in the lowest quantum state of the system which belongs to the antisymmetric eigenfunction. The force between the atoms is then, according to (1.1.5), equal to $-dE_1/dR$ and a stable molecule will be formed if, at a certain distance R_0 the energy E_1 has a minimum.

In general it is only possible to solve problems of this type by the use of approximation methods. As an example we shall consider here one of the simplest cases occurring in the quantum theory of molecules, namely the homoeopolar bond between two hydrogen atoms at such distances R that the interaction energy between the atoms is small compared with the internal energies of the isolated atoms. For this purpose we shall make use of the perturbation method introduced in section 7.2.

Let $\mathbf{R}_1$ and $\mathbf{R}_2$ be the position vectors of the two nuclei, and $\mathbf{r}_1$ and $\mathbf{r}_2$, as before, the position vectors of the two electrons; then $R = |\mathbf{R}_1 - \mathbf{R}_2|$. According to (7.2.38) the expectation value of the energy E_1 of the system in the ground state is

$$\bar{E}_1(R) = \int \cdots \int \Psi_1^* \mathscr{H} \Psi_1 \, dV_1 \, dV_2, \tag{8.2.9}$$

where $\Psi_1(\mathbf{r}_1, \mathbf{r}_2)$ is the eigenfunction of the unperturbed system in the ground state, that is when the two atoms are completely separated ($R = \infty$), and $\mathscr{H}$ the Hamiltonian operator of the perturbed system.

As exchanges of electrons between the two atoms can take place, electron 1 may be attached to atom 1 or atom 2. Let the corresponding single-particle wave functions be $\psi_1^1 = \psi_1(\mathbf{r}_1)$ and $\psi_2^1 = \psi_2(\mathbf{r}_1)$. Similarly, for electron 2 the single-particle wave functions are $\psi_1^2 = \psi_1(\mathbf{r}_2)$ and $\psi_2^2 = \psi_2(\mathbf{r}_2)$. We know that ψ must be antisymmetric in the complete set of position coordinates and spins of the electrons. There are, therefore, two possibilities: either the spins of the two electrons are *parallel*, in which case Ψ is symmetric in the spin parameter and therefore must be *antisymmetric* in the coordinates of the electrons, or the spins of the electrons are *antiparallel* when Ψ is antisymmetric in the spin parameter and consequently must be *symmetric* in the coordinates of the electrons. It thus follows from (8.1.10) and (8.1.11) that Ψ must have the form

$$\Psi = C(\overset{1}{\psi}_1\overset{2}{\psi}_2 \pm \overset{2}{\psi}_1\overset{1}{\psi}_2), \tag{8.2.10}$$

where C is a normalization factor and the upper sign is to be used if Ψ is to be symmetric in the position vectors $\mathbf{r}_1, \mathbf{r}_2$ and the lower sign if it is to be antisymmetric in $\mathbf{r}_1, \mathbf{r}_2$.

From the expression (5.3.25) for the eigenfunction of the ground state of the hydrogen atom we obtain immediately (leaving out the normalization factors)

$$\left.\begin{aligned} \overset{1}{\psi}_1 &= e^{-|\mathbf{r}_1-\mathbf{R}_1|/a}, & \overset{1}{\psi}_2 &= e^{-|\mathbf{r}_1-\mathbf{R}_2|/a}, \\ \overset{2}{\psi}_1 &= e^{-|\mathbf{r}_2-\mathbf{R}_1|/a}, & \overset{2}{\psi}_2 &= e^{-|\mathbf{r}_2-\mathbf{R}_2|/a}, \end{aligned}\right\} \tag{8.2.11}$$

where a is the Bohr radius for the hydrogen atom which follows from (5.3.18) for $Z = 1$.

The Hamiltonian of the system is equal to the sum of the kinetic energies of the two electrons and the electrostatic mutual potential energies of the electrons of charge $-e$ and the nuclei of charge $+e$. Hence the Hamiltonian operator is

$$\mathscr{H} = -\frac{\hbar^2}{2m}(\nabla_1^2 + \nabla_2^2)$$

$$+ e^2\left(\frac{1}{R} + \frac{1}{|\mathbf{r}_1 - \mathbf{r}_2|} - \frac{1}{|\mathbf{r}_1 - \mathbf{R}_1|} - \frac{1}{|\mathbf{r}_1 - \mathbf{R}_2|} - \frac{1}{|\mathbf{r}_2 - \mathbf{R}_1|} - \frac{1}{|\mathbf{r}_2 - \mathbf{R}_2|}\right). \tag{8.2.12}$$

The interaction energy E' as a function of the distance R between the nuclei is now obtained from (8.2.9) wih the help of (8.2.10), (8.2.11), and (8.2.12). It is easily verified that E' tends to zero for large R and to e^2/R for small R. It is not possible to evaluate the integrals in a closed form but they can, of course, be computed numerically. The resulting function $E'(R)$ is represented by the two curves of Fig. 11 for the symmetric and the antisymmetric case.

The function $E'(R)$ corresponding to a wave function which is *antisymmetric* in the coordinates of the electrons shows a monotonic decrease with increasing distance R; thus $dE'/dR < 0$ which indicates a *repulsive* force between the atoms. The curve belonging to a *symmetric* wave function, on the other hand, has a minimum for a certain distance R_0, which means that the two atoms are in *stable equilibrium* at this distance, the force between the atoms being *attractive* for $R > R_0$ and *repulsive* for $R < R_0$. It is further seen that, although these results have been derived by means of a first-order perturbation method under the assumption that the interaction energy is small enough, the value of E' obtained in this way for $R = R_0$ is only about 10 per cent of the self-energies of the separated atoms while R_0 is actually smaller than $2a$. Thus the application of the perturbation method to this problem is seen to be justified.

It appears from this consideration that it is indeed possible to account for the formation of hydrogen molecules and the existence of homoeopolar bonds between the atoms on the basis of the idea of exchange forces, and that a necessary condition for a homoeopolar chemical reaction between two atoms to take place is that their wave functions must be antisymmetric in the spins of the exchanging electrons or, in other words, that the spins of these electrons must be opposite in direction to each other. Thus in a saturated homoeopolar chemical compound the electronic spins compensate each other.

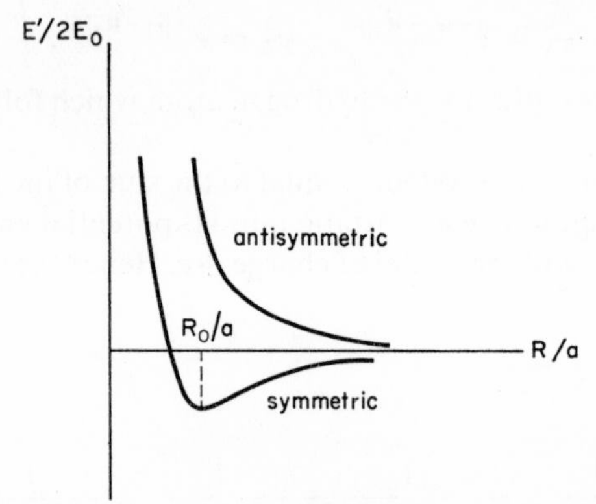

FIG. 11. Interaction energy of two hydrogen atoms for symmetric and antisymmetric wave functions, demonstrating the formation of a hydrogen molecule.

Another method for dealing with problems of the exchange of electrons between atoms consists in considering the wave function ψ of *one* of the electrons concerned only which represents the movement of that electron while it is shared between the two nuclei. For this reason it is called the method of "*molecular orbitals*" (Lennard-Jones, 1929). It has the advantage that it makes it possible to visualize the movement of the electron in the form of a continuous space charge distribution according to the ideas developed in section 6.2. The joint space charge density ϱ_e of all the electrons involved may then be constructed as a superposition of their individual charge densities. This in turn enables one to evaluate the exchange forces between the two atoms as the result of the Coulomb interaction between the negatively charged "electronic cloud", on the one hand, and the positively charged nuclei. This procedure cannot be followed in the previously discussed method since the wave function Ψ can only be depicted in the form of a continuous distribution in a *six-dimensional* space, and hence it is not immediately apparent that the quantum mechanical exchange forces are only a special case of electrostatic interaction between systems of charged particles.

Let us again consider two atoms between which an exchange of electrons may take place. We know that the wave function Ψ of the whole system must have definite symmetry properties regarding the exchange of electrons between the atoms. We may equally well describe these symmetry properties in terms of exchanges of the two nuclei with respect to the electronic system. Thus we shall postulate that the one-particle wave function ψ of an electron used in the method of molecular orbitals shall be either symmetric or antisymmetric with respect to an exchange of the two nuclei to which it may be attached.

We shall again assume that the distance between the atoms is large enough so as to make the interaction energy small compared with the energies of the completely separated atoms. Let ψ_1 be the eigenfunction of the electron in the lowest quantum state when attached to nucleus 1 and ψ_2 the same function referring to nucleus 2. A wave function ψ of the form

$$\psi_s(\mathbf{r}) = C_s(\psi_1 + \psi_2) \tag{8.2.13}$$

will then evidently be symmetric with respect to an exchange of the two nuclei and a wave function of the form

$$\psi_a(\mathbf{r}) = C_a(\psi_1 - \psi_2) \tag{8.2.14}$$

will be antisymmetric with respect to this process. The corresponding charge density distribution is in both cases, according to (6.2.3) given by

$$\varrho_e(\mathbf{r}) \equiv -e\psi\psi^*. \tag{8.2.15}$$

In order to see how this method works in detail we shall consider as a simple one-dimensional example the exchange interaction of two linear harmonic oscillators, each consisting of a heavy particle with charge $+e$ and an electron with charge $-e$. We assume that the heavy particles are fixed at the points $R/2$ and $-R/2$ respectively of the x-axis and that the electrons can perform oscillations along the x-axis with these points as centres.

From (5.1.20) we obtain at once, apart from an arbitrary factor,

$$\psi_1 = e^{-\frac{1}{2}(\xi-\delta)^2}, \quad \psi_2 = e^{-\frac{1}{2}(\xi+\delta)^2}, \tag{8.2.16}$$

where we have introduced the dimensionless quantities

$$\xi = x/a, \quad \delta = R/2a, \tag{8.2.17}$$

a being the classical amplitude of the oscillations of the dipoles. Thus according to (8.2.13) and (8.2.14) we have in first approximation

$$\left.\begin{aligned} \psi_s(\xi) &= C_s\left(e^{-\frac{1}{2}(\xi-\delta)^2} + e^{-\frac{1}{2}(\xi+\delta)^2}\right), \\ \psi_a(\xi) &= C_a\left(e^{-\frac{1}{2}(\xi-\delta)^2} - e^{-\frac{1}{2}(\xi+\delta)^2}\right). \end{aligned}\right\} \tag{8.2.18}$$

The constants C_s and C_a in these expressions are determined by the normalization condition

$$\int_{-\infty}^{+\infty} \psi\psi^* \, d\xi = 1, \tag{8.2.19}$$

which yields

$$C_s^2 = \frac{1}{2\sqrt{\pi}(1 + e^{-\delta^2})}, \quad C_a^2 = \frac{1}{2\sqrt{\pi}(1 - e^{-\delta^2})}. \tag{8.2.20}$$

The shape of the functions (8.2.18) for $\delta = 1{\cdot}5$ is shown in Fig. 12.

The electric density distribution associated with the pair of electrons shared between the two oscillators, which in the present case is a "linear density" (electric charge per unit length) is, according to (8.2.15), proportional to $\psi\psi^* = \psi^2$ and thus proportional to

$$\varrho_e(\xi) \sim \{e^{-(\xi-\delta)^2} + e^{-(\xi+\delta)^2} \pm 2e^{-(\xi^2+\delta^2)}\}, \tag{8.2.21}$$

where the upper sign holds for the symmetric and the lower sign for the antisymmetric case. in Fig. 13 curves representing this function for two distances corresponding to $\delta = 0{\cdot}5$ and $\delta = 1{\cdot}5$ are shown for both cases.

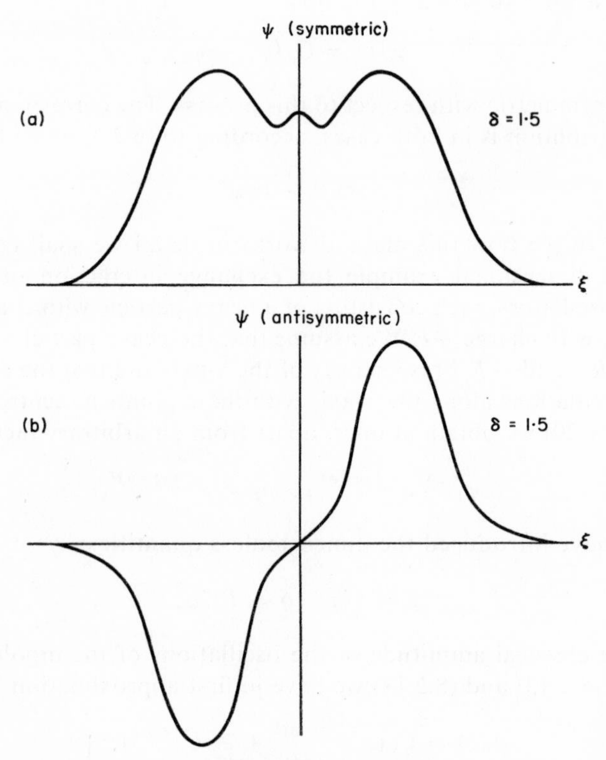

FIG. 12. Symmetric and antisymmetric orbital wave functions for two one-dimensional harmonic oscillators.

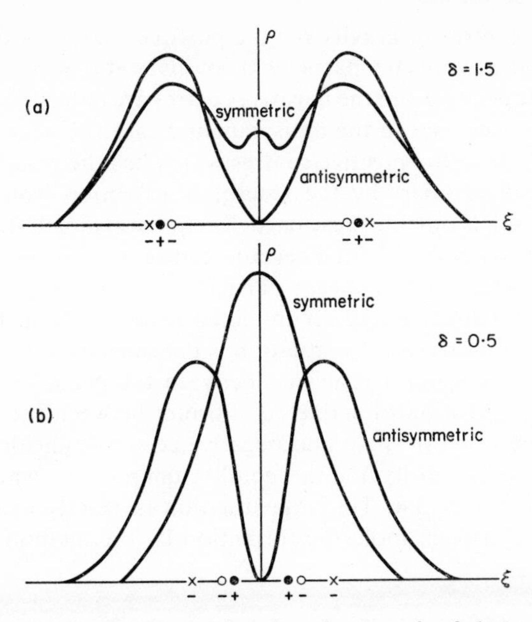

FIG. 13. Electric density distribution functions for the pair of electrons shared between two linear harmonic oscillators for symmetric and antisymmetric orbital wave functions at two distances of the positive centres, illustrating the mechanism of the exchange interaction.

Instead of explicitly evaluating the interaction forces between the negative electronic cloud and the positive centres of oscillations we shall restrict ourselves to calculating the positions $\bar{\xi}$ of the "electric centres of gravity" of the right-hand and of the left-hand parts of the distribution (8.2.21) in which we may imagine the charges $-e$ of the two electrons to be concentrated. $\bar{\xi}$ is defined by

$$\bar{\xi} = \pm 2 \int_0^{\infty} \xi \psi^2(\xi)\, d\xi. \tag{8.2.22}$$

Substituting from (8.2.18) into (8.2.22) we obtain, after evaluating the simple integral s,

$$\left.\begin{aligned} \bar{\xi}_s &= \pm 2C_s^2 \left(\delta \sqrt{\pi} \operatorname{erf} \delta + 2e^{-\delta^2}\right) \\ \bar{\xi}_a &= \pm 2C_a^2 \delta \sqrt{\pi} \operatorname{erf} \delta \end{aligned}\right\} \tag{8.2.23}$$

and using the values (8.2.20) for the constants C_s and C_a we obtain finally

$$\bar{\xi}_s = \pm \delta \frac{\operatorname{erf} \delta + \dfrac{2}{\delta \sqrt{\pi}} e^{-\delta^2}}{1 + e^{-\delta^2}}, \qquad \bar{\xi}_a = \pm \delta \frac{\operatorname{erf} \delta}{1 - e^{-\delta^2}}. \tag{8.2.24}$$

The electric centres of gravity for the positive charges $+e$ are, of course, identical with the two heavy particles themselves at $\xi = \delta$ and $\xi = -\delta$. The positions of the positive and the negative centres are indicated in the diagrams of Fig. 13. It is seen that in the antisymmetric case the negative centres ($\times$) are on the *outside* of the positive centres which has the result that the heavy particles are pulled apart by the Coulomb attraction from the electronic cloud, whose bulk is outside these particles. In the symmetric case we see that for the larger distance $\delta = 1{\cdot}5$ the negative centres ($\bigcirc$) are on the *inside* of the positives ones, and hence the heavy particles are drawn together by virtue of the negative electronic charge accumulated between them. For the smaller distance $\delta = 0{\cdot}5$, however, the situation is the same as in the antisymmetric case, and we have again a repulsion between the particles. It follows that there must be a position of stable equilibrium between the oscillators at a certain distance when positive and negative centres coincide. By numerical computation one can easily find this equilibrium position which corresponds to approximately $\delta \simeq 3/4$. Thus the situation is exactly analogous to that found for the hydrogen molecule formation by the method of Heitler and London.

Problems and exercises

P. 8.1. Compile a table of the combinations of quantum numbers n, l, u, s for the electrons of the Ne-atom and show that they satisfy Pauli's exclusion principle.

P. 8.2. Prove that the wave function belonging to an assembly of atoms, consisting of an even number of protons, neutrons, and electrons, is always symmetric in the coordinates of the atoms as a whole, and that it is antisymmetric if the total number of protons, neutrons, and electrons is odd.

P. 8.3. Using equations (8.2.7) and (8.2.8) calculate the energies necessary to produce an H^+- and an H^{++}-ion respectively.

P. 8.4. Use equations (8.2.9) and (8.2.12) to derive an explicit expression for the interaction energy E' of the atoms in a hydrogen molecule as a function of the distance R between the nuclei, and show that $E' \to 0$ for $R \to \infty$, and $E' \simeq e^2/R$ for very small R.

PART II

Relativity Theory

CHAPTER 9

The Classical Principle of Relativity

IT WAS pointed out in the introduction that the belief in a universal preferential frame of reference for the mathematical formulation of the laws of physics was a basic principle of classical physics. This frame of reference was identified with a physical substance, the "ether", supposed to fill the whole universe and to act as medium for the propagation of light and as the "carrier" of the electromagnetic field. Nevertheless, the principle of relativity, according to which the laws of physics should be independent of the frame of reference, has a certain restricted validity in classical physics. In this chapter it will be demonstrated to what degree of accuracy the "classical principle of relativity" holds in mechanics, optics, and electrodynamics, and it will be shown that experimental evidence demands the principle of relativity to be precisely valid which is incompatible with classical physics.

9.1. Relativity in Newtonian dynamics

In Chapter 1 we already formulated the Newtonian laws of motion for a system of N material particles in the form of the vector equations

$$m_i\mathbf{a}_i \equiv m_i\ddot{\mathbf{r}}_i \equiv m_i\dot{\mathbf{v}}_i = \mathbf{F}_i \quad (i = 1, 2, \ldots, N). \tag{9.1.1}$$

These equations clearly have a definite meaning only if the position vectors $\mathbf{r}_i$ and the velocity vectors $\mathbf{v}_i$ of the various particles are defined with respect to a specified reference system. In order to make it possible to measure the positions and velocities of the particles at any time, it is necessary to use a system of coordinates whose origin and axes are in some way connected with material bodies. The first task is therefore to see whether a material frame of reference can indeed be found with respect to which the equations (9.1.1) are precisely valid. Such a system of reference is called an "*inertial system*" because the law of inertia will be valid with respect to such a system, that is, in the absence of forces ($\mathbf{F}_i = 0$) the velocities $\mathbf{v}_i$ of the particles remain constant.

The simplest choice for a frame of reference would be a system of rigid bodies fixed with respect to the earth. Indeed, Newton's laws of motion were suggested by the outcome of mechanical experiments in terrestrial laboratories, and they were subsequently found to be valid to a high degree of accuracy with respect to such "laboratory systems of reference". However, from various experiments, like the famous Foucault pendulum experiment, it became clear that the Newtonian equations of motion were not precisely valid for such terrestrial reference systems but for celestial ones, defined by the positions of "fixed stars". In actual fact the stars are known to be not really "fixed" but to be in motion with respect to each other with very considerable speeds. However, on a human time scale the displacements of the stars are negligibly small and we may therefore state (though with a certain lack of precision) that the system of the fixed stars "as a whole" constitutes a valid inertial frame of reference.

Newton himself was of the opinion that it must be possible to define a preferential inertial system without reference to material bodies, and he identified it with the "absolute space" which, in his own words, he described as "remaining always similar and immovable without regard to any thing external". From this metaphysical point of view the notion of "absolute movement" of a body, as distinct from its relative motion with respect to other bodies, has a definite meaning. From the point of view of the physicist, however, absolute space has no meaning unless it is endowed with some physical properties which make it possible to ascertain by physical measurements whether a body is in "absolute movement" or at "absolute rest". The concept of "ether" fulfils this requirement, and we may therefore also think of the ether as the inertial frame of reference. We shall return to this concept in sections 9.2 and 9.3.

A further requirement for the equations of motion (9.1.1) to have a definite meaning is that it must be specifically stated in what way the quantities "length" and "time" are to be measured which appear in these equations. The usual measurement of length is based on the use of "measuring rods" made of "rigid" bodies, that is of bodies whose dimensions have invariable relationships to each other. The usual measurement of time is based on the rotational movement of the earth, which is practically not influenced by external forces, and the fact that a great variety of mechanisms, called "clocks", can be devised in such a way that they can be "synchronized" with each other and the rotational movement of the earth, that is that the displacement of their "pointers" can be made to keep in precise proportions to each other. Experience has indeed shown that the Newtonian equations of motion are valid if length and time measurements are performed with rods and clocks of this description.

Again, Newton was of the opinion that there existed what he called "absolute time", namely something which "from its own nature flows equally without regard to any thing external", and that the variable t appearing in his equations was meant to represent absolute time. Whether one accepts this metaphysical thesis or not the fact remains that the above described method of time measurement was supposed to be valid for the whole realm of classical physics, irrespective of the frame of reference used.

Let us now consider two systems of reference S and S' such that the system S' as a whole moves with respect to S in such a way that a point P with position vector $\mathbf{r}$ and *fixed* in S' moves with velocity $\mathbf{V}(\mathbf{r}, t)$. Call $\mathbf{v}$ the velocity of a particle with respect to S and $\mathbf{v}'$ its velocity with respect to S'. The law of kinematics which is incorporated in classical physics, sometimes called the "*law of superposition of velocities*", maintains that if a point performs two simultaneous movements the corresponding velocities add up according to the rule of vector addition. Thus in the present case the velocities $\mathbf{v}$, $\mathbf{v}'$, and $\mathbf{V}$ are related by

$$\mathbf{v} = \mathbf{v}' + \mathbf{V}. \tag{9.1.2}$$

One may be inclined to regard the relationship (9.1.2) as self-evident. But this is a fallacy, for the law of superposition of velocities is not a geometrical theorem which can be logically deduced from the Euclidean axioms, since it must necessarily depend on the methods used for measuring length and time. Thus the rule (9.1.2) for the vector addition of velocities is an empirical law which has been found to be valid for velocities which are small compared with the velocity of light.

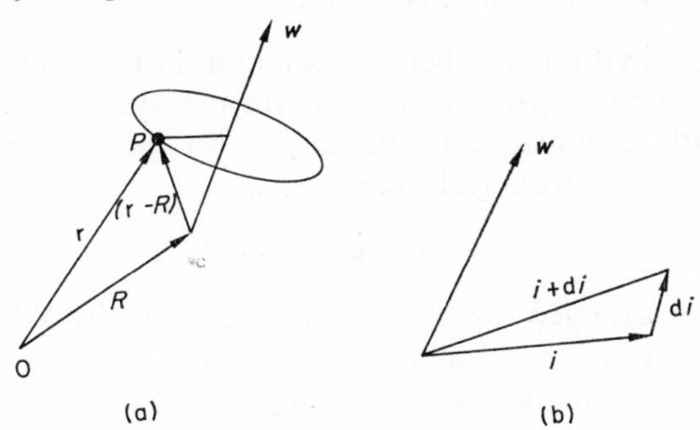

FIG. 14. Diagram illustrating the derivation of the relationship between the accelerations of a mass point in two different frames of reference.

We now wish to establish a relationship between the accelerations $\mathbf{a}$ and $\mathbf{a}'$ of our mass point with respect to S and S'. It is evidently possible to split the velocity $\mathbf{V}$ into a translational and a rotational component within a short time interval dt. Let $\mathbf{u}(\mathbf{r}, t)$ be the translational velocity, $\mathbf{R}(t)$ the position vector of the instantaneous centre of rotation, and $\mathbf{w}(t)$ the rotation vector whose direction is that of the axis of rotation and whose magnitude is equal to the angular velocity ω. It follows that (Fig. 14a)

$$\mathbf{V}(\mathbf{r}, t) = \mathbf{u}(\mathbf{r}, t) + \mathbf{w}(t) \wedge (\mathbf{r} - \mathbf{R}). \tag{9.1.3}$$

Thus, with the help of (9.1.2),

$$\dot{\mathbf{V}} = \dot{\mathbf{u}} + \dot{\mathbf{w}} \wedge (\mathbf{r} - \mathbf{R}) + \mathbf{w} \wedge (\mathbf{v} - \dot{\mathbf{R}})$$

$$= \dot{\mathbf{u}} + \dot{\mathbf{w}} \wedge (\mathbf{r} - \mathbf{R}) + \mathbf{w} \wedge \mathbf{v}' + \mathbf{w} \wedge \mathbf{u} + \mathbf{w} \wedge [\mathbf{w} \wedge (\mathbf{r} - \mathbf{R})] - \mathbf{w} \wedge \dot{\mathbf{R}}. \tag{9.1.4}$$

Let further **i, j, k** be unit vectors in the directions of the axes of a coordinate system defining S'. We can then write

$$\mathbf{v}' = \mathbf{i}v_x' + \mathbf{j}v_y' + \mathbf{k}v_z'. \tag{9.1.5}$$

Differentiating (9.1.5) with respect to time results in

$$\dot{\mathbf{v}}' = \mathbf{a}' + \dot{\mathbf{i}}v_x' + \dot{\mathbf{j}}v_y' + \dot{\mathbf{k}}v_z'.$$

But (Fig. 14b)

$$\dot{\mathbf{i}} = \mathbf{w} \wedge \mathbf{i}, \quad \dot{\mathbf{j}} = \mathbf{w} \wedge \mathbf{j}, \quad \dot{\mathbf{k}} = \mathbf{w} \wedge \mathbf{k}, \tag{9.1.6}$$

thus

$$\dot{\mathbf{v}}' = \mathbf{a}' + \mathbf{w} \wedge \mathbf{v}'. \tag{9.1.7}$$

From (9.1.2), (9.1.4), and (9.1.7) we now obtain

$$\mathbf{a} = \dot{\mathbf{v}}' + \dot{\mathbf{V}} = \mathbf{a}' + \dot{\mathbf{u}} + \mathbf{w} \wedge \mathbf{u} + 2(\mathbf{w} \wedge \mathbf{v}') + \mathbf{w} \wedge [\mathbf{w} \wedge (\mathbf{r} - \mathbf{R})]$$

$$- \mathbf{w} \wedge \dot{\mathbf{R}} + \dot{\mathbf{w}} \wedge (\mathbf{r} - \mathbf{R}), \tag{9.1.8}$$

which is the required relation between **a** and **a**′ in terms of **r**, **R**, **u**, and **w**.

We first consider the special case where the two systems S and S' perform a pure *translational* movement with respect to each other, that is when $\mathbf{w} = 0$. We then have from (9.1.8) and (9.1.3)

$$\mathbf{a}' = \mathbf{a} - \dot{\mathbf{u}} = \mathbf{a} - \dot{\mathbf{V}}. \tag{9.1.9}$$

If it is further assumed that the translational velocity **V** is *constant*, that is that the origin of S' moves along a straight line with constant speed with respect to S, we see that $\mathbf{a} = \mathbf{a}'$ or, for any system of particles

$$\dot{\mathbf{v}}_i = \dot{\mathbf{v}}_i' \quad (\mathbf{V} = \text{const}). \tag{9.1.10}$$

If therefore, the Newtonian equations of motion (9.1.1) are valid in S they are also valid in S'; in other words, all reference systems which perform a uniform translational movement with respect to the inertial system, as defined by the system of fixed stars, are also inertial systems. This theorem is called the "*classical principle of relativity in dynamics*" and was, of course, well known to Newton. Alternatively, we may express it in the following form: *An observer in a laboratory which performs a uniform translational movement with respect to the fixed stars cannot make any statement concerning the velocity of this movement from the outcome of mechanical experiments within the laboratory alone.*

The relationship (9.1.10) can be expressed in the form of a *coordinate transformation.* Let $\mathbf{r}\,(x, y, z)$ be the position vector of a point with respect to a coordinate system S and $\mathbf{r}'(x', y', z')$ the position vector of the same point with respect to a coordinate system S'. Integration of (9.1.2) with respect to time then yields the "*transformation equations*"

$$\mathbf{r} = \mathbf{r}_0 + \mathbf{r}' + \mathbf{V}t \tag{9.1.11}$$

in vector form (where $\mathbf{r}_0(x_0, y_0, z_0)$ is the position vector of the origin of S' at $t = 0$), or

$$\left.\begin{aligned} x &= x_0 + x' + \mathbf{V}_x t, \\ y &= y_0 + y' + \mathbf{V}_y t, \\ z &= z_0 + z' + \mathbf{V}_z t, \end{aligned}\right\} \tag{9.1.12}$$

in coordinate form. The transformation which is defined by (9.1.11) or (9.1.12) is called the "*Galileo transformation*"†, and the classical principle of relativity may further be expressed in the form: *The Newtonian equations of motion (9.1.1) are* invariant *with respect to the Galileo transformation.* This statement includes the fact that the law of inertia is valid in all coordinate systems related to each other by a Galileo transformation, which justifies the nomenclature.

We can, of course, always choose our coordinate systems in such a way that they coincide at $t = 0$ and that the x-axis has the direction of $\mathbf{V}$. In that case the Galileo transformations reduce to

$$x' = x - Vt, \quad y' = y, \quad z' = z, \quad t' = t, \tag{9.1.13}$$

where the last equation has been added in order to indicate that, according to the principle of classical physics previously stated, the time measurements by clocks are not affected by the frame of reference used by the observer.

Let us now consider the more general case of a translational movement of S' with respect to S which is *non-uniform,* that is where $\mathbf{V}$ is a function of time. It then follows from (9.1.1) and (9.1.9) that

$$m_i \mathbf{a}_i' = \mathbf{F}_i - m_i \dot{\mathbf{V}}, \tag{9.1.14}$$

which shows that the Newtonian equations of motion are *not* valid for the frame of reference S' if $\mathbf{V} \gtrless 0$. We may, however, write these equations formally as if the Newtonian equations *were* satisfied, namely in the form

$$m_i \mathbf{a}_i' = \mathbf{F}_i - \mathbf{F}_i^* \tag{9.1.15}$$

by introducing an "apparent" force

$$\mathbf{F}_i^* = -m_i \dot{\mathbf{V}}. \tag{9.1.16}$$

† This generally accepted term was first suggested by Ph. Frank (1911).

In fact, an observer making mechanical experiments in a laboratory moving with non-uniform velocity with respect to an inertial system, may quite legitimately describe the outcome of his measurements with respect to the walls of the laboratory as frame of reference as an indication for the existence of forces $\mathbf{F}_i^*$ which are proportional to the masses m_i of the particles concerned. These forces are therefore usually called "*inertial forces*".

As a further application of the general relationship (9.1.8) we now assume that S' performs a purely *rotational* movement with respect to S with fixed centre $\mathbf{R}$ of rotation; thus in this case $\mathbf{u} = 0$ and $\dot{\mathbf{R}} = 0$. We then obtain

$$\mathbf{a}' = \mathbf{a} - \mathbf{w} \wedge [\mathbf{w} \wedge (\mathbf{r} - \mathbf{R})] - \dot{\mathbf{w}} \wedge (\mathbf{r} - \mathbf{R}) - 2(\mathbf{w} \wedge \mathbf{v}'). \quad (9.1.17)$$

Again, we may use the equation of motion in the form (9.1.15) in which the inertial forces $\mathbf{F}_i^*$ are now given by

$$\mathbf{F}_i^* = -m_i \{\mathbf{w} \wedge [\mathbf{w} \wedge (\mathbf{r}_i - \mathbf{R})] + \dot{\mathbf{w}} \wedge (\mathbf{r}_i - \mathbf{R}) + 2(\mathbf{w} \wedge \mathbf{v}_i')\} \quad (9.1.18)$$

which, as in the previous case, are proportional to the masses of the particles. The first term in curly brackets gives rise to the "*centrifugal force*" which is easily seen to be perpendicular to the axis of rotation and directed away from it and to have a magnitude proportional to ω^2 and to the distance of the mass point from the axis. The second term is due to a change in time of the rotation vector $\mathbf{w}$, and the corresponding inertial force is only present if either the angular velocity of the rotation is not constant or the axis of rotation changes its direction in space. The third term gives rise to the so-called "*Coriolis force*" which, amongst other things, is responsible for the phenomenon of rotation of the Foucault pendulum in the terrestrial frame of reference. It occurs only if $\mathbf{v}_i' \lessgtr 0$, that is if the particle is in motion with respect to S'; its direction lies in a plane normal to the rotation axis of S' and is itself normal to $\mathbf{v}_i'$.

Newton thought that the inertial forces had their origin in the non-uniformity of the *absolute* movement of a body and he maintained that it was possible to measure the absolute acceleration of material bodies from observations of the effects of inertial forces. If instead of the hypothetical absolute space the system of the fixed stars is accepted as the preferential frame of reference, it appears that an observer in a closed laboratory can make definite statements about the acceleration of the laboratory with respect to the system of fixed stars from mechanical experiments within the laboratory. Likewise, observations of effects due to centrifugal and Coriolis forces by terrestrial observers lead to the conclusion that the earth is performing a uniform rotation relative to the fixed stars. This, of course, is no more than can be ascertained directly by observing the changes in the positions of the stars in the sky; but the inference to be drawn is that the phenomenon of inertia has its origin in the presence of the external material bodies in the universe, as first suggested by E. Mach.

This view is strongly supported by the previously emphasized fact that the *inertial* forces on a body are proportional to its mass. For according to Newton's law of gravitation the *gravitational* forces, which also act between distant bodies, are equally proportional to their masses. This suggests an intimate relationship between the phenomena of inertia and gravitation which is the basis of Einstein's theory of gravitation and will be discussed in Chapter 12.

9.2. The classical concept of ether in optics

In the preceding section it was shown that all frames of reference which move with constant velocity with respect to the system of the fixed stars are inertial systems and that it is impossible to detect or measure such a translational movement by means of observations of mechanical phenomena within that frame of reference. The question arises whether this also holds for optical phenomena.

According to Newton's particle theory light was supposed to consist of some kind of material particles which obeyed the laws of dynamics and were capable of moving through material bodies without resistance. If this theory is accepted it follows at once that the classical principle of relativity should hold in the realm of geometrical optics as well as in that of dynamics. In particular, if it is observed that the rays of light are straight lines in "homogeneous media" in one frame of reference they will also appear as straight lines in any other reference system which performs a uniform translational movement with respect to the first. But the speed and the direction of a ray relative to the axes ought to differ from one such frame of reference to another.

Numerous experiments which were undertaken in order to find out whether the orbital movement of the earth affected the laws of geometrical optics in a terrestrial frame of reference showed clearly that this was not the case and that therefore the principle of relativity seems to hold in the realm of geometrical optics. Furthermore, the phenomenon of "*astronomical aberration*" of light, discovered by Bradley in 1725, also confirms the above-mentioned predictions from the particle theory of light.

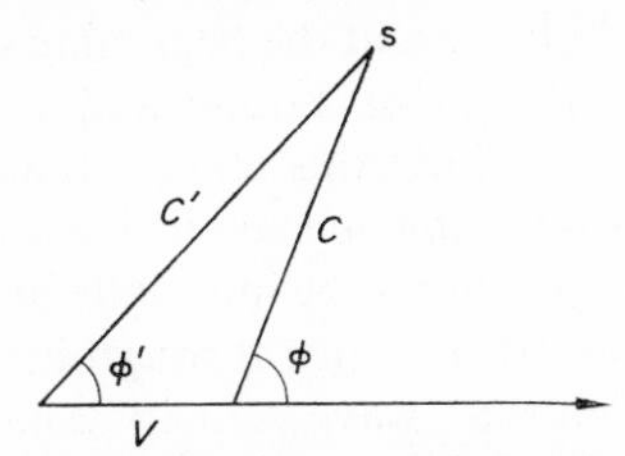

FIG. 15. Diagram illustrating the astronomical aberration of light according to the classical particle theory.

Consider a star s (Fig. 15) at an elevation φ above the horizon as measured by an observer at rest with respect to s. This means that the velocity vector $\mathbf{c}$ of the light ray entering his observing instrument makes an angle φ with the horizontal. If now the observer moves with velocity $\mathbf{V}$ in a horizontal direction towards s, the vector $\mathbf{c}'$ of the relative velocity of the ray with respect to him is, according to (9.1.2) $\mathbf{c}' = \mathbf{c} - \mathbf{V}$ and includes an angle φ' with the horizontal, related to φ, φ', V, and c by

$$\frac{V}{c} = \frac{\sin(\varphi - \varphi')}{\sin \varphi'}. \tag{9.2.1}$$

Thus the angular position of the star will now appear to be altered, the elevation φ being *lowered* in the direction of $\mathbf{V}$ by an amount $\alpha = \varphi - \varphi'$. This is the phenomenon of aberration and α is called the "*aberration angle*".

In terms of α the relation (9.2.1) can be written

$$\frac{V}{c} = \frac{\sin \alpha}{\sin(\varphi - \alpha)}. \tag{9.2.2}$$

For velocities V that are small compared with the velocity c of light α becomes small, and from (9.2.2) follows the simple approximate formula for the aberration angle,

$$\alpha \simeq \frac{V}{c} \sin \varphi \quad (V \ll c). \tag{9.2.3}$$

The aberration effect due to a strictly uniform translational movement of an observer with respect to a star can obviously not be detected because the observer can only measure φ' but not φ. However, in the case of a terrestrial frame of reference the vector $\mathbf{V}$ of the orbital velocity of the earth changes its direction continuously and performs a complete cycle of 360° in the course of a year. It is easy to see that, owing to this movement, the apparent position of a star should, according to (9.2.3), perform a harmonic oscillation with amplitude V/c in the direction parallel to the plane of the ecliptic, and with amplitude $V \sin \varphi/c$ normal to that direction when φ is the astronomical latitude of the star. Very precise measurements of the aberration effect are indeed in complete agreement with these theoretical predictions.

The particle theory of light was, at the beginning of the nineteenth century, superseded by the wave theory of Young and Fresnel which is based on Huygens's concept of a "luminiferous ether", that is a continuous material medium, capable of transmitting mechanical disturbances in the form of waves. Young supposed the ether to be immobile in space and penetrating all material bodies which can freely move through it. According to this theory the velocity of light, which is a "wave velocity", has the universal value c in empty space or "pure ether", irrespective of whether the light source is at rest or in motion with respect to the ether, whereas according to the particle theory the velocity of a light ray, emitted from a source moving with velocity $\mathbf{v}$, should be the vector sum of $\mathbf{c}$ and $\mathbf{v}$.

The mechanical wave theory of light was subsequently replaced by the electromagnetic wave theory of Maxwell. According to this theory the ether is not supposed to have any mechanical properties and serves only as a "carrier" for the propagation of the electromagnetic waves. We shall deal with this electromagnetic aspect of the ether in the following section 9.3. The considerations in this section apply to both mechanical and electromagnetic ether waves.

If the concept of the immobile ether is correct, the ether provides a unique preferential frame of reference for the description of optical phenomena, and it is to be expected that observable effects will be caused by the movement of light sources and observers through the ether. Let us first examine the influence of such movements on the frequency of monochromatic light waves which constitute the "*Doppler effect*".

Assume first the light source to be at rest and the observer to move with velocity V away from the light source. The measured wave velocity will then have the value $u = c - V$ instead of c, while the wavelength λ of the wave will obviously not be affected by the movement of the observer. Hence the frequency ν', as measured by the moving observer, is, according to (1.2.9),

$$\nu' = \frac{u}{\lambda} = \frac{c - V}{\lambda}, \tag{9.2.4}$$

while the frequency measured by a resting observer is

$$\nu = \frac{c}{\lambda}. \tag{9.2.5}$$

The "*Doppler shift*" $\Delta\nu = \nu' - \nu'$ is therefore given by

$$\frac{\Delta\nu}{\nu} = \frac{c - V}{c} - 1 = -\frac{V}{c}. \tag{9.2.6}$$

Assume next that the observer remains stationary and the light source moves away from him with velocity v. The measurement of the wave velocity will now give the value c, but the wavelength of the emitted light, that is the distance between two consecutive points of equal phase, is now $\lambda' = \lambda + vT$, where $T = 1/\nu$ is the periodic time of the oscillation. Thus

$$\lambda' = \lambda + v/\nu, \tag{9.2.7}$$

and hence the Doppler shift is

$$\frac{\Delta\nu}{\nu} = \frac{\lambda}{\lambda'} - 1 = \frac{1}{1 + v/c} - 1 = -\frac{v}{c + v}. \tag{9.2.8}$$

In the more general case, where both the light source and the observer are moving, one has, instead of (9.2.4),

$$\nu' = \frac{c - V}{\lambda'} = \frac{c - V}{\lambda + v/\nu}, \tag{9.2.9}$$

and hence the Doppler shift becomes

$$\frac{\Delta\nu}{\nu} = \frac{c - V}{c + v} - 1 = -\frac{V + v}{c + v} = -\frac{v'}{c + v} = -\frac{v'}{c + v' - V}, \tag{9.2.10}$$

where $v' = v + V$ is the *relative* velocity of the light source with respect to the observer.

If $v' = 0$, that is if light source and observer form a system which as a whole performs a translational movement through the ether, it follows from (9.2.10) that $\Delta\nu = 0$. Thus no information concerning the velocity V of that movement can be obtained from observations of the frequency of a light source within the system, for example from measurements of that frequency from different directions.

Further, as long as V and v, and hence also v', are small compared with c, all three formulae (9.2.6), (9.2.8), and (9.2.10) give the same result, namely

$$\frac{\Delta\nu}{\nu} = -\frac{v'}{c} \quad (v' \ll c). \tag{9.2.11}$$

Thus the Doppler effect depends on the *relative* velocity v' only, and its observation allows only v' to be determined but not V. The principle of relativity therefore holds under these conditions for this type of optical phenomena. We shall return to the question as to what happens when v/c and V/c are not negligibly small later on in this section.

Whereas in the particle theory of light the notion of "light rays" as the paths of the light particles has a prime significance, in the wave theory this place is taken by the plane monochromatic waves. The question arises whether a wave which appears to be plane from the point of view of an observer who is at rest in the ether will remain to be a plane wave in a reference system that moves with constant velocity V with respect to the ether.

According to (1.2.10) the wave function for a plane wave has the form

$$\Phi(\mathbf{r}, t) = \gamma\, e^{2\pi i(\nu t - \mathbf{kr})}, \tag{9.2.12}$$

where $\mathbf{k}$ is the wave vector of the wave, with components k_x, k_y, k_z. For simplicity we assume $\mathbf{V}$ to have the direction of the x-axis of the coordinate system used. We can then transform Φ from the resting to the moving reference system by means of the Galileo transformations (9.1.13) and thus obtain for the exponent in (9.2.12)

$$2\pi i(\nu t - k_x x' - k_y y' - k_z z' - k_x V t) = 2\pi i(\nu' t - \mathbf{kr}'),$$

where we have written

$$\nu' = \nu - k_x V. \tag{9.2.13}$$

Thus the plane wave (9.2.12) as seen from the resting observer in the system S will also appear to be a plane wave in the reference system S', *progressing in the same direction* but with a different frequency ν' given by (9.2.13).

Let A be the direction cosine of **k** with respect to the x-axis. Then

$$k_x = \frac{A\nu}{c}, \tag{9.2.14}$$

and, from (9.2.13) and (9.2.14), we obtain for the relative change of frequency

$$\frac{\Delta\nu}{\nu} = \frac{\nu' - \nu}{\nu} = -\frac{AV}{c}. \tag{9.2.15}$$

$\Delta\nu$ is evidently again the Doppler shift (derived here from a different consideration) and the expression (9.2.15) is a generalization of (9.2.6) with which it coincides if $A = 1$, that is if the observer moves in the direction of the light wave. $\Delta\nu$ becomes zero if $A = 0$, that is if the observer moves at right angles to the direction of the wave propagation. Thus no "transverse Doppler effect" is expected on the basis of the classical wave theory if the principle of superposition of velocities and hence the Galileo transformations (9.1.13) are supposed to be valid.

The result, derived above, that the direction of a plane wave in empty space is not affected by the movement of the frame of reference, seems to be in contradiction to the observed astronomical aberration effect which, as we have shown before, is well accounted for by the particle theory. In actual fact, however, the angular position of a star is not measured by determining the direction of a plane wave, issuing from the star, but by the measurement of the position of a *star image* in the focal plane of a telescope. It was first

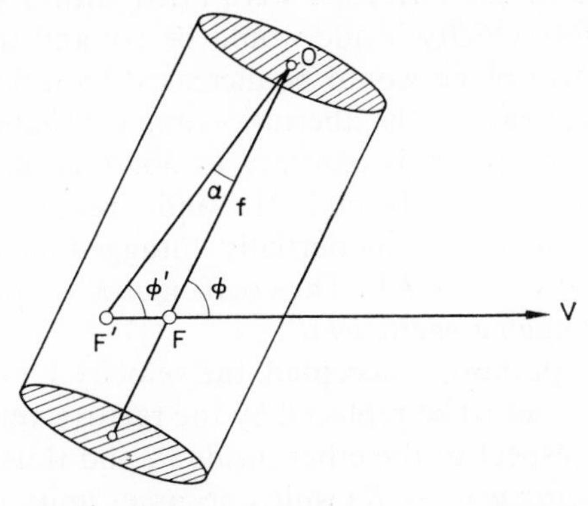

FIG. 16. Diagram illustrating the astronomical aberration of light according to the classical wave theory.

pointed out by Young that this fact provides the explanation of the observed aberration effect on the basis of the wave theory.

Let O (Fig. 16) be the objective lens of a telescope and F its focal point in which the image of a star of elevation φ above the horizon would be formed if the telescope were stationary in the ether. Let the focal length of the telescope be f. If the telescope is now moving in the horizontal direction with velocity V the wave emerging at a certain moment from O will still be moving in the same direction φ, but when it reaches the focal plane after the time $t = f/c$ the focus will have moved from F to F' over a distance $Vt = Vf/c$ in the direction of V. Thus the star image appears to the observer to be displaced from F in the opposite direction by the same amount, and the observer will conclude that the star's angular position has been changed by an angle $\alpha = \varphi - \varphi'$ given by

$$\frac{\sin \alpha}{\sin (\varphi - \alpha)} = \frac{Vt}{ct} = \frac{V}{c},$$

which is identical with (9.2.2).

There is, however, an essential difference between the explanations of the aberration effect by the particle and by the wave theory. For according to the former the effect is due to the *relative* motion between object and observer whereas according to the latter it is due to the *absolute* movement of the observing instrument through the ether. But again, as already emphasized before, the velocity V of this movement cannot be measured as long as it remains unaltered. The observed aberration effect can thus only be used for determining the velocity of the *orbital* movement of the earth but not for ascertaining whether the whole solar system performs a translational movement through the ether.

If the aberration effect has indeed its origin in the movement of the observing instrument through the ether, the effect ought to be modified by the refractive properties of the substances through which the light has to pass. For if the tube of the telescope were filled with a substance of refractive index n the light velocity inside would be c/n and the displacement of the image in the focal plane would be increased by a factor n, and this would amount to an increase of the aberration angle α (outside the telescope) by a factor n^2. This, however, is contrary to observation (which was decisively demonstrated by Airy only in 1871), and Fresnel therefore proposed the hypothesis that the ether was partially "dragged along" by material bodies with a fractional velocity KV. The coefficient K is now generally referred to as Fresnel's "*dragging coefficient*".

If Fresnel's hypothesis is accepted, the velocity V in Young's theory of the aberration effect has to be replaced by the relative velocity $V(1 - K)$ of the telescope with respect to the ether inside it, and thus α would in fact be increased by a factor $n^2(1 - K)$ which becomes unity if one assumes that

$$K = 1 - 1/n^2. \tag{9.2.16}$$

The laws of geometrical optics, as derived from Huygens' principle, are valid within the framework of the wave theory with respect to the ether as preferential system. It is thus to be expected that they will be modified if the reference system used performs a uniform translational movement with respect to the ether. In order to find out whether this is in fact the case we formulate the laws of geometrical optics in the form of Fermat's variational principle which we have already discussed in section 1.3 and which stipulates that the time T taken for a light signal to pass from a point P_1 to a point P_2 in an inhomogeneous medium always takes the path which makes T a minimum.

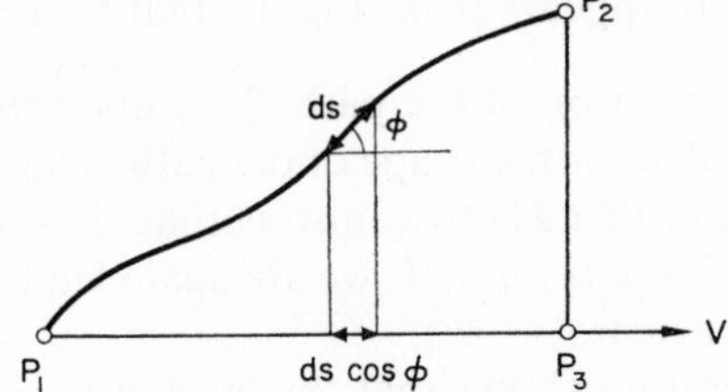

FIG. 17. Application of Fermat's principle for the derivation of the expression for Fresnel's dragging coefficient.

In a system S, at rest with respect to the ether, T is given by the expression (1.3.24) where the wave velocity is $u = c/n(\mathbf{r})$. Thus Fermat's principle takes the form

$$\delta \int_{P_1}^{P_2} \frac{ds}{u} = \delta \int_{P_1}^{P_2} n\, ds = 0. \tag{9.2.17}$$

In the system S', which moves with respect to S with velocity V, the wave velocity is $u' = u - V(1 - K)$ if the ether dragging effect is taken into account and thus we have now

$$T' = \int_{P_1}^{P_2} \frac{ds}{u'} = \int_{P_1}^{P_2} \frac{ds}{|u - V(1 - K)|}. \tag{9.2.18}$$

If the ordinary laws of geometrical optics are to be valid in S' as well as in S, the expression (9.2.18) must be a minimum or

$$\delta \int_{P_1}^{P_2} \frac{ds}{|u - V(1 - K)|} = 0. \tag{9.2.19}$$

Denoting the angle between ds and $\mathbf{V}$ by φ (Fig. 17) and neglecting terms of higher order than the first in V/c, we find

$$|u - V(1 - K)|^{-1} \simeq \left\{\frac{c^2}{n^2} - \frac{2cV}{n}(1 - K)\cos\varphi\right\}^{-1/2}$$

$$= \frac{1}{c}\left\{n + (1 - K)\frac{n^2 V}{c}\cos\varphi\right\},$$

and thus (9.2.19) takes the form

$$\delta \int_{P_1}^{P_2} \left\{ n + (1 - K)\, n^2 \frac{V}{c} \cos\varphi \right\} ds = 0. \tag{9.2.20}$$

If now (9.2.17) and (9.2.20) are to hold simultaneously the integral

$$\Delta T = \int_{P_1}^{P_2} (1 - K) \frac{n^2 V}{c^2} \cos\varphi \, ds \tag{9.2.21}$$

must be a constant. It appears from Fig. 17 that $\int_{P_1}^{P_2} \cos\varphi \, ds$ is simply the length P_1P_3 of the projection of the path P_1P_2 upon the direction of $\mathbf{V}$ and it is therefore independent of the shape of the path. Thus the necessary condition for the variation of (9.2.21) to vanish is that $(1 - K)\, n^2$ must be a constant and, as clearly $K = 0$ for $n = 1$, we are again led to the relation (9.2.16) for the dragging coefficient.

The above consideration shows that the laws of geometrical optics ought to be valid in all frames of reference which move with *constant* velocity V with respect to the ether; and that it therefore should be impossible for an observer in such a system to measure V from geometrical–optical experiments within his frame of reference, provided that V/c is small and that Fresnel's dragging hypothesis with the expression (9.2.16) for the dragging coefficient is accepted. Experimental evidence proves, indeed, conclusively that the classical principle of relativity does in fact hold as far as the laws of geometrical optics are concerned, and thus Fresnel's hypothesis is vindicated. There are, however, grave objections against it, for it is not based on any physical theory of interaction between ether and matter and, furthermore, as the refractive index depends on frequency, the relationship (9.2.16) cannot hold independently of frequency. The concept of "ether drag" thus cannot be taken at its face value.

It follows from the same consideration that it ought to be equally impossible to determine V by sending two light beams from the same source in P_1 in a suitable apparatus to P_2 along different paths and by observing whether a change of interference pattern occurs when the apparatus is rotated with respect to the direction of the orbital velocity of the earth. For according to (9.2.16) and (9.2.21) ΔT must have the same value along the two paths, and hence the phase difference between the two wave trains is not affected by the change in the magnitude or the direction of V as long as V/c is small. Again, experiments of this type have confirmed these conclusions, so that it follows that the classical principle of relativity also holds for typical wave-optical phenomena, like interference, to a first approximation, that is if effects of the second or higher order in V/c are not taken into account.

It ought, nevertheless, to be possible to measure the velocity of the movement of material bodies through the ether by means of optical experiments where either different parts of the apparatus used move with different velocities or where the system as a whole performs a *non-uniform* movement.

Experiments of the first type are actually mainly used for directly measuring the velocity of light in moving transparent bodies. The first of these was carried out by Fizeau in 1851. The underlying principle of these experiments is illustrated by the schematic Fig. 18.

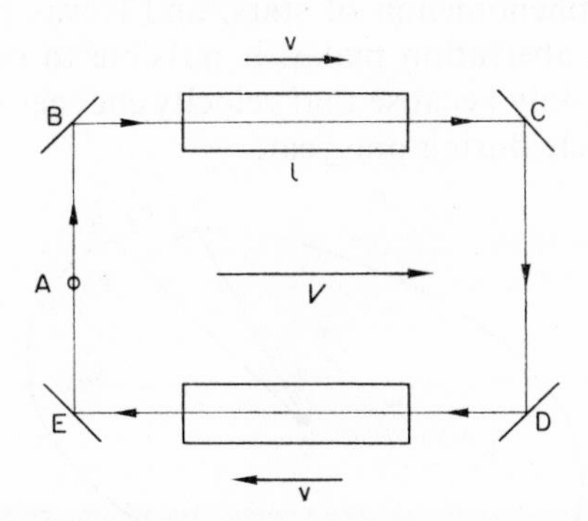

FIG. 18. Fizeau's experiment for measuring the velocity of light in moving bodies.

A monochromatic light beam, originating in A, is sent along the closed path $ABCDEA$ so that an interference pattern is produced in A. On its way it passes through two tubes of length l through which a liquid is flowing with velocity v with respect to the laboratory frame of reference in the direction of the arrows. If now the direction of the flow is reversed a fractional shift Δ of the fringe pattern is observed which can be calculated as follows.

Suppose the laboratory to be moving through the ether with a velocity whose component in the direction of the tubes is V. The *relative* light velocity with respect to the observer in the upper tube is then, according to the superposition principle, $c/n + K(v+V)$, and in the lower tube $c/n + K(v-V)$ as indicated in the figure. Thus the time t taken by the beam to traverse the two tubes is, for $v/c \ll 1$ and $V/c \ll 1$,

$$t_1 = \frac{l}{c/n + K(v+V)} + \frac{l}{c/n + K(v-V)} \simeq \frac{2ln}{c}\left(1 - \frac{nKv}{c}\right)$$

independently of V. Similarly, the time t_2 when the flow is reversed is

$$t_2 \simeq \frac{2ln}{c}\left(1 + \frac{nKv}{c}\right).$$

Thus the fringe shift is

$$\Delta = \frac{(t_2 - t_1)}{\lambda}\, c \simeq \frac{4n^2Kl}{\lambda}\,\frac{v}{c}. \tag{9.2.22}$$

This formula may now be used to determine the dragging coefficient K direct, and the results of Fizeau's and subsequent investigations have fully confirmed Fresnel's relation (9.2.16).

One example of experiments of the second type when the observer's frame of reference moves as a whole with non-uniform velocity through the ether has already been described before. It consists in the observation of the astronomical aberration phenomenon of stars, and it was pointed out that the measurement of the aberration makes it possible to determine the orbital velocity of the earth only because that velocity changes its direction and performs a complete cycle during one year.

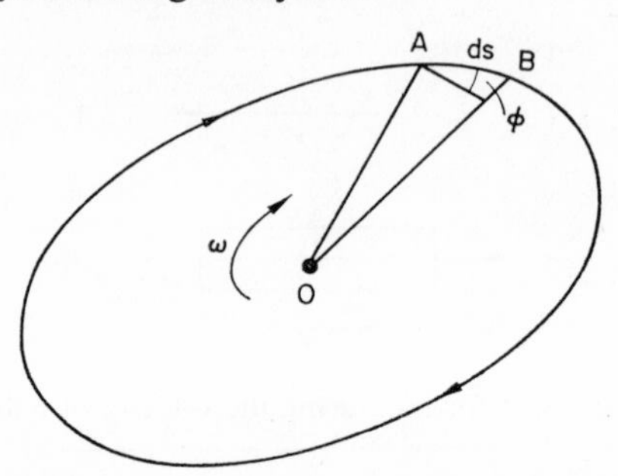

FIG. 19. Light moving in a closed path in a rotating frame of reference.

Another example is provided by an arrangement whereby a beam of light travels in a closed path in a plane normal to the axis of a system that rotates with constant angular velocity ω with respect to the ether (Fig. 19). In this case we have $V = \omega r$, and from (9.2.21) we obtain for the difference ΔT between the times taken by a light signal to travel round the path when the system is rotating and when it is at rest the expression

$$\Delta T = \frac{\omega}{c^2} \oint r \cos \varphi \, ds. \tag{9.2.23}$$

Here the integrand is equal to twice the area of the narrow triangle OAB, and thus the integral is equal to twice the area a enclosed by the path. Thus if two light beams are made to travel along the same path in opposite directions and then superposed, the phase difference produced by the rotation of the system is

$$\Delta = \frac{4\omega a}{c\lambda}. \tag{9.2.24}$$

Consequently a measurable shift in the interference pattern should be produced by a sufficiently large change of either ω or r. The former effect was indeed shown up in laboratory experiments where the apparatus was placed on a rotating disk and found to be in agreement with (9.2.24) (Sagnac, 1913). The latter procedure forms the basis of a famous experiment by Michelson and Gale (1925) which was undertaken in order to demonstrate the influence of the *rotational* motion of the earth with respect to the ether on the propagation of light waves.

The effect of the rotation of a frame of reference expressed by the equation (9.2.24) may be regarded as the optical analogue to the effect of the Coriolis forces in mechanics which was discussed in section 9.1, and the Michelson and Gale experiment is the optical analogue of the Foucault pendulum experiment. But whereas the appearence of Coriolis forces indicate that the frame of reference used rotates with respect to the *inertial* system of mechanics, that is the system of the fixed stars, the optical effect (9.2.24) is supposed to occur when the frame of reference rotates with respect to the *ether*, the preferential reference system of optics. The experimental evidence referred to above indicates that the inertial system of mechanics and the ether of classical optics are identical.

So far we have shown that the classical principle of relativity applies to optics as well as to mechanics as long as second-order effects in V/c are too small to be detected. It is clear, however, that such effects must exist if the ether hypothesis is correct, for according to the superposition principle (9.1.2) the velocity c' of light in empty space within a reference system S' that moves with velocity $\mathbf{V}$ with respect to the ether ought to be different from $\mathbf{c}$, namely

$$\mathbf{c}' = \mathbf{c} - \mathbf{V}. \tag{9.2.25}$$

Thus an observer in S' ought to have the impression that the light waves were affected by an "*ether wind*" or "*ether drift*" of velocity $-\mathbf{V}$. It is to be expected that effects of this type are of the second order in V/c.

The first ether drift experiment capable of detecting effects of this order of magnitude was carried out by Michelson in 1881, following a suggestion by Maxwell. The basic idea of this and the many subsequent experiments of a similar type is to send two light beams from the same monochromatic source at O (Fig. 20a) along two arms of lengths l at right angles to each other

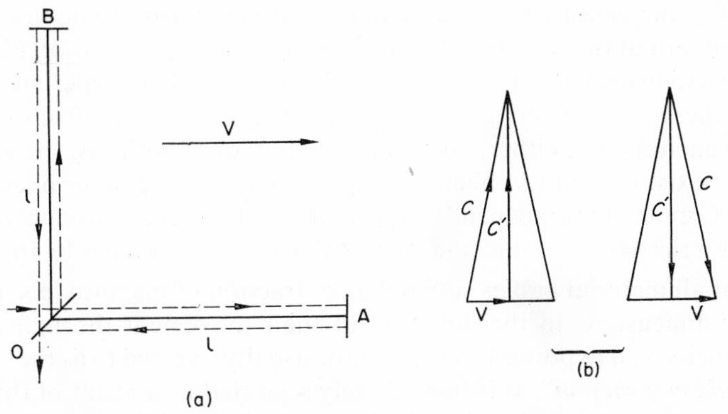

FIG. 20. Diagram illustrating the classical theory of Michelson's experiment.

forwards and backwards in a horizontal plane, and to observe whether the interference pattern produced by the superposition of the two beams at O is affected by a rotation of the apparatus about a vertical axis.

Suppose that the apparatus is first orientated in such a way that OA has the direction of the orbital velocity of the earth. In this case, according to (9.2.25), the time taken for the light beam travelling along the arm OA to return to O is

$$t_1 = \frac{l}{c - V} + \frac{l}{c + V} = \frac{2l}{c\,(1 - V^2/c^2)}$$

and the time taken along the arm OB is (Fig. 20b)

$$t_2 = \frac{2l}{\sqrt{c^2 - V^2}} = \frac{2l}{c\sqrt{1 - V^2/c^2}}\,.$$

Thus neglecting terms of higher than the second order in V/c we get for the time difference

$$t_1 - t_2 = \frac{2l}{c}\left(\frac{1}{1 - V^2/c^2} - \frac{1}{\sqrt{1 - V^2/c^2}}\right)$$

$$\simeq \frac{l}{c}\frac{V^2}{c^2}\,. \tag{9.2.26}$$

If now the apparatus is rotated through 90°, $t_1 - t_2$ changes its sign and correspondingly a shift of the fringe pattern of magnitude

$$\Delta = \frac{2\,(t_1 - t_2)\,c}{\lambda} = \frac{2l}{\lambda}\frac{V^2}{c^2} \tag{9.2.27}$$

ought to occur as a result of the ether drift.

Although the second factor on the right-hand side of (9.2.27) is very small, the first factor can be made large enough and the accuracy of the measurement of Δ increased to such an extent that ether drift velocities of about one-twentieth of that due to the orbital motion of the earth could be detected in some sophisticated versions of the original Michelson experiment. But in fact all these experiments gave completely negative results, from which one has to conclude that either, contrary to the ether hypothesis, the ether drift does not exist or that its effect is compensated by some other phenomenon.

H. A. Lorentz and (independently) J. Fitzgerald, indeed, attempted in 1893 to explain the negative outcome of the Michelson experiment by the assumption that all material bodies suffered a contraction of magnitude $\sqrt{1 - V^2/c}$ of their dimensions in the direction of their movement through the ether with velocity V, a hypothesis which is now usually referred to as the "*Lorentz–Fitzgerald contraction*". It is immediately seen that as a result of this phenomenon the first term in brackets in the expression (9.2.26) becomes equal to

$1/\sqrt{1 - V^2/c}$ and thus is exactly cancelled by the second term, so that $t_1 - t_2 = 0$ for any position of the apparatus. The Lorentz–Fitzgerald contraction is clearly an *ad hoc* hypothesis, just as Fresnel's dragging hypothesis, and although Lorentz made an attempt to explain it on a physical basis, as an outcome of the electromagnetic interaction forces between moving electrically charged particles (see section 9.3), it is difficult to understand why the contraction should be independent of the atomic constitution of the bodies concerned. Moreover, it appears that the Lorentz–Fitzgerald contraction is not in itself sufficient to enforce the precise validity of the principle of relativity in classical optics.

In order to demonstrate this let us assume that the two arms OA and OB in Fig. 20 have different lengths l_1 and l_2. In this case the expression (9.2.26) for the time difference $t_1 - t_2$ is modified to

$$t_1 - t_2 = \frac{2}{c}\left(\frac{l_1}{1 - V^2/c^2} - \frac{l_2}{\sqrt{1 - V^2/c^2}}\right), \tag{9.2.28}$$

and if the Lorentz contraction of the arm OA is taken into account, one obtains for the expected phase difference between the two beams

$$\Delta = \frac{2\nu\,(l_1 - l_2)}{c\sqrt{1 - V^2/c^2}}. \tag{9.2.29}$$

This is still independent of the *direction* of $\mathbf{V}$, and thus a rotation of the apparatus could not produce a shift in the fringe pattern, but it depends on the *magnitude* of $\mathbf{V}$. Now it is known that the solar system as a whole is performing a translational movement with respect to the system of the stars of the galaxy and the superposition of the velocity of this movement and that of the orbital movement of the earth about the sun should result in a periodic annual change of V. Thus observation with a suitably constructed instrument of this type should reveal a corresponding small periodic fringe shift. This experiment was actually carried out by Kennedy and Thorndike in 1932 and showed a complete absence of this effect.

In order to explain the negative outcome of the Kennedy–Thorndike experiment one can make another *ad hoc* hypothesis by postulating that the frequency ν of a spectral line emitted by an atom that moves with velocity V through the ether is changed to a frequency

$$\nu' = \nu\sqrt{1 - V^2/c^2}\,. \tag{9.2.30}$$

For if ν' is substituted into (9.2.29) instead of ν the variation of Δ with V, which we considered above, would not occur. The effect, expressed by (9.2.30) would constitute a kind of "second-order Doppler effect" which would be superposed over the ordinary, previously discussed first-order Doppler effect, and would be *independent of the direction* of V.

The vibrations of monochromatic light emitted by an atom can and are, in fact, nowadays used for measuring time intervals. The relation (9.2.30) indicates that an "atomic clock" would slow down by a relative amount $\sqrt{1 - V^2/c^2}$ when it moves with velocity V through the ether. It was first suggested by H. A. Lorentz (1899) that any mechanism suitable for measuring time might be slowed down on account of its movement through the ether by a factor $\sqrt{1 - V^2/c^2}$. Thus for a time interval of length Δt, as measured by a resting clock, the moving clock would register a time interval

$$\Delta t' = \Delta t \sqrt{1 - V^2/c^2} \,. \tag{9.2.31}$$

We shall refer to this hypothetical phenomenon as the "*Lorentz clock retardation*".

It can be shown (H. Littman-Fürth, 1954) that an ether drift should also influence the frequencies of the normal modes of stationary, electromagnetic waves in cavities, and that the frequency ν' for a position of the cavity where the standing wave has the direction of the ether drift, and the frequency ν for a direction normal to it should be related by a formula of the form (9.2.30). But recent experiments with microwaves and coherent light waves (generated in "lasers") have given no indication of such an effect which would, of course, be compensated by a Lorentz contraction of the dimensions of the cavity in the direction of the ether drift.

We finally return to the Doppler effect and the question whether its observation could not be used for detecting an ether drift in cases where the relative velocity v' of the light source with respect to the observer is large enough so that the quantity $v'V/c^2$ is not negligibly small compared with unity, although v'^2/c^2 still is. One then finds from the general formula (9.2.10) that

$$\frac{\Delta\nu}{\nu} \simeq -\frac{v'}{c}\left(1 + \frac{V}{c}\right). \tag{9.2.32}$$

If therefore an observer looks at the light source moving away from him with velocity v' and measures the frequency of the emitted light first when $\mathbf{v}'$ has the same direction as $\mathbf{V}$ and then when it has the opposite direction to $\mathbf{V}$, he should observe a frequency shift equal to $\frac{\Delta\nu}{\nu} = \frac{2v'V}{c^2}$. But again, optical experiments of this type and recent experiments with microwaves in "masers" (Cedarholm and Townes, 1958) have shown a complete absence of this effect.

It appears, therefore, that the principle of relativity is *strictly* valid in the realm of wave optics which is incompatible with the notion of the immobile universal ether unless some highly dubious and *ad hoc* invented hypotheses are introduced into the theory. This state of affairs calls for a complete revision of the ether concept and, indeed, of the fundamental principles on which classical continuum physics is based, as briefly indicated in the introduction.

9.3. The classical ether concept in electrodynamics

Electrodynamics deals with the mutual interactions of material bodies by electric and magnetic forces. Classical electrodynamics is based on the idea that this interaction takes place in two stages, namely that material bodies which carry electric charges and currents or are magnetically polarized "generate" electromagnetic fields and that these fields in turn influence the electromagnetic state of material bodies situated in them and exert forces on these bodies.

The mathematical description of the field-generating processes is given by the phenomenological *equations of Maxwell* which are typical continuum laws as described in section 1.2. For the present purpose we write them in the form†

$$\operatorname{div} \mathbf{E} = 4\pi(\varrho - \operatorname{div} \mathbf{P}), \tag{9.3.1}$$

$$\operatorname{div} \mathbf{B} = 0, \tag{9.3.2}$$

$$\operatorname{curl} \mathbf{B} - \frac{1}{c}\dot{\mathbf{E}} = \frac{4\pi}{c}(\mathbf{j} + \dot{\mathbf{P}} + c \operatorname{curl} \mathbf{M}), \tag{9.3.3}$$

$$\operatorname{curl} \dot{\mathbf{E}} + \frac{1}{c}\dot{\mathbf{B}} = 0. \tag{9.3.4}$$

The vector quantities $\mathbf{E}$ (electric field strength) and $\mathbf{B}$ (magnetic induction) on the left-hand sides of these equations describe the electromagnetic field. The scalar quantity ϱ (electric charge density) and the vector quantities $\mathbf{j}$ (electric current density), $\mathbf{P}$ (electric polarization), and $\mathbf{M}$ (magnetic polarization) on the right-hand sides describe the electromagnetic state of the field generating bodies. They are not independent because, as can easily be seen, they are connected by the continuity equation [see eqn. (6.2.14)]

$$\dot{\varrho} + \operatorname{div} \mathbf{j} = 0. \tag{9.3.5}$$

The vector quantity $\mathbf{H}$ (magnetic field strength) is in this presentation not primarily defined but connected to $\mathbf{B}$ and $\mathbf{M}$ by the equation

$$\mathbf{H} = \mathbf{B} - 4\pi\mathbf{M}, \tag{9.3.6}$$

and similarly the quantity $\mathbf{D}$ (dielectric displacement) is defined in terms of $\mathbf{E}$ and $\mathbf{P}$ by

$$\mathbf{D} = \mathbf{E} + 4\pi\mathbf{P}. \tag{9.3.7}$$

† The system of electric and magnetic units used here is the classical "c.g.s. system", which in the author's opinion is best suited for the present purpose.

The influence of the electromagnetic field upon the state of *stationary* isotropic bodies is expressed by the equations

$$\mathbf{j} = \sigma \mathbf{E}, \tag{9.3.8}$$

$$\mathbf{P} = \varkappa \mathbf{E}, \tag{9.3.9}$$

$$\mathbf{M} = \chi \mathbf{H}. \tag{9.3.10}$$

Here the scalar quantities σ (electric conductivity), $\varkappa$ (dielectric susceptibility), and χ (magnetic susceptibility) are, within the framework of the phenomenological theory, properties of the materials of the bodies and thus, in general, known functions of space.

Finally, the force $\mathbf{f}$ per unit volume, exerted by the electromagnetic field upon the material bodies, consists of four parts:

$$\mathbf{f}_1 = \varrho \mathbf{E}, \tag{9.3.11}$$

$$\mathbf{f}_2 = \frac{1}{c}(\mathbf{j} \wedge \mathbf{B}), \tag{9.3.12}$$

$$\mathbf{f}_3 = -\mathbf{E} \cdot \operatorname{div} \mathbf{P}, \tag{9.3.13}$$

$$\mathbf{f}_4 = -\mathbf{H} \cdot \operatorname{div} \mathbf{M}. \tag{9.3.14}$$

$\mathbf{f}_1$ and $\mathbf{f}_2$ represent the forces exerted on charges and currents respectively and $\mathbf{f}_3$ and $\mathbf{f}_4$ the forces on electrically and magnetically polarized bodies.

Obviously the electromagnetic field equations have a definite meaning only if they refer to a specific frame of reference for the coordinates. Owing to the continuous character of these equations the reference system ought to be defined by a *continuous medium*, and as electromagnetic fields are known to exist in empty space, the "luminiferous ether" of optics suggests itself as the "carrier" of the electromagnetic field and so as the preferential system of reference in electrodynamics as well.

If this is so, one has to expect that the *movement* of material bodies through the ether will give rise to additional electromagnetic fields, and that additional forces will be exerted by the field on such moving bodies. The laws of the "*electrodynamics of moving bodies*" cannot, of course, be deduced from the set of equations (9.3.1) to (9.3.14) as these apply only to systems of bodies at *rest* with respect to each other. But under plausible assumptions tentative mathematical formulations of these laws may be made which can then be put to the test by suitable experiments.

Faraday had already suggested in 1838 that the bodily movement of an electric charge might be equivalent to an electric current and thus produce a magnetic field. Consider a cylindrically shaped body of unit cross-sectional area and length v, which is filled with electric charge of density ϱ. If this body moves with constant velocity $\mathbf{v}$ in the direction of the cylinder axis, the total charge ϱv within the body will be carried through the unit area of a fixed plane normal to $\mathbf{v}$ in unit time, which is equivalent to a "*convection current*" of density

$$\mathbf{j}_c = \varrho\mathbf{v}. \tag{9.3.15}$$

According to (9.3.3) this should, in the absence of electric and magnetic polarization, produce a magnetic field, given by

$$\text{curl } \mathbf{B}_c = \frac{4\pi}{c}\mathbf{j}_c = \frac{4\pi}{c}\varrho\mathbf{v}. \tag{9.3.16}$$

This relation was formulated by Maxwell in 1873 and verified in famous experiments with rotating charged discs by Rowland, Eichenwald, and others which prove that the electric charge on a material body is carried along with it.

Equation (9.3.1) further indicates that in addition to the "true charges" ϱ, additional charges ϱ_p as sources of the electric field are generated by a non-uniform electric polarization of a medium according to

$$\varrho_p = -\text{div } \mathbf{P}. \tag{9.3.17}$$

This suggests that another additional convection current of density $\mathbf{j}_p$ is produced by the movement of a non-uniformly polarized dielectric which, in analogy to (9.3.15), is given by

$$\mathbf{j}_p = \varrho_p\mathbf{v} = -\mathbf{v}\,\text{div } \mathbf{P} \tag{9.3.18}$$

and should also give rise to a magnetic field.

The existence of such an effect was first suggested by Roentgen in 1888 and verified by himself and later by Eichenwald (1903) by means of experiments with dielectric disks rotating in an electric field which thus became electrically polarized according to the relation (9.3.9). The experimental results prove that the electric polarization of a dielectric body is also carried along with it in its movement through the ether; we shall come back to this important fact later on in this section.

It can be shown that the magnetic field $\mathbf{B}$ produced by the convection current $\mathbf{j}_p$ is the same as if the body would have acquired a magnetic polarization $\mathbf{M}_p$. For in the absence of magnetic polarization the expression in brackets on the right-hand side of (9.3.3) becomes $\mathbf{j} + \mathbf{j}_p + \dot{\mathbf{P}}$, and in the absence of electric polarization it is $\mathbf{j} + c\,\text{curl } \mathbf{M}_p$. Thus we have, with the help of (9.3.18)

$$\text{curl } \mathbf{M}_p = \frac{1}{c}(\dot{\mathbf{P}} - \mathbf{v}\,\text{div } \mathbf{P}). \tag{9.3.19}$$

The x-component of the right-hand member in (9.3.19) is

$$\frac{1}{c}\left(\frac{\partial \mathsf{P}_x}{\partial x}v_x + \frac{\partial \mathsf{P}_x}{\partial y}v_y + \frac{\partial \mathsf{P}_x}{\partial z}v_z - v_x\frac{\partial \mathsf{P}_x}{\partial x} - v_x\frac{\partial \mathsf{P}_y}{\partial y} - v_x\frac{\partial \mathsf{P}_z}{\partial z}\right),$$

and it can easily be verified that the expression in the brackets is equal to the x-component of curl $(\mathbf{P} \wedge \mathbf{v})$. The same consideration applies to the other components and it follows that

$$\mathbf{M}_p = \frac{1}{c}(\mathbf{P} \wedge \mathbf{v}). \qquad (9.3.20)$$

Thus it is indeed proved that an electrically polarized body which performs a uniform translational movement with respect to the ether appears to a stationary observer to have acquired a magnetic moment $\mathbf{M}_p$ given by the relation (9.3.20).

The relationship (9.3.20) suggests that, similarly, the movement of a magnetic body through the ether might produce an electric polarization $\mathbf{P}_m$ of the body which, as measured by a stationary observer, were connected with $\mathbf{M}$ and $\mathbf{v}$ by a relation of the form

$$\mathbf{P}_m = \frac{1}{c}(\mathbf{v} \wedge \mathbf{M}). \qquad (9.3.21)$$

It will be proved later in section 10.4 from the basic principle of relativistic electrodynamics that this tentative relationship is indeed valid.

This indicates that the movement of magnets through the ether ought to produce electric fields which in turn generate electric currents in conducting circuits. This fact is, of course, well known and widely utilized for the generation of electric current. But, according to Maxwell's formulation of Faraday's law of induction, this phenomenon should occur only if the movement of the magnets produces a *change in time of the magnetic flux* through the circuit. Such a change of flux does, however, not take place if, for example, an axially symmetrical, permanent magnet rotates about its own axis, and it therefore ought not to be possible to use such a device as an electric generator. But Faraday himself discovered that an induction current is in fact induced in a wire whose ends make sliding contacts with the axis and a circle on the surface of a rotating cylindrical magnet, a phenomenon which is known under the name of "*unipolar induction*".

The phenomenon of unipolar induction cannot be deduced from the system of Maxwell's equations, but it follows as a direct consequence of the tentative equation (9.3.21). For according to this relation a rotating magnet acquires, from the point of view of a resting observer, an electric polarization P_m and thus a potential difference must be created between the end points of the wire in Faraday's mentioned experiment. It has indeed been proved by a whole series of ingenious experiments that unipolar induction is not due to a direct action of the magnetic field of the rotating magnet on the circuit but to the acquisition of an electric polarization of the magnet, and that the relationship (9.3.21) is indeed correct. These experiments finally also prove that, like the electric polarization, the magnetic polarization of a material body is carried along with its movement through the ether.

We now turn to the discussion of the forces exerted by electromagnetic fields on moving bodies. We have shown that the electromagnetic field generated by a "convection current" is in every respect the same as that generated by a "conduction current" of the same magnitude. It is therefore to be expected that a magnetic field $\mathbf{B}$ will exert a force $\mathbf{f}_2$ on a moving body carrying electric charge and polarization which can be calculated from (9.3.12) if the total convection current $\mathbf{j}_c + \mathbf{j}_p$ is inserted for $\mathbf{j}$. Adding to these forces the sum of the forces $\mathbf{f}_1$ and $\mathbf{f}_3$ which constitute the action of the electric field $\mathbf{E}$ on the charge and polarization of the body we obtain for the force per unit volume

$$\mathbf{f} = (\varrho - \operatorname{div} \mathbf{P})\left[\mathbf{E} + \frac{1}{c}(\mathbf{v} \wedge \mathbf{B})\right]. \tag{9.3.22}$$

This formula shows that the action of a *magnetic* field on a moving body is equivalent to that of an additional *electric* field $\mathbf{E}_b$ given by

$$\mathbf{E}_b = \frac{1}{c}(\mathbf{v} \wedge \mathbf{B}). \tag{9.3.23}$$

Proof of the validity of the relationship (9.3.23) is provided by a great amount of experimental evidence from which the following examples are selected.

(a) A particle charged with an electric charge e will in addition to the force $\mathbf{F}_e = e\mathbf{E}$, exerted by an electric field $\mathbf{E}$ also experience an additional force

$$\mathbf{F}_m = \frac{e}{c}(\mathbf{v} \wedge \mathbf{B}) \tag{9.3.24}$$

exerted by a magnetic field $\mathbf{B}$.

The existence of this force was predicted by H. A. Lorentz from his "theory of electrons" (see later) and is therefore called "*Lorentz force*". It plays a most important part in experiments for the determination of either the "specific charge" e/m or the velocity of elementary particles, in "mass spectrography", and in the function of particle accelerators.

(b) If a metal cylinder of thickness a and radius R rotates with angular velocity ω in a homogeneous magnetic field with field strength $\mathbf{H}$ parallel to its axis a radial electric field $\mathbf{E}_b$ of magnitude $\frac{1}{c}\mathbf{H}R\omega$ is produced in the cylinder, which gives rise to a potential difference $(a\omega \mathbf{H} R)/c$ between two points across the cylinder which can be used for generating a current in an external circuit between these two points. Similarly, potential differences will occur in a conducting liquid flowing through a tube in a magnetic field.

(c) If the experiment described above is carried out with a dielectric instead of a conducting cylinder, the field $\mathbf{E}_b$ will, according to (9.3.9), gvie rise to a polarization $\mathbf{P}_b = \varkappa \mathbf{E}_b$ which can be measured, and thus the relation (9.3.23) can be verified (H. A. Wilson, 1904).

(d) If excited atoms in an atomic ray move across a magnetic field they should according to (9.3.23) experience an electric field $\mathbf{E}_b$ which ought to produce a "Stark effect" splitting of the spectral lines emitted by the atoms. This experiment was carried out by W. Wien in 1914–16 and the expected effect was indeed discovered and shown to be in quantitative agreement with the relationship (9.3.23).

We have now shown that moving bodies carrying electric charges and electric and magnetic polarization give rise to electromagnetic fields and that these fields in turn must exert forces on the bodies. Thus one should expect that a system of such bodies, which is in equilibrium when it is stationary in the ether, will not remain in equilibrium if it moves as a whole with uniform speed V through the ether; in other words that the principle of relativity will *not* hold in the realm of classical electrodynamics.

However, eqns. (9.3.16), (9.3.19), (9.3.20), and (9.3.21) show that the field quantities are all proportional to the ratio V/c, and it seen from eqns. (9.3.23) and (9.3.24) that the forces exerted by the fields on the bodies are also proportional to V/c. Hence the effect to be expected is of second order in V/c and the principle of relativity should thus be valid to a *first-order approximation* in electrodynamics as well as in optics. This is, indeed, to be expected, for the Maxwellian equations are based on experiments with electromagnetic systems on earth which are evidently not at rest with respect to the ether.

It is, of course, not surprising to find that the principle of relativity should hold to the same approximation in electrodynamics and in optics, for according to the electromagnetic theory of light the laws of optics follow from those of electrodynamics as special cases. But whereas the mechanical ether theory made it necessary to introduce Fresnel's dragging hypothesis with all its shortcomings in order to explain the validity of the principle of relativity in optics to a first-order approximation, as pointed out in section 9.2, no such hypothesis was introduced in this section. This indicates that it must be possible to account for the appearance of the "dragging coefficient" K in the laws of light propagation in moving bodies direct from the basic principles of the electrodynamics of moving bodies without any further special hypothesis.

This can indeed be shown to be the case by the following simple consideration. An electromagnetic wave in a dielectric is described by the wave vectors $\mathbf{D}$ and $\mathbf{H}$. Now we have seen that only the second part of $\mathbf{D}$ in the expression (9.3.7) is carried along with the moving dielectric, whereas the first part refers to the stationary ether. Hence only the fraction

$$K = \frac{4\pi\mathbf{P}}{\mathbf{E} + 4\pi\mathbf{P}} = \frac{4\pi\varkappa}{1 + 4\pi\varkappa} = \frac{\varepsilon - 1}{\varepsilon} = 1 - \frac{1}{\varepsilon} \tag{9.3.25}$$

(where ε is the "*permittivity*" of the dielectric) takes part in the movement of the body. But according to the electromagnetic wave theory the refractive index n of the body for a wave of given frequency is equal to the square root of its permittivity for this frequency and thus the expression (9.3.25) for K becomes indeed equal to Fresnel's expression (9.2.16).

So far our considerations have been entirely based on the system of the phenomenological Maxwell's equations (9.3.1) to (9.3.14) and the tentative equations (9.3.15) to (9.3.24) for moving bodies suggested by them. The first successful attempt to derive the laws of the electrodynamics of moving bodies from first principles is due to H. A. Lorentz on the basis of his *Theory of Electrons* (1895). This theory assumes that there exists an *immovable* non-material ether which is the carrier of electromagnetic fields described by the field vectors $\mathbf{E}$ and $\mathbf{H}$, and that all electromagnetic phenomena in material bodies are due to the presence of "electrons" within them. In particular *all* electric currents are supposed to be due to the movement of the electrons, electric polarization due to their displacement from equilibrium positions, and magnetic polarization due to their periodic orbital movement inside the atoms.

The fundamental equations of this theory, called the "*Lorentz–Maxwell equations*" which correspond to the equations (9.3.1) to (9.3.4), are

$$\left.\begin{aligned} &\operatorname{div} \mathbf{E} = 4\pi\varrho, \quad \operatorname{div} \mathbf{H} = 0, \\ &\operatorname{curl} \mathbf{H} = \frac{4\pi}{c}(\varrho \mathbf{v}) + \frac{1}{c}\dot{\mathbf{E}}, \quad \operatorname{curl} \mathbf{E} = -\frac{1}{c}\dot{\mathbf{H}}, \end{aligned}\right\} \tag{9.3.26}$$

where ϱ is equal to the electronic charge times the number density of the electrons and $\mathbf{v}$ their velocity. To these equations is to be added the continuity equation in the form

$$\dot{\varrho} + \operatorname{div}(\varrho \mathbf{v}) = 0 \tag{9.3.27}$$

and the equations (9.3.11) to (9.3.14) are replaced by the single equation

$$\mathbf{F} = e\left[\mathbf{E} + \frac{1}{c}(\mathbf{v} \wedge \mathbf{H})\right], \tag{9.3.28}$$

representing the force exerted by the field on an electron.

From the point of view of this theory *all* currents are convection currents, and (9.3.15) holds automatically for "conduction" currents as well. On the basis of the theory one can calculate the electromagnetic field surrounding an electron at the origin of coordinates moving through the ether with velocity $\mathbf{v}$ in the following plausible way.

A so-called "current element" $I d\mathbf{s}$ in a wire of cross-sectional area a is, according to (9.3.15), equal to

$$I\, d\mathbf{s} = \mathbf{j} a\, ds = \mathbf{j}\, dV = \varrho\, dV \cdot \mathbf{v} = Q\mathbf{v}, \tag{9.3.29}$$

where Q is the total charge contained in the volume element dV. The magnetic field $\mathbf{H}$ which it generates at a point with position vector $\mathbf{r}$, is, by

Biot–Savart's law,

$$\mathbf{H} = \frac{I}{c}\frac{d\mathbf{s} \wedge \mathbf{r}}{r^3}. \tag{9.3.30}$$

Thus combining (9.3.29) and (9.3.30) and identifying Q with the electronic charge e, we obtain

$$\mathbf{H} = \frac{e}{c}\frac{\mathbf{v} \wedge \mathbf{r}}{r^3}. \tag{9.3.31}$$

The electric field $\mathbf{E}$ of the electron is given by Coulomb's law, irrespective of the fact that the electron is moving, and thus

$$\mathbf{E} = e\frac{\mathbf{r}}{r^3}. \tag{9.3.32}$$

Combining (9.3.31) and (9.3.32) we can finally set up the following relationship between $\mathbf{E}$, $\mathbf{H}$, and $\mathbf{v}$:

$$\mathbf{H} = \frac{1}{c}(\mathbf{v} \wedge \mathbf{E}). \tag{9.3.33}$$

From the point of view of the theory of electrons all macroscopic charges and currents are superpositions of moving electrons, and the phenomenological magnetic induction vector $\mathbf{B}$ is identified with the Lorentz vector $\mathbf{H}$. Hence the relation

$$\mathbf{B} = \frac{1}{c}(\mathbf{v} \wedge \mathbf{E}) \tag{9.3.34}$$

ought to hold generally. It can, indeed, be shown that the experimentally verified relationship (9.3.20) between $\mathbf{M}_p$ and $\mathbf{P}$ follows directly from (9.3.34). For the particular case of a charged plate condenser this is illustrated by Fig. 21 which is self-explanatory.

Lorentz was able to deduce from his theory the macroscopic equations of the electrodynamics of stationary and moving bodies and to prove the

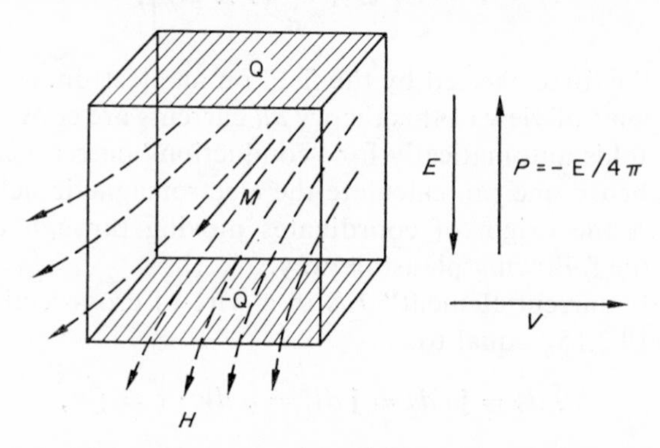

FIG. 21. Magnetic field created by the movement of a charged plate condenser through the ether, according to classical theory.

validity of the principle of relativity for small values of V/c, but as his equations are *not* invariant with respect to the Galileo transformations (9.1.13), he had to modify the last of them by introducing (1904) the concept of a so-called "local time" t' by setting

$$t' = t - \frac{Vx}{c^2}. \tag{9.3.35}$$

This is clearly another "*ad hoc* hypothesis" which had to be introduced into the theory in order to account for the observed phenomena, in particular the phenomenon of unipolar induction, while, on the other hand, Fresnel's hypothesis could be abandoned. We shall not discuss the significance of the transformation equation (9.3.35) here because we shall show in the following chapter 10 that the concept of local time follows logically from the basic principles of the theory of relativity, and it will be discussed there in section 10.1.

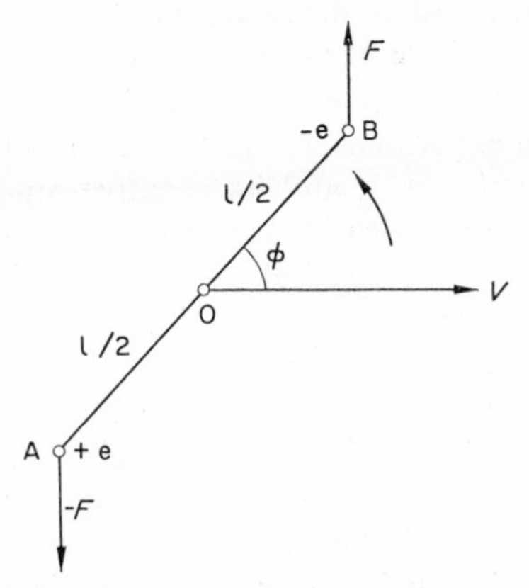

FIG. 22. Diagram illustrating the creation of a couple acting on a pair of electric charges, due to its movement through the ether, according to classical theory.

We now return to the question of the second-order effects to be expected to occur in an electromagnetic system when it moves as a whole with uniform velocity **V** through the ether. As a simple example let us consider two point charges $+e$ and $-e$, fixed at the end points A and B (Fig. 22) of a horizontal rigid beam of length l which is suspended in its centre O and is inclined at an angle φ to the direction of the earth's orbital velocity **V**.

It follows from (9.3.31) that the moving charge $+e$ creates a magnetic field of magnitude

$$\mathsf{H} = eV \sin\varphi / l^2 c$$

at B in a vertical direction and this in turn, according to (9.3.28), exerts a force of magnitude

$$F = eV\mathsf{H}/c = e^2V^2 \sin\varphi / l^2c^2$$

on the charge $-e$ in a horizontal direction, as indicated in the figure. This force, together with the equal and opposite force exerted by the charge $-e$ on the charge $+e$ at A, forms a couple acting on the beam with magnitude

$$T = Fl \cos \varphi = \frac{e^2}{2l} \frac{V^2}{c^2} \sin 2\varphi, \tag{9.3.36}$$

which tends to turn the beam in a direction perpendicular to $\mathbf{V}$ and has its maximum for $\varphi = 45°$. The effect is indeed seen to be of the second order in V/c.

It was first suggested by Fitzgerald to put this to the test, and an experiment of this type was in fact carried out by Trouton and Noble in 1903 in which the beam with the two charges was replaced by a plate condenser suspended with the plates hanging down vertically. A charging up of the condenser should then produce a small but measurable angular deflection. But no such deflection could be detected, and later experiments with much greater accuracy, by which "ether drifts" of less than 1 km/sec would produce a measurable deflection, proved equally negative.

The Trouton–Noble experiment is the analogue of the Michelson experiment in optics (see section 9.2), and its negative outcome proves that the principle of relativity is *exactly* valid in electrodynamics as well as in optics—contrary to the predictions of classical electromagnetic theory.

It could, of course, be argued that the failure to detect a positive effect in the Trouton–Noble experiment is due to the same hypothetical Lorentz contraction which we have shown in section 9.2 to be capable of explaining the negative outcome of the Michelson experiment. This can indeed be shown to be the case as we shall now demonstrate for the simplified model of this experiment considered above.

The Lorentz contraction which operates in the direction of $\mathbf{V}$ will reduce the length l of the beam to

$$l' = [l^2 \sin^2 \varphi + l^2 \cos^2 \varphi\,(1 - V^2/c^2)]^{1/2} = l\left(1 - \frac{V^2}{c^2} \cos^2 \varphi\right)^{1/2}$$

and thus increase the electrostatic energy $W = -e^2/l$ of the system to

$$W' = -\frac{e^2}{l'} = -\frac{e^2}{l}\left(1 - \frac{V^2}{c^2} \cos^2 \varphi\right)^{-1/2} \simeq -\frac{e^2}{l}\left(1 + \frac{V^2}{2c^2} \cos^2 \varphi\right)$$

$$= W - \frac{e^2 V^2}{2lc^2} \cos^2 \varphi. \tag{9.3.37}$$

The additional energy $\Delta W = W' - W$ must give rise to an additional couple T'

$$T' = -\frac{d(\Delta W)}{d\varphi} = -\frac{e^2 V^2}{lc^2} \cos \varphi \sin \varphi = -\frac{e^2 V^2}{2lc^2} \sin 2\varphi, \tag{9.3.38}$$

which is equal and opposite to T and thus compensates it, so that, indeed, no deflection would be caused by the movement of the system through the ether.

Problems and exercises

P.9.1. Prove that the plane of oscillation of a Foucault pendulum, located at latitude φ, will appear to rotate with an angular velocity $-\omega \sin \varphi$, where ω is the angular velocity of the earth's rotation.

P.9.2. A rigid body of mass M can rotate freely about a fixed centre at $\mathbf{R}$. Use eqn. (9.1.18) for calculating the effect of a change of direction of the rotation vector $\mathbf{w}$ on the movement of the body.

P.9.3. According to (9.1.18) the Coriolis force acting on a particle of mass m, moving with velocity $\mathbf{v}'$ at a point of latitude φ on the earth's surface can be split up into two parts:

$$\mathbf{F}_1^* = -2m\omega \sin \varphi \, (\mathbf{i} \wedge \mathbf{v}'),$$
$$\mathbf{F}_2^* = -2m\omega \cos \varphi \, (\mathbf{j} \wedge \mathbf{v}'),$$

where $\mathbf{i}$ and $\mathbf{j}$ are unit vectors pointing in the vertical and horizontal northerly directions respectively, and ω is the angular velocity of the earth's rotation. Discuss fully the effects of $\mathbf{F}_1^*$ and $\mathbf{F}_2^*$ on m for the cases when (a) the particle moves in a horizontal, and (b) in a vertical direction.

P.9.4. Prove that, as a result of the astronomical aberration, the apparent position of a star of stellar latitude φ describes an ellipse in the course of a year with the major axis parallel to the plane of the ecliptic, and that the ratio of the axes is equal to $\sin \varphi$, assuming the earth's orbit to be circular.

P.9.5. In Sagnac's experiment four mirrors are placed at the corners of a square of side l, mounted on a disk which is made to rotate with angular velocity ω. Two rays of light are travelling around the square, one in the direction of rotation and the other in the opposite direction. Calculate the difference between the travelling times t_1 and t_2, and show that formula (9.2.24) for the phase difference between the two rays, produced by the rotation, is satisfied.

P.9.6. A plate A acts as emitter of plane electromagnetic waves of frequency ν in the direction of the positive x-axis, and another plate B, parallel to A, emits waves of the same frequency in the opposite direction. The whole system moves with speed V with respect to the ether in the positive x-direction. Show that, from the point of view of an observer moving with the system, the wave function Φ will appear to be zero at certain "nodal" planes at mutual distances $d = \dfrac{\lambda}{2}\left(1 - \dfrac{V^2}{c^2}\right)$, where λ is the wavelength of the waves measured for $V = 0$.

P.9.7. Show that in the Michelson experiment no second-order effect will be observed for *any* orientation of the arms with respect to the velocity $\mathbf{V}$ of the earth if it is assumed that the length l of an arm is reduced to $l' = l\sqrt{1 - \dfrac{V^2 \cos^2 \alpha}{c^2}}$, where α is the angle between $\mathbf{V}$ and the arm.

P.9.8. A dielectric disc of radius R carries on its surface a uniformly distributed electric charge Q and rotates at n revolutions per second. Calculate the magnetic field strength H, generated by the convection current, at a point on the axis of rotation at a distance l from the disk, in particular for $l \ll R$.

P.9.9. A very long steel bar with rectangular cross-section, placed parallel to the x-axis with its sides parallel to the xy- and the xz-plane respectively, is uniformly magnetized in the y-direction and moves with velocity V in the x-direction. Show that an electric potential difference φ will be set up by unipolar induction between two sliding contacts which are a distance a apart in the z-direction. Use eqn. (9.3.21) to calculate the magnitude of φ.

P.9.10. A beam of particles of mass m, carrying electric charges e, is issuing from $x = 0$ with initial velocity v in the positive x-direction. It is acted upon simultaneously by a homogeneous electric field **E** and a homogeneous magnetic field **H**, both in the z-direction. It produces a spot with coordinates y, z on a fluorescent screen normal to the x-axis at $x = l$. Derive expressions for y and z as functions of v and show that a beam with particles of various velocities will trace out a parabola on the screen.

P.9.11. Show that a charged particle, originally moving with velocity v at right angles to the direction of a homogeneous magnetic field **H**, performs a circular orbital movement in a plane normal to **H**. Calculate the radius R of the circle as a function of v and H and prove that the number n of revolutions per second is independent of v.

P.9.12. Evaluate the couple **T** which, according to classical electrodynamics, should be exerted on the condenser of capacitance C in the Trouton–Noble experiment (described in section 9.3) when the condenser is charged up to a potential Φ, making use of the relations (9.3.24) and (9.3.34).

CHAPTER 10

The Special Theory of Relativity

IN CHAPTER 9 it was shown that according to the laws of classical physics the classical principle of relativity ought to be valid precisely in the realm of Newtonian dynamics, but only to a first-order approximation for sufficiently small velocities in the realms of optics and electrodynamics. But overwhelming experimental evidence suggests that the principle of relativity should in fact hold precisely for *all* physical phenomena. It was further shown that this would indeed follow from classical theory if certain *ad hoc* hypotheses were introduced.

However, Einstein (1905c) for the first time declared the principle of relativity as an *epistomological* principle of physics and developed the so-called "special theory of relativity" on its basis which was to replace the classical theory. Einstein's "*restricted principle of relativity*" maintains that the laws of physics should be valid for all *inertial* frames of reference (see section 9.1) and should therefore only contain the *relative* velocities of particles with respect to the observer. Hence the notion of a *preferential* frame of reference, like the Newtonian absolute space or the Lorentzian ether has no place in a theory based on this principle and an "ether drift" does not exist. Einstein further postulated that the speed of light should have the same universal value for all observers using inertial frames of reference and should be independent of the velocity of the light source.

It is evidently not possible to reconcile these principles with the classical notions of space and time, and hence classical kinematics has first to be modified. In this chapter we shall first develop the main principles of "relativistic kinematics" and then proceed to apply these new principles to the derivation of the laws of relativistic mechanics, optics, and electrodynamics.

10.1. Relativistic kinematics

Just as in classical kinematics (section 9.1) the measurement of lengths and times in relativistic kinematics is supposed to be carried out by means of measuring rods made of rigid materials, and clocks performing periodic movements. But in classical kinematics it was thought to be self-evident that two measuring rods, which appear to be of the same length to an observer with respect to whom they are both at rest, should remain to be equal in length if one of them is in motion with respect to the observer. Similarly, it was considered as obvious that two clocks, which appear to be perfectly synchronous to an observer with respect to whom they are both at rest, should remain to keep time with one another if one of the clocks moves with respect to the observer. The classical law (9.1.2) of the superposition of velocities follows direct from these assumptions and also the Galileo transformations (9.1.11), (9.1.12), and (9.1.13).

It was pointed out by Einstein for the first time that these assumptions are in fact by no means evident, because in order to compare lengths of measuring rods and time rates of clocks, which are moving relative to each other, length and time measurements have always to be combined, and that the assumptions of classical kinematics need to be verified by experiment. The classical law of superposition of velocities and the Galileo transformations are clearly incompatible with the experimentally established principle of the constancy of the velocity of light, as was shown in Chapter 9, and this fact may further be taken as a proof for the incorrectness of the basic assumptions of classical kinematics which therefore must be appropriately modified.

As in classical kinematics, coordinate systems have to be used for describing the position of a point with respect to a frame of reference. In principle these can be quite arbitrary, but the most suitable ones are Cartesian coordinate systems (rectangular or oblique) in which straight lines are represented by linear equations between the coordinates. In the following we shall use rectangular coordinates x, y, z throughout. It is sometimes useful to represent *kinematic* relationships in a four-dimensional space with coordinates x, y, z, ct in which they appear as *geometrical* relations. Another representation in an abstract four-dimensional space with coordinates x, y, z, ict, to be explained later on in this section, will be made use of repeatedly in this chapter.

The simplest way to deduce the relativistic coordinate transformations, which are to replace the classical Galileo transformations, is to consider the propagation of light in the form of spherical waves from a point source. Assume this source to be at the origin of a coordinate system S. Then a signal sent out at $t = 0$ will reach all points of a spherical surface of radius ct at time t whose equation is thus

$$x^2 + y^2 + z^2 - c^2t^2 = 0. \tag{10.1.1}$$

If now another observer makes measurements in another coordinate system S' that coincides with S at $t = 0$ and moves with a velocity V with respect to S in the x-direction, then, if the principle of the constancy of light velocity is correct, his observations should give precisely the same result, that is the coordinates x', y', z' of points on the wave surface, measured with his measuring rods, and the time t' measured with his clock must be related by the equation

$$x'^2 + y'^2 + z'^2 - c^2 t'^2 = 0, \tag{10.1.2}$$

provided that for $t = 0$ also $t' = 0$.

We now have to find a relationship between the variables x, y, z, t, on the one hand, and the variables x', y', z', t' on the other hand, which will transform eqn. (10.1.1) into (10.1.2) or vice versa, or in other words, we have to find a coordinate transformation for which the left-hand side of (10.1.1) remains invariant, that is

$$x^2 + y^2 + z^2 - c^2 t^2 = x'^2 + y'^2 + z'^2 - c^2 t'^2. \tag{10.1.3}$$

As space and time must be supposed to be homogeneous, the transformation equations must be independent of the particular choice of the zero point for the measurement of space and time and must therefore necessarily be *linear*. Furthermore, as the x-direction (and x'-direction) was chosen to be the direction of the relative movement of the two systems, the transformation equations must, for reasons of symmetry, about that direction, have the form

$$x' = \alpha t + \beta x, \quad y' = \varepsilon y, \quad z' = \varepsilon z, \quad t' = \gamma t + \delta x, \tag{10.1.4}$$

where $\alpha, \beta, \gamma, \delta, \varepsilon$ are constants which depend on the velocity V.

Since the origin of S' moves with velocity V in the x-direction, $x = Vt$ must correspond to $x' = 0$ and hence we must have

$$\alpha = -\beta V. \tag{10.1.5}$$

Since, further, the transformations (10.1.4) have to coincide with the Galileo transformations (9.1.13) for sufficiently small values of V (under which condition the latter are known to be valid from experiment), we must further have

$$\beta = 1, \gamma = 1, \delta = 0, \varepsilon = 1 \quad \text{for} \quad V/c \ll 1. \tag{10.1.6}$$

We now substitute (10.1.4) into (10.1.3) and obtain

$$x^2 (1 - \beta^2 + c^2 \delta^2) + t^2 (c^2 \gamma^2 - c^2 - \alpha^2) + 2xt (c^2 \gamma \delta - \alpha\beta) + y^2 (1 - \varepsilon^2) + z^2 (1 - \varepsilon^2) = 0.$$

As this equation must be valid for all values of x, y, z, t, the coefficients of the various powers of these quantities must vanish separately so that we obtain the following equations:

$$\left.\begin{aligned} c^2 \delta^2 &= \beta^2 - 1, \\ c^2 \gamma \delta &= \alpha\beta, \\ c^2 \gamma^2 &= c^2 + \alpha^2, \\ \varepsilon &= 1. \end{aligned}\right\} \tag{10.1.7}$$

Elimination of δ from the first two of these equations yields

$$\alpha^2 = c^2 (\beta^2 - 1) \quad \text{and} \quad \gamma^2 = \beta^2,$$

from which, with the help of (10.1.5) and by making use of the conditions (10.1.6), we obtain

$$\beta = \gamma = \frac{1}{\sqrt{1 - V^2/c^2}}. \tag{10.1.8}$$

From the third of the equations (10.1.7) one has then

$$\delta = \frac{\alpha}{c^2} = -\frac{\beta V}{c^2}, \tag{10.1.9}$$

which automatically satisfies (10.1.6).

Thus, finally, the relativistic transformation equations (10.1.4) become

$$\left.\begin{aligned} x' &= \beta (x - Vt), \quad y' = y, \quad z' = z, \\ t' &= \beta \left(t - \frac{Vx}{c^2} \right), \quad \beta = \frac{1}{\sqrt{1 - V^2/c^2}}. \end{aligned}\right\} \tag{10.1.10}$$

These equations are called the "*Lorentz transformations*" because they were first derived by H. A. Lorentz in 1904 (before Einstein) but not on the basis of relativistic considerations.

We now proceed to show that the Lorentz transformations satisfy the principle of relativity, and that Einstein's two basic principles are thus consistent with each other. For if the equations (10.1.10) are solved with respect to x, y, z, t, one obtains from an elementary calculation

$$\left.\begin{aligned} x &= \beta (x' + Vt'), \quad y = y', \quad z = z', \\ t &= \beta \left(t' + \frac{Vx'}{c^2} \right). \end{aligned}\right\} \tag{10.1.11}$$

These equations differ from the equations (10.1.10) only in so far as V is replaced by $-V$. Since $-V$ is the relative velocity of S with respect to S', one sees that the equations (10.1.10) and (10.1.11) are indeed completely equivalent and, complying with the principle of relativity, contain only *relative velocities*.

In order to understand the physical meaning of the Lorentz transformations we shall now discuss some special cases in greater detail. Let us first assume that an observer in S wishes to measure the length l' of a rod fixed on the x'-axis of the system S' (which in the following will be called the "moving rod") in terms of the length l of a "resting" rod, fixed in S. This can, for example, be achieved if an operator in S' simultaneously, that is at the same time t', sends out two light signals from the endpoints of l' towards the observer. If x_1 and x_2 are the positions of the end points of l and x'_1 and x'_2 those of l', then one has from the first of the equations (10.1.11)

$$x_1 = \beta (x'_1 + Vt'), \quad x_2 = \beta (x'_2 + Vt'),$$

and hence

$$l = \beta l'$$

or

$$l' = l \sqrt{1 - V^2/c^2}. \tag{10.1.12}$$

This means that the dimensions of material bodies in the direction of their movement with velocity V with respect to an observer appear to be reduced by a factor $\sqrt{1 - V^2/c^2}$, whereas, according to (10.1.11) no change in the dimensions at right angles to the direction of V is observed.

This is precisely the "Lorentz–Fitzgerald contraction" which, as was pointed out in section 9.2, was introduced as an *ad hoc* hypothesis into classical optics in order to account for the negative outcome of the Michelson experiment. It is here deduced logically as a special consequence of the relativistic kinematic equations without any further assumption. This is, of course, not so surprising as these equations are an outcome of the principle of the constancy of the velocity of light. But whereas in Lorentz's theory the contraction of a material body is supposed to be caused by its *absolute* movement through the ether, according to the theory of relativity the contraction should be the result of the *relative* movement between body and observer.

This can indeed be proved easily. For if the observer is now in the system S' and wishes to measure the length of the rod l, the measurements must be taken at the same time t (and not t'), and thus from the first of the equations (10.1.10) one has

$$x'_1 = \beta (x_1 - Vt), \quad x'_2 = \beta (x_2 - Vt),$$

from which follows

$$l' = \beta l$$

or

$$l = l' \sqrt{1 - V^2/c^2}. \tag{10.1.13}$$

This seems at first sight incompatible with (10.1.12), but as now l is the length of the rod moving with respect to the observer and l' that of the rod at rest with respect to the observer, we see again that the moving rod is contracted in comparison with the resting rod in the ratio $\sqrt{1 - V^2/c^2}$. In Lorentz's theory, on the contrary, a rod moving through the ether would *always* be found to be compressed in the direction of that movement compared with a body at rest in the ether.

Since l and l' must necessarily be *real* quantities, eqns. (10.1.12) and (10.1.13) become meaningless if $V > c$. We must therefore conclude that two observers cannot move with relative velocities greater than that of light, and that the light velocity c in relativity theory plays the part of infinite velocity in classical theory. We shall discuss this statement later from various other points of view.

Let us now assume that an observer in S wants to compare the rate of a clock, fixed at some point x' in S', with that of a clock stationary in S. If measurements are made at two times t_1 and t_2, corresponding to times t_1' and t_2' in S', one obtains from the fourth of the equations (10.1.11)

$$t_1 = \beta\left(t_1' + \frac{Vx'}{c^2}\right), \quad t_2 = \beta\left(t_2' + \frac{Vx'}{c^2}\right),$$

from which follows

$$\Delta t \equiv t_2 - t_1 = \beta\,(t_2' - t_1') = \beta\,\Delta t'$$

or

$$\Delta t' = \Delta t\,\sqrt{1 - V^2/c^2}. \tag{10.1.14}$$

The observer will therefore conclude that the "moving" clock is *retarded* in comparison with the "resting" clock in the ratio $\sqrt{1 - V^2/c^2}$. This is the same phenomenon which we called "Lorentz clock retardation" in section 9.2, and formula (10.1.14) is indeed identical with formula (9.2.31). Lorentz had introduced the clock retardation into classical optics as a special hypothesis in order to account for the negative outcome of experiments to detect an ether drift. But whereas in Lorentz's view the clock retardation was caused by the *absolute* movement of the clock mechanism through the ether, the theory of relativity ascribes the retardation to the *relative* movement of clock and observer.

This can again be shown to be the case by placing the observer in the system S'. The observed clock is now fixed in S and thus is again moving with relative velocity V with respect to the observer. From the fourth of the equations (10.1.10) one now has

$$t_1' = \beta\left(t_1 - \frac{Vx}{c^2}\right), \quad t_2' = \beta\left(t_2 - \frac{Vx}{c^2}\right),$$

and hence

$$\Delta t' = \beta\,\Delta t$$

or

$$\Delta t = \Delta t'\,\sqrt{1 - V^2/c^2}. \tag{10.1.15}$$

This is not in contradiction to (10.1.14) because now Δt is the time interval on the "moving clock" and $\Delta t'$ the corresponding time interval on the "resting" clock, so that the former is again shorter than the latter and the moving clock retarded.

We notice here again that, since Δt and $\Delta t'$ must necessarily be real, eqns. (10.1.14) and (10.1.15) become meaningless if $V > c$. This is another way of demonstrating that in relativity theory c is the upper limit for any velocity.

From the point of view of the theory of relativity the time retardation phenomenon should not only affect the rate of periodic mechanisms but quite generally the rate of any physical process. In particular the decay time of a radioactive process occurring in a fast-moving particle should appear to be longer than that of a stationary particle of the same type. This is indeed strikingly demonstrated by the observation of *mesons* which are unstable elementary particles. Measurements of the life time of so-called "μ-mesons" in the cosmic radiation first showed that they approximately obeyed the relationship (10.1.14), and more recent measurements of the life times of artificially produced "π-mesons" (by means of the measurement of their path lengths in cloud chambers) have proved that the relationship (10.1.14) is indeed precisely satisfied.

Another problem, which is intimately connected with relativistic kinematics, is that of the *simultaneity* of two events. Intuitively one would think that this term has an absolute meaning and that the fact that two events occur at the same time in two different regions of space should hold for all observers of these events, independently of their relative movement with respect to one another. However, it has to be kept in mind that simultaneity can only be established by the observation of clocks which are placed close to the location of the events. Now in classical kinematics the measurement of time by clocks is not affected by the movement of the observer with respect to the clock, as expressed mathematically by the Galileo transformation $t = t'$ (9.1.13). But Lorentz had already found it necessary to introduce the notion of "local time" as an *ad hoc* hypothesis, as was explained in section 9.3. His original relation (9.3.35) is indeed, apart from the missing factor β, identical with the fourth of the Lorentz transformations (10.1.10). From the point of view of the theory of relativity the situation is as follows.

Suppose that the clocks are placed at two points x_1 and x_2 on the x-axis of the system S and found to keep time perfectly by an observer in S. Two events at x_1 and x_2 respectively will then be said to occur simultaneously by this observer if the two clocks at the moments of the events show the same time: $t_1 = t_2 = t$. It then follows from the fourth of the equations (10.1.11) that for an observer in S'

$$t = \beta\left(t_1' + \frac{Vx_1'}{c^2}\right) = \beta\left(t_2' + \frac{Vx_2'}{c^2}\right)$$

so that

$$\Delta t' = -\frac{l'V}{c^2}. \tag{10.1.16}$$

This means that the two events are *not* simultaneous with respect to the observer in S' but that the event at the point x_1' appears to occur *earlier* than that in x_2' by the amount $l'V/c^2$, where l' is the distance between the locations of the events as measured by that observer.

Vice versa, if the two events appear to occur simultaneously to the observer in S', that is if $t_1' = t_2' = t'$, then, from the fourth of the equations (10.1.10),

$$t' = \beta\left(t_1 - \frac{Vx_1}{c^2}\right) = \beta\left(t_2 - \frac{Vx_2}{c^2}\right),$$

and hence

$$\Delta t = \frac{lV}{c^2}. \qquad (10.1.17)$$

That is to say, for the observer who is at rest with respect to the locations of the events the one at x_2 will appear to occur *later* than the one at x_1.

These results clearly show that simultaneity of two events at different points in space has only a relative meaning, and that two events which appear to occur simultaneously to one observer will in general not appear to be simultaneous to another observer, one of the events occurring either earlier or later than the other. However, this is only true within certain limits. For $T = l/c$ in (10.1.17) is evidently the time a light signal takes to travel from x_1 to x_2, thus $\Delta t/T = V/c$. Hence as V is always smaller than c, one must have $\Delta t < T$. In other words, if an event E_2 occurs at a point P_2 a time Δt later than an event E_1 at P_1 so that Δt is longer than the time T taken for a light signal to travel from P_1 to P_2, then this statement is *always* true, irrespective of the frame of reference chosen. The event E_2 may then have been *caused* by the event E_1. But if $\Delta t < T$ then E_1 may appear to be earlier or later than E_2 according to the frame of reference used, and thus a causal relationship between the two events cannot exist. This means that a physical interaction of any kind between two systems cannot proceed with a velocity greater than that of light.

So far no direct experimental proof of the relation (10.1.16) or (10.1.17) has been given although an experiment to this effect should not be too difficult to perform (Fürth, 1965). However, there is a number of indirect proofs, among them the Cedarholm–Townes experiment, mentioned in section 9.2. For it follows from (10.1.16) that then a relative time shift $\dfrac{\Delta t'}{t'}$ will produce a relative frequency shift of "atomic clocks" of magnitude $\dfrac{\Delta \nu}{\nu} = -\dfrac{\Delta t'}{t'} = \dfrac{v'V}{c^2}$ which exactly compensates the term $-\dfrac{v'V}{c^2}$ in (9.2.32), so that the effect predicted on the basis of classical optics should in fact not take place if relativistic kinematics is valid.

We now proceed to derive the transformation equations for velocity in relativistic kinematics. As usual the components of velocity in the system S are defined by

$$v_x = \frac{dx}{dt}, \quad v_y = \frac{dy}{dt}, \quad v_z = \frac{dz}{dt}, \tag{10.1.18}$$

and similarly we define for the system S'

$$v'_x = \frac{dx'}{dt'}, \quad v'_y = \frac{dy'}{dt'}, \quad v'_z = \frac{dz'}{dt'}. \tag{10.1.19}$$

Taking differentials on both sides of the Lorentz equations (10.1.10) one obtains for constant V

$$\left.\begin{aligned} &dx' = \beta\,(dx - V\,dt), \quad dy' = dy, \quad dz' = dz, \\ &dt' = \beta\left(dt - \frac{V\,dx}{c^2}\right). \end{aligned}\right\} \tag{10.1.20}$$

From the last of these equations follows the relationship

$$\frac{dt'}{dt} = \beta\left(1 - \frac{Vv_x}{c^2}\right), \tag{10.1.21}$$

which will be made use of later. Further we obtain with the help of (10.1.18) and (10.1.19) the desired velocity transformation equations in the form

$$v'_x = \frac{v_x - V}{1 - Vv_x/c^2}, \quad v'_y = \frac{v_y\sqrt{1 - V^2/c^2}}{1 - Vv_x/c^2}, \quad v'_z = \frac{v_z\sqrt{1 - V^2/c^2}}{1 - Vv_x/c^2}. \tag{10.1.22}$$

These equations are clearly different from the equations (9.1.2) which express the classical law of superposition of velocities; for these, in the particular case of V having the direction of the x-axis, would demand that $v'_x = v_x - V$, $v'_y = v_y$, $v'_z = v_z$. However, for very small values of V/c the equations (10.1.22) assume indeed this empirically established form, in confirmation of the statement made in section 9.1.

Starting from the inverse Lorentz transformations (10.1.11) one can derive, in a completely analogous way, the following relations:

$$\frac{dt}{dt'} = \beta\left(1 + \frac{Vv'_x}{c^2}\right) \tag{10.1.23}$$

and

$$\left.\begin{aligned} &v_x = \frac{v'_x + V}{1 + Vv'_x/c^2}, \quad v_y = \frac{v'_y\sqrt{1 - V^2/c^2}}{1 + Vv'_x/c^2}, \\ &v_z = \frac{v'_z\sqrt{1 - V^2/c^2}}{1 + Vv'_x/c^2}, \end{aligned}\right\} \tag{10.1.24}$$

which differ from the relations (10.1.21) and (10.1.22) respectively only in so far as the primed symbols are interchanged with the unprimed ones and V is replaced by $-V$, in accordance with the principle of relativity.

We notice that, if one puts $V = c$ in (10.1.24) one obtains

$$v_x = c, \quad v_y = v_z = 0,$$

which indicates that if any velocity $\mathbf{v}' > \mathbf{c}$ is added to a velocity $\mathbf{c}$ the result is again a velocity $\mathbf{c}$ in the same direction. This again shows that in relativistic kinematics c plays the part of an infinite velocity.

From (10.1.22) one can also derive an expression for the magnitude v' of the velocity in S' in terms of the velocity components v_x, v_y, v_z in S. From

$$v'^2 = v_x'^2 + v_y'^2 + v_z'^2 = \frac{(v_x - V)^2 + (1 - V^2/c^2)(v_y^2 + v_z^2)}{(1 - Vv_x/c^2)^2}$$

$$= \frac{v^2(1 - V^2/c^2) + v_x^2 V^2/c^2 - 2v_x V + V^2}{(1 - Vv_x/c^2)^2}$$

$$= \frac{v^2(1 - V^2/c^2) + (Vv_x/c^2 - 1)^2 c^2 + (V^2 - c^2)}{(1 - Vv_x/c^2)^2}$$

one has

$$1 - \frac{v'^2}{c^2} = \frac{(c^2 - V^2) - v^2(1 - V^2/c^2)}{c^2(1 - Vv_x/c^2)^2}$$

$$= \left(1 - \frac{v^2}{c^2}\right)\frac{1 - V^2/c^2}{(1 - Vv_x/c^2)^2}. \tag{10.1.25}$$

This relation may be used to transform the expression (10.1.21) for dt'/dt into

$$\frac{dt'}{dt} = \frac{1 - Vv_x/c^2}{\sqrt{1 - V^2/c^2}} = \frac{\sqrt{1 - v^2/c^2}}{\sqrt{1 - v'^2/c^2}},$$

from which follows

$$\frac{1}{\sqrt{1 - v^2/c^2}}\frac{d}{dt} = \frac{1}{\sqrt{1 - v'^2/c^2}}\frac{d}{dt'}. \tag{10.1.26}$$

The relationship (10.1.26) shows that the operator $\dfrac{1}{\sqrt{1 - v^2/c^2}}\dfrac{d}{dt}$ is invariant with respect to the transformation from S to S', or if we introduce a variable τ such that

$$d\tau = \sqrt{1 - v^2/c^2}\, dt \tag{10.1.27}$$

it follows that

$$\frac{d}{d\tau} = \frac{d}{d\tau'}. \tag{10.1.28}$$

This means that differentiation with respect to τ is an invariant operation with respect to the Lorentz transformations. τ may be considered to be a time measure which is carried along with a particle that moves with a velocity $\mathbf{v}$. It was therefore called "*proper time*" by Minkowski who introduced this concept into relativity theory.

To Minkowski (1908) we also owe the idea of representing point events, that is events taking place at a point x, y, z at time t, in an abstract four-dimensional space with coordinates

$$x_1 = x, \quad x_2 = y, \quad x_3 = z, \quad x_4 = ict. \tag{10.1.29}$$

A continuous sequence of events will then be represented by a line in this space. A point P with coordinates x_1, x_2, x_3, x_4 will, for convenience, be called a "*world point*" and a line a "*world line*". The distance D between two world points in Minkowski space is called their "*world distance*" and represents both the spatial and temporal distance between two events.

One can always choose the four-dimensional coordinate frame in such a way that one of the points, say O, is in the origin and the other P has coordinates x_1, x_2, x_3, x_4. Their world distance is defined, in analogy to distance in ordinary three-dimensional space, by

$$D^2 = x_1^2 + x_2^2 + x_3^2 + x_4^2 = x^2 + y^2 + z^2 - c^2t^2. \tag{10.1.30}$$

We have shown that the expression on the right-hand side of (10.1.30) is invariant with respect to a Lorentz transformation. Thus the world distances between any two points in Minkowski space are invariant with respect to the transition from one to another inertial frame of reference. The Lorentz transformations are thus seen to be linear, homogeneous transformations which have the property to leave the distances in Minkowski space invariant and therefore represent *four-dimensional rotations* in that space.

The world distance between two neighbouring points, or the line element of a world line is, according to (10.1.30),

$$\begin{aligned} d\sigma^2 &= (dx_1)^2 + (dx_2)^2 + (dx_3)^2 + (dx_4)^2 \\ &= (dx)^2 + (dy)^2 + (dz)^2 - c^2\,(dt)^2 \end{aligned} \tag{10.1.31}$$

and must be invariant with respect to a Lorentz transformation and thus also the quantity

$$d\tau^2 \equiv -\frac{d\sigma^2}{c^2} = dt^2\,(1 - v^2/c^2). \tag{10.1.32}$$

Comparing this with the expression (10.1.27) one sees that τ is identical with the "proper time", and this proves again that $d\tau$ is indeed an invariant with respect to the Lorentz transformations.

10.2. Relativistic mechanics

It was shown in section 9.1 that the classical principle of relativity was valid in Newtonian mechanics or that the Newtonian equations of motion were invariant with respect to the Galileo transformations. In relativistic kinematics these transformations are replaced by the Lorentz transformations with respect to which the Newtonian equations of motion are *not* invariant. Hence these equations cannot be valid in relativistic mechanics and have to be replaced by suitably modified equations of motion which are invariant with respect to the Lorentz transformations.

There is no unique guidance for setting up the laws of relativistic dynamics except for the condition that they must go over into those of classical dynamics in the limiting case of velocities v of particles that are small compared with c. It is, however, plausible to demand that the laws of conservation of momentum and energy should remain valid in relativistic mechanics.

It will be assumed that the relativistic expression for the momentum $\mathbf{p}$ of a particle differs from the classical expression (1.1.2) only in so far as the differentiation with respect to t is replaced by differentiation with respect to proper time τ, the latter being defined by (10.1.27). Thus we put

$$\mathbf{p} = m^0 \frac{d\mathbf{r}}{d\tau} \tag{10.2.1}$$

or in components

$$p_x = m^0 \frac{dx}{d\tau}, \quad p_y = m^0 \frac{dy}{d\tau}, \quad p_z = m^0 \frac{dz}{d\tau}, \tag{10.2.2}$$

where m^0 is the classical mass of the particle as defined by its inertial behaviour at very small speeds. The expressions (10.2.1) and (10.2.2) evidently become identical with the classical expression (1.1.2) for $v \ll c$.

The components of the total momentum of a closed system of N particles are then

$$P_x = \sum_i m_i^0 \frac{dx_i}{d\tau}, \quad P_y = \sum_i m_i^0 \frac{dy_i}{d\tau}, \quad P_z = \sum_i m_i^0 \frac{dz_i}{d\tau}. \tag{10.2.3}$$

If one now goes over to another frame of reference S' which moves relative to the previously used frame S with velocity V, the components of the total momentum become

$$P'_x = \sum_i m_i^0 \frac{dx'_i}{d\tau}, \quad P'_y = \sum_i m_i^0 \frac{dy'_i}{d\tau}, \quad P'_z = \sum_i m_i^0 \frac{dz'_i}{d\tau}, \tag{10.2.4}$$

since $d\tau$ is invariant with respect to a transformation from S to S'. Making use of the Lorentz transformations (10.1.10) one can re-write (10.2.4) in the form

$$\left.\begin{aligned} P'_x &= \beta\left(\sum_i m_i^0 \frac{dx_i}{d\tau} - V\sum_i \frac{m_i^0}{\sqrt{1-v_i^2/c^2}}\right) \\ &= \beta P_x - \beta V \sum_i \frac{m_i^0}{\sqrt{1-v_i^2/c^2}}, \\ P'_y &= P_y, \quad P'_z = P_z. \end{aligned}\right\} \tag{10.2.5}$$

Thus, if the law of conservation of momentum for the system of particles holds in S, that is if P_x, P_y, P_z are constants, the conservation law will also hold in the system S' provided that the quantity $\sum_i \frac{m_i^0}{\sqrt{1-v_i^2/c^2}}$ also remains constant.

It is plausible to interpret this condition as an expression for the conservation of the total kinetic energy K of the particles in the system S in the absence of interaction forces between the particles. If this is to be the case the relativistic expression for the kinetic energy K_i of the ith particle must be a linear function of $\frac{m_i^0}{\sqrt{1-v_i^2/c^2}}$; thus

$$K_i = \frac{a_i m_i^0}{\sqrt{1-v_i^2/c^2}} + b_i,$$

where a_i and b_i are constants. For small values of v_i this reduces to

$$K_i = a_i m_i^0\left(1 + \frac{v_i^2}{2c^2}\right) + b_i \quad \left(\frac{v_i}{c} \ll 1\right)$$

which must be equal to the classical expression (1.1.2)

$$K_i = m_i^0 \frac{v_i^2}{2}.$$

Thus it appears that $a_i = c^2$ and $b_i = -m_i^0 c^2$ so that the relativistic expression for the kinetic energy of a particle of mass m^0 is

$$K = m^0 c^2\left(\frac{1}{\sqrt{1-v^2/c^2}} - 1\right). \tag{10.2.6}$$

The laws of conservation of momentum and energy are therefore satisfied in accordance with the general principles of the theory of relativity if the expression (10.2.6) is taken to be the relativistic formula for the kinetic energy of a particle and, from (10.2.1) and (10.1.27), the expression

$$\mathbf{p} = \frac{m^0 \mathbf{v}}{\sqrt{1-v^2/c^2}} \tag{10.2.7}$$

as the "*relativistic momentum vector*" of a particle.

It follows from these relations that **p** and E would become infinitely large for $v = c$, which means that the velocity of a particle with finite mass can never reach the velocity of light. We may regard this as the *physical* reason for the kinematic fact, emphasized in 10.1, that c in the theory of relativity plays the part of an infinite velocity.

The expression (10.2.7) can formally be written in the classical form (1.1.2)

$$\mathbf{p} = m\mathbf{v} \tag{10.2.8}$$

if one defines a quantity

$$m = \frac{m^0}{\sqrt{1 - v^2/c^2}}. \tag{10.2.9}$$

m is called the "*relativistic mass*" of the particle, which is a function of its speed v, and m^0 is then referred to as its "*rest mass*", for m becomes equal to m^0 for $v \to 0$. The expression (10.2.6) for the kinetic energy can then also be re-written in the simple form

$$K = (m - m^0)\, c^2. \tag{10.2.10}$$

It appears from (10.2.10) that the relativistic mass of a particle is increased by an amount δm if its kinetic energy is increased by an amount δK so that

$$\delta m = \delta K/c^2. \tag{10.2.11}$$

It seems reasonable to generalize this relationship between mass and kinetic energy to other forms of energy, since all forms of energy can be transformed into one another, so that *any* increase of internal energy δE of a material body would increase its mass by

$$\delta m = \delta E/c^2. \tag{10.2.12}$$

This implies that also the rest mass m^0 of a body ought to increase by an amount $\delta E/c^2$ if its potential energy in a conservative field of force or its internal elastic energy is increased by δE.

This prediction was first made by Einstein from the general principles of his theory and has since then become of the utmost importance in the field of nuclear physics. For it stipulates that the rest mass of an atomic nucleus should not be equal to the sum of the rest masses of the constituent nucleons (protons and neutrons) but *smaller* than the latter ("*mass defect*") by an amount $\delta m = W/c^2$ if W is the work which has to be done in order to split the nucleus into its constituents, and *larger* than the latter ("*mass excess*") if on the contrary energy can be gained when the nucleus is split.

It has indeed been established experimentally that all stable nuclei show mass defects, given by the above-mentioned relationship, and this fact is made use of widely for the determination of the binding energies of nuclei from their measured mass defects. The unstable nuclei, on the other hand, show a mass excess, and it is often maintained in textbooks that this fact is responsible for the possibility of utilizing nuclear energy by the process of "fission". This is true in so far as a mass excess is a measure for a surplus internal energy; but this holds not only for nuclear but also for atomic, chemical reactions. The physical cause for the nuclear surplus potential energy is the existence of the very strong repulsive forces between the protons in the nucleus which are released in the process of fission, giving rise to the transformation of the potential into kinetic energy of the fission products.

If the argument leading to the relationship (10.2.12) is carried further to its logical conclusion one is led to assume that the rest mass m^0 of a particle itself is a measure of its total internal energy E_0 so that quite generally

$$E_0 = m^0 c^2. \tag{10.2.13}$$

The relationship embodied in this equation is usually called the "*principle of equivalence of mass and energy*", according to which the classical law of conservation of mass is not an independent principle but is incorporated in the principle of conservation of energy.

The total energy E of a free particle of rest mass m^0 moving with velocity v is the sum of its kinetic energy K (10.2.6) and its internal energy E_0 (10.2.13) and thus equal to

$$E = K + E_0 = \frac{m^0 c^2}{\sqrt{1 - v^2/c^2}}\,. \tag{10.2.14}$$

We shall call E the "*relativistic energy*" of the particle.

Combining (10.2.14) with (10.2.9) we obtain the relation

$$E = mc^2 \tag{10.2.15}$$

of which (10.2.13) is a special case for $v = 0$ and hence $m = m^0$, and which holds *generally* also for particles subject to external forces because, according to (10.2.12), an additional potential energy U of the particle is equivalent to an increase U/c^2 of mass.

The mass–energy relationship (10.2.15) is the most important outcome of the theory of relativity as it implies that matter may be "created" at the expense of another form of energy, say electromagnetic radiation, and that matter can be "annihilated" whereby energy is released in another form. There is abundant experimental evidence for processes of this kind, examples of which are the creation of pairs of electrons and positrons by γ-radiation and the annihilation of such pairs resulting in the emission of photons. Especially in recent times the methods of high energy physics have made it possible to study the interaction of the various types of fundamental particles in which some of them disappear while others emerge. The study of these processes is entirely based on Einstein's mass–energy relation (10.2.15).

The relations (10.2.7) and (10.2.14) can be combined to give an equation between p, E, and m^0, which is independent of v, namely

$$p^2 - \frac{E^2}{c^2} = -(m^0 c)^2. \tag{10.2.16}$$

For velocities small compared with c the kinetic energy K is small compared with $E_0 = m^0c^2$, and under this condition (10.2.16) goes over into

$$p \simeq \sqrt{2Km_0} \qquad (K \ll E_0), \tag{10.2.17}$$

in agreement with the classical relation (1.1.2). For the opposite limit of extremely energetic particles, in the so-called "ultra-relativistic" region, one has

$$p \simeq E/c \qquad (K \gg E_0), \tag{10.2.18}$$

which also follows direct from (10.2.7) and (10.2.14) for $v \to c$.

The relationship (10.2.16) further suggests to define a four-dimensional vector, or in short a "*four-vector*" $\mathbf{J}$, whose first three components J_1, J_2, J_3 are the three components of the momentum vector $\mathbf{p}$, and the fourth component is equal to the energy E multiplied by i/c; thus

$$J_1 = p_x, \; J_2 = p_y, \; J_3 = p_z, \; J_4 = iE/c. \tag{10.2.19}$$

$\mathbf{J}$ is called the "*momentum–energy vector*".

The left-hand member of (10.2.16) is then evidently the square of the magnitude of $\mathbf{J}$ which is thus equal to $-(m^0c)^2$ and therefore invariant with respect to the Lorentz transformations, and these, as we have seen in section 10.1, are equivalent to rotations in Minkowski space. Further, the supposed validity of the laws of conservation of momentum and energy in all inertial systems of coordinates demands that all four components of the sum of the energy–momentum vectors for an isolated system of particles be invariant with respect to transformations from one system to another; an elegant way of expressing the validity of the conservation laws in relativity theory is therefore to say that the total momentum–energy vector is invariant with respect to the Lorentz transformations.

We now proceed to derive the equations of motion of relativistic mechanics. There is again *a priori* no unique way towards the solution of this problem. Einstein chose the apparently most natural one, namely simply to take over the formulation of the Newtonian law in the form (1.1.4) and to replace the classical momentum by the relativistic momentum. Thus the relativistic equation of motion for a particle is

$$\frac{d\mathbf{p}}{dt} = \mathbf{F}, \tag{10.2.20}$$

where $\mathbf{F}$ is the Newtonian force, defined by the acceleration of a resting or *slowly* moving particle. We shall first prove that this law of motion is consistent with the general theorem of conservation of energy.

The work W done by the force $\mathbf{F}$ upon a particle of rest mass m^0 along a path leading from a point P_1 to a point P_2 is by definition

$$W = \int_{P_1}^{P_2} \mathbf{F}\, d\mathbf{s}. \tag{10.2.21}$$

Substituting from (10.2.20) for $\mathbf{F}$ and making use of the formula (10.2.7) for $\mathbf{p}$ we obtain, by means of partial integration,

$$\begin{aligned} W &= \int_{P_1}^{P_2} \frac{d\mathbf{p}}{dt}\, d\mathbf{s} = \int_{P_1}^{P_2} \mathbf{v}\, d\mathbf{p} = \int_{P_1}^{P_2} \mathbf{v}\, d\left(\frac{m^0\mathbf{v}}{\sqrt{1 - v^2/c^2}}\right) \\ &= \left[\frac{m^0 v^2}{\sqrt{1 - v^2/c^2}}\right]_{v_1}^{v_2} - \frac{m^0}{2}\int_{P_1}^{P_2} \frac{d(v^2)}{\sqrt{1 - v^2/c^2}} \\ &= \left[\frac{m^0 v^2}{\sqrt{1 - v^2/c^2}} + m^0 c^2 \sqrt{1 - v^2/c^2}\right]_{v_1}^{v_2} = \left[\frac{m^0 c^2}{\sqrt{1 - v^2/c^2}}\right]_{v_1}^{v_2}. \end{aligned}$$

Thus we find from (10.2.6) that

$$W = K_2 - K_1, \tag{10.2.22}$$

which means that the increase of kinetic energy of the particle is equal to the work done by the force to produce this increase, or equal to the decrease of potential energy; and this is indeed the law of conservation of energy.

To illustrate the use of the relativistic equation of motion (10.2.20) we shall give two examples of its application to problems which are of particular practical importance.

Let us first assume that a particle carrying an electric charge e, and whose initial velocity $\mathbf{v}$ is in the xy-plane is acted upon by a homogeneous magnetic field of strength H in the direction of the z-axis. It will then experience a "Lorentz force" (9.3.24) of magnitude evH/c in a direction perpendicular to $\mathbf{v}$ in the xy-plane. As the component of $\mathbf{F}$ in the $\mathbf{v}$-direction is zero the tangential component of $\mathbf{p}$, and hence the speed v of the particle remains unchanged. The radial component p_r satisfies the differential equation

$$\frac{dp_r}{dt} = \frac{evH}{c},$$

and its time derivative is thus seen to be constant. Hence the path must be a circle in the xy-plane with radius R. Figure 23 shows that

$$\frac{dp_r}{dt} = \frac{dp_r}{d\varphi}\,\omega = p\omega = \frac{pv}{R};$$

thus one obtains

$$p = eRH/c. \tag{10.2.23}$$

This relation is widely used for the determination of the momenta of fast-moving charged fundamental particles by measuring the curvature of their traces in a cloud- or bubble-chamber in a magnetic field.

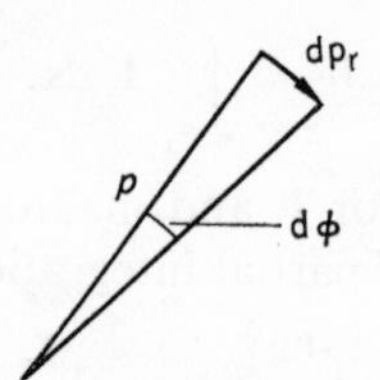

FIG. 23. Change of the radial component p_r of the linear momentum of a particle moving in a circular path.

Alternatively one can put (10.2.23) in the form

$$p = mv = evH/\omega c,$$

where m is the relativistic mass (10.2.9), so that one obtains a formula for the angular velocity of the movement of the particle

$$\omega = eH/mc. \tag{10.2.24}$$

This formula is the basis for the function of the cyclotron for the acceleration of charged particles up to velocities small compared with c, for then m is practically equal to m^0 and ω is therefore independent of v. But for very high speeds m changes with v during the process of acceleration, and hence either ω or H has to be appropriately altered as is done in the so-called synchro-cyclotrons and synchrotrons.

If at $t = 0$ the particle is at the origin of coordinates and moves in the x-direction then its path will show a deflection y in the y-direction

$$y = R - \sqrt{R^2 - x^2}$$

which for small y, according to (10.2.8) and (10.2.23), is approximately equal to

$$y = \frac{x^2}{2R} = \frac{eHx^2}{2mvc}. \tag{10.2.25}$$

Let us now assume that the same charged particle is subjected to a homogeneous electric field $\mathbf{E}$ in the z-direction. It will then in time t be deflected in that direction by an amount z, and if z is small enough, the original speed v of the particle will remain practically the same, and also its relativistic mass m. We therefore can integrate the equation of motion

$$\frac{dp_z}{dt} = m\frac{d^2z}{dt^2} = eE$$

at once, which results in

$$z = \frac{eE}{2m} t^2 = \frac{eEx^2}{2mv^2}. \tag{10.2.26}$$

If the particle is finally acted upon by the electric and the magnetic field simultaneously one obtains from (10.2.25) and (10.2.26)

$$\left.\begin{aligned} m &= \frac{eH^2}{2Ec^2} \cdot \frac{zx^2}{y^2}, \\ v &= \frac{Ec}{H} \cdot \frac{y}{z}. \end{aligned}\right\} \tag{10.2.27}$$

Thus by making measurements of y and z for a given x, under the condition that $y \ll x$ and $z \ll x$, for various initial speeds v of one and the same type of particle, one can determine the dependence of m on v. In this way it was first discovered by W. Kaufmann (1901) that the mass m of fast-moving electrons was not constant, that is not identical with their rest mass m^0, and later experiments, using this and various other methods, have shown that the relation (10.2.9) for the relativistic mass is precisely valid for all types of charged particles.

It remains to derive the transformation equations for momentum, energy, and force from a frame of reference S to another S' which moves relative to S with velocity V in the common x-direction.

For this purpose we start from the definition of momentum (10.2.1). Since $d\tau$ there is invariant with respect to such a transformation, we have

$$\mathbf{p} = m^0 \frac{d\mathbf{r}}{d\tau}, \quad \mathbf{p}' = m^0 \frac{d\mathbf{r}'}{d\tau} \tag{10.2.28}$$

from which follows

$$\frac{p'_x}{p_x} = \frac{dx'}{dx}, \quad \frac{p'_y}{p_y} = \frac{dy'}{dy}, \quad \frac{p'_z}{p_z} = \frac{dz'}{dz}. \tag{10.2.29}$$

Now from (10.1.20) one has

$$\frac{dx'}{dx} = \beta(1 - V/v_x), \quad \frac{dy'}{dy} = 1, \quad \frac{dz'}{dz} = 1.$$

Hence the equations (10.2.29) yield the required *momentum transformation equations*

$$p'_x = p_x \frac{1 - V/v_x}{\sqrt{1 - V^2/c^2}}, \quad p'_y = p_y = p'_z = p_z. \tag{10.2.30}$$

The expressions for the energy in the two coordinate systems are from (10.2.14)

$$E = \frac{m^0 c^2}{\sqrt{1 - v^2/c^2}}, \quad E' = \frac{m^0 c^2}{\sqrt{1 - v'^2/c^2}}, \tag{10.2.31}$$

from which follows

$$\frac{E'}{E} = \frac{\sqrt{1 - v^2/c^2}}{\sqrt{1 - v'^2/c^2}}. \tag{10.2.32}$$

We now make use of the equation (10.1.25) which leads to

$$\frac{1 - v^2/c^2}{1 - v'^2/c^2} = \frac{(1 - Vv_x/c^2)^2}{1 - V^2/c^2},$$

and hence we obtain from (10.2.32)

$$E' = E\,\frac{1 - Vv_x/c^2}{\sqrt{1 - V^2/c^2}}, \tag{10.2.33}$$

which is the required *energy transformation equation.*

Finally, in order to derive the transformation equations for the force components we consider a particle of rest mass m^0 which moves *very slowly* within the system S. Under these circumstances, as was stated at the beginning of this section, the classical laws of dynamics must hold. Hence for $v \to 0$ one has

$$\mathbf{p} = m^0 \mathbf{v} \quad \text{and} \quad \mathbf{F} = m^0 \frac{d\mathbf{v}}{dt} \tag{10.2.34}$$

and from (10.1.21)

$$dt'/dt = \beta. \tag{10.2.35}$$

Thus with the help of (10.2.30) and (10.2.34) one obtains

$$p'_x = \beta m^0 v_x (1 - V/v_x) = \beta m^0 v_x - \beta m^0 V, \; p'_y = p_y, \; p'_z = p_z,$$

and further by means of (10.2.35)

$$\left.\begin{aligned} \frac{dp'_x}{dt'} = \frac{dp'_x}{dt}\frac{dt}{dt'} = m^0 \frac{dv_x}{dt}, \quad \frac{dp'_y}{dt'} = m^0 \frac{dv_y}{dt}\sqrt{1 - V^2/c^2}, \\ \frac{dp'_z}{dt'} = m^0 \frac{dv_z}{dt}\sqrt{1 - V^2/c^2}. \end{aligned}\right\} \tag{10.2.36}$$

Finally, substituting from (10.2.34) and (10.2.36) into the equation of motion (10.2.20), we obtain the required *force transformation equations*

$$F'_x = F_x, \quad F'_y = F_y\sqrt{1 - V^2/c^2}, \quad F'_z = F_z\sqrt{1 - V^2/c^2}. \tag{10.2.37}$$

If the relativistic equation of motion (10.2.20) is to be valid universally, the transformation equations (10.2.37) must hold for *any* type of force, for example for the forces exerted on a charged particle in an electromagnetic field. If such a particle is accelerated from rest in a system S of field generating bodies, then an observer, also at rest in S, will measure the force components F_x, F_y, F_z. But if he is at rest with respect to a system S', which moves with speed V in the x-direction relative to the field generating bodies, his measurements will give the values F_x', F_y', F_z' of the force components. We shall show later, in section 10.4, that the relations (10.2.37) can indeed be deduced from the general transformation equations for the electromagnetic field.

10.3. Relativistic optics

It was pointed out in the introduction to this chapter that according to the basic principles of the special theory of relativity, in particular the principle of the constancy of the velocity of light in empty space, all experiments designed to detect differences in the light velocity between frames of reference, moving with constant velocity relative to one another, ought to give negative results. This is indeed borne out by the failure of all attempts to detect an "ether drift" which were discussed at some length in Chapter 9.

In this section we shall develop the main laws of relativistic optics, which are to replace those of classical optics as set out in section 9.2, on the basis of relativistic kinematics and mechanics. We first discuss the propagation of light in moving bodies.

The velocity transformation equations (10.1.24) must also apply to the propagation of light, irrespective of whether we regard this as a phenomenon of wave propagation or of the movement of photons. Thus if a beam of light propagates in the x-direction within a transparent body of refractive index n, which also moves in the x-direction with velocity V, one has $v_x' = u' = c/n$, and hence the speed of light $v_x = u$, as measured by a resting observer, is

$$u = \frac{c/n + V}{1 + V/nc}. \tag{10.3.1}$$

For velocities V, which are small compared with c, u becomes approximately equal to

$$u \simeq \left(\frac{c}{n} + V\right)\left(1 - \frac{V}{nc}\right) \simeq u' + V(1 - 1/n^2) = u' + KV \tag{10.3.2}$$

with

$$K = 1 - 1/u^2. \tag{10.3.3}$$

This is seen to be identical with Fresnel's formula (9.2.16) for the "dragging coefficient" K, which is here shown to be simply a consequence of relativistic kinematics, without the need to make special assumptions about the partial carrying along of an ether (section 9.2), or about the dielectric displacement D in a material body (section 9.3).

We now consider the propagation of light in empty space from the point of view of the wave theory. In section 9.2 we studied the propagation of a plane, monochromatic wave on the assumption of the validity of the Galileo transformations. We shall now treat the same problem by using the Lorentz transformations in the form (10.1.11) instead.

We write the wave function Φ of the wave in the system S in the form (9.2.12), namely

$$\Phi(\mathbf{r}_1 t) = \gamma\, e^{2\pi i(\gamma t - \mathbf{k}.\mathbf{r})} \tag{10.3.4}$$

and obtain for the exponent in the coordinate system x', y', z', t'

$$\begin{aligned} 2\pi i(\nu t - k_x x - k_y y - k_z z) &\\ = 2\pi i\,[\nu\beta(t' + x'V/c^2) - k_x\beta(x' + Vt') - k_y y' - k_z z'] &\\ = 2\pi i(\nu' t' - k_x' x' - k_y' y' - k_z' z'), & \end{aligned}$$

where we have put

$$\left.\begin{aligned} \nu' &= \beta(\nu - k_x V), \\ k_x' &= \beta(k_x - \nu V/c^2),\ k_y' = k_y,\ k_z' = k_z. \end{aligned}\right\} \tag{10.3.5}$$

The first of these equations replaces the classical relation (9.2.13) and represents the *relativistic Doppler effect;* the second equation indicates that the component of the wave vector in the direction of the movement of S' with respect to S is altered from k_x to k_x', whereas classically no change at all in the wave vectors should take place.

As before, we introduce the direction cosines A, B, C of the wave normal in S and the corresponding quantities A', B', C' in S' so that

$$\left.\begin{aligned} k_x &= A\nu/c, \quad k_y = B\nu/c, \quad k_z = C\nu/c, \\ k_x' &= A'\nu'/c,\ k_y' = B'\nu'/c,\ k_z' = C'\nu'/c. \end{aligned}\right\} \tag{10.3.6}$$

From (10.3.5) and (10.3.6) we then obtain easily

$$\nu' = \beta\nu(1 - VA/c) \tag{10.3.7}$$

and

$$A' = \frac{A - V/c}{1 - AV/c}, \quad B' = \frac{B}{\beta(1 - AV/c)}, \quad C' = \frac{C}{\beta(1 - AV/c)}. \tag{10.3.8}$$

Since A, B, C are direction cosines of a vector in S they must satisfy the condition

$$A^2 + B^2 + C^2 = 1.$$

We shall now prove that A', B', C' satisfy the same condition. For it follows from (10.3.8) that

$$A'^2 + B'^2 + C'^2 = \frac{(A - V/c)^2 + (B^2 + C^2)(1 - V^2/c')}{(1 - AV/c)^2}$$

$$= \frac{(A - V/c)^2 + (1 - A^2)(1 - V^2/c^2)}{(1 - AV/c)^2} = 1.$$

Thus A', B', C' are also direction cosines in the system S' and it follows that, just as in classical optics, a plane wave in a system S remains a plane wave in the system S'. But whereas, according to the considerations in section 9.2, no change of direction of such a wave should take place in classical optics, relativistic optics demands that such a change of direction should occur, which constitutes the *aberration* effect.

The question as to whether the aberration of light takes place in the space between light source and observer or within the observer's instrument was shown to be of importance for the interpretation of the observations of this effect in classical optics. This was discussed in section 9.2, and if the classical wave theory is supposed to be valid, the latter mechanism has to be assumed. But from the point of view of relativistic wave optics the aberration effect is solely the outcome of relativistic kinematics, and hence only dependent on the relative movement of observer and light source, irrespective of the type of instrument used for the measurement.

The equations (10.3.8) are the relativistic expressions for the aberration effect. It is easily shown that they give the same result as the classical formula (9.2.2) for velocities V that are *small* compared with c. Assuming the wave normal to be in the xy-plane, that is $C = C' = 0$, and using the same symbols as in section 9.2, we can write the equations (10.3.8) in the form

$$\cos\varphi' = \frac{\cos\varphi - V/c}{1 - \cos\varphi \cdot V/c}, \quad \sin\varphi' = \frac{\sin\varphi}{1 - \cos\varphi \cdot V/c},$$

which show that the aberration angle $\alpha = \varphi - \varphi'$ is small. Taking this fact into account we obtain from either of the above relations approximately

$$|\alpha| \simeq \frac{V}{c}\sin\varphi \quad (V \ll c),$$

which is indeed identical with (9.2.3).

We now return to the formula (10.3.7) for the Doppler effect. It evidently differs from the classical formula (9.2.15) only by the presence of the factor β which is practically equal to unity for small velocities V but becomes significant for velocities which are comparable with that of light. We shall illustrate this by two special examples.

Assume first that $A = 1$, that is that the observer moves in the direction of the wave normal away from the light source. We then have from (10.3.7)

$$\frac{\nu'}{\nu} = \frac{1 - V/c}{\sqrt{1 - V^2/c^2}} = \sqrt{\frac{1 - V/c}{1 + V/c}}. \tag{10.3.9}$$

Here V is, of course, the *relative* velocity of the observer with respect to the light source, and the relativistic expression for the Doppler shift

$$\frac{\Delta\nu}{\nu} = \sqrt{\frac{1 - V/c}{1 + V/c}} - 1 \tag{10.3.10}$$

differs significantly from both classical expression (9.2.6) and (9.2.8) in which V is the *absolute* velocity of the observer and v that of the light source.

Expanding the right-hand side of (10.3.10) in powers of V/c one obtains

$$\frac{\Delta\nu}{\nu} = -\frac{V}{c} + \frac{1}{2}\frac{V^2}{c^2}\cdots, \tag{10.3.11}$$

whereas, according to the classical formula (9.2.8), the coefficient of the quadratic term is equal to 1, and in the formula (9.2.6) this term is missing altogether. One way of testing the validity of the expression (10.3.10) is to observe the spectrum of light emitted by a fast-moving excited atom or molecule in and opposite to the direction of its movement, and to compare the Doppler shifts of the spectral lines. Such experiments were carried out by Ives and Stillwell (1941), and the results obtained show that the two Doppler shifts are indeed asymmetric and in agreement with the formula (10.3.11).

Next we assume that $A = 0$, which means that the observer is at rest in S while the light source is at rest in S' and observations are made in a direction at right angles to $\mathbf{V}$. Under these conditions equation (10.3.7) reduces to

$$\frac{\nu'}{\nu} = \beta = \frac{1}{\sqrt{1 - V^2/c^2}} \quad (A = 0). \tag{10.3.12}$$

This indicates that the frequency ν, as measured by the "resting" observer, is smaller than the frequency ν', measured by an observer moving *with* the light source, by a factor $\sqrt{1 - V^2/c^2}$. This phenomenon is called the "*transverse Doppler effect*". It is a characteristic feature of relativistic optics which is alien to classical optics (see section 9.2).

If, on the other hand, the observer is at rest in S' and the light source in S, one obtains the expression for the transverse Doppler effect by putting $A' = 0$, or, according to (10.3.8), $A = V/c$. Substituting this into (10.3.7) one gets

$$\nu' = \beta\nu\,(1 - V^2/c^2) = \nu\sqrt{1 - V^2/c^2} \quad (A' = 0). \tag{10.3.13}$$

This relation expresses the same fact as (10.3.12), for now ν is the frequency measured by an observer at rest with respect to the light source and ν' the frequency measured by the "moving" observer, and it is again seen that the latter is smaller than the former by a factor $\sqrt{1 - V^2/c^2}$.

The relations (10.3.12) and (10.3.13) could, of course, have been directly deduced from the expressions (10.1.14) and (10.1.15) for the Lorentz clock retardation, since the mechanisms responsible for the emission of spectral lines fall into the same category as clocks. This point of view had already been taken in section 9.2, and the expression (9.2.30) for what we then called "second order Doppler effect", suggested by the negative outcome of the Kennedy–Thorndike experiment, is indeed identical with the expression (10.3.13).

In recent years it has become possible to demonstrate the existence of the transverse Doppler effect directly and to verify formula (10.3.12) by making use of the fact that certain substances emit soft γ-rays of extremely narrow frequency width and also act as extremely selective absorbers of such rays (Mössbauer effect). The source is placed at the centre and the absorber near the rim of a fast-rotating rotor, and the intensity of the radiation penetrating through the absorber is measured by means of a counter; from this the Doppler displacement between source and absorber can be calculated indirectly (Fig. 24).

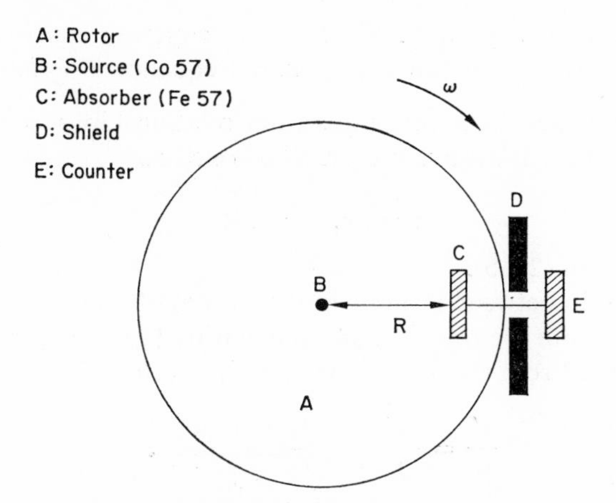

FIG 24. Experiment for measuring the transverse Doppler effect by means of the Mössbauer effect.

If the position vector of an atom in the plane normal to the rotation axis is $\mathbf{r}_1$ and that of an atom of the absorber is $\mathbf{r}_2$, and if $\mathbf{i}$ is a unit vector in the direction of the axis, then the velocity vectors of these two atoms are evidently

$$\mathbf{v}_1 = \omega (\mathbf{r}_1 \wedge \mathbf{i}) \quad \text{and} \quad \mathbf{v}_2 = \omega (\mathbf{r}_2 \wedge \mathbf{i}),$$

where ω is the angular velocity of the rotation. Hence the relative velocity between absorbing and emitting atoms is

$$\mathbf{v}_1 - \mathbf{v}_2 = \omega\,(\mathbf{r}_1 - \mathbf{r}_2 \wedge \mathbf{i}).$$

As $\mathbf{r}_1 - \mathbf{r}_2$ is the vector joining the two atoms one sees that, irrespective of the positions of the atoms on the rotor, the relative velocity is always precisely at right angles to the line joining the atoms, and the exhibited line shift is therefore indeed the outcome of the transverse Doppler effect. The dependence of $\Delta\nu$ on $V = R\omega$, measured in the way just described, fully confirms the expression (10.3.12).

The results derived so far in this section were based on the concept of light propagation in the form of *waves*. We shall now show that the same results can be derived by regarding a light beam as a stream of separate *photons* which move along straight lines with speed c.

If the general laws of relativistic mechanics also apply to photons it follows from the energy expression (10.2.14) that the rest mass of a photon must be zero, for otherwise its energy E would be infinitely large. The relation (10.2.16) then shows that a momentum of magnitude

$$p = \frac{E}{c} \tag{10.3.14}$$

must be assigned to a photon. Einstein consequently concluded that the quantum relationship between energy and frequency $E = h\nu$ for photons must be accompanied by another quantum relationship $p = \frac{h\nu}{c}$ or $\mathbf{p} = h\mathbf{k}$ between momentum and wave vector. Thus one has

$$E = h\nu, \quad \mathbf{p} = h\mathbf{k} \tag{10.3.15}$$

in conformity with (2.2.5).

The relationship between ν' and ν in the frames of reference S' and S can now be derived from the transformation equation (10.2.33) for the energy if one sets $v_x = cA$. From the first of the equations (10.3.15) one has indeed

$$\frac{\nu'}{\nu} = \frac{E'}{E} = \frac{1 - AV/c}{\sqrt{1 - V^2/c^2}}, \tag{10.3.16}$$

which is identical with the relativistic Doppler formula (10.3.7), previously obtained from wave optical considerations.

The formula for the aberration effect follow similarly from the transformation equations (10.1.22) for velocity if one puts

$$v_x = Ac, \quad v_y = Bc, \quad v_z = Cc,$$

$$v_x' = A'c, \quad v_y' = B'c, \quad v_z' = C'c.$$

Thus one obtains the relations

$$\left.\begin{aligned} A' &= \frac{A - V/c}{1 - VA/c}, \quad B' = \frac{B\sqrt{1 - V^2/c^2}}{1 - VA/c}, \\ C' &= \frac{C\sqrt{1 - V^2/c^2}}{1 - VA/c}, \end{aligned}\right\} \tag{10.3.17}$$

which are again identical with the relations (10.3.8) that follow from the application of the Lorentz transformations to a plane wave.

Various problems concerning the interaction between photons and particles with finite rest mass can also be solved by the application of relativistic mechanics. As an example we shall deal here with the so-called "*Compton effect*", arising in the process of scattering of light by free electrons, for which only the laws of conservation of momentum and energy are required.

Let us therefore consider the process of interaction of a photon of frequency ν with a stationary free electron of rest-mass m^0, resulting in the "absorption" of the original photon and the "emission" of a new photon of frequency ν' and the acceleration of the electron to a velocity v. Call θ the angle between the directions of flight of the secondary and the primary photons and φ the angle between the velocity vector $\mathbf{v}$ of the electron and the direction of the primary photon (Fig. 25).

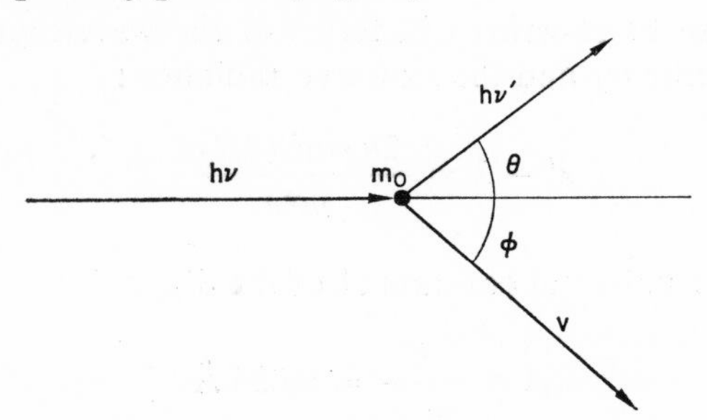

FIG. 25. Diagram illustrating the Compton effect.

The law of conservation of energy demands that the total energy of the system before the "collision" be equal to that after the collision. From (10.2.14) and the first of the equations (10.3.15) one has, therefore,

$$h\nu + m^0c^2 = h\nu' + \frac{m^0c^2}{\sqrt{1 - v^2/c^2}}. \tag{10.3.18}$$

From the law of conservation of momentum follows that the component of the total momentum of the system in the direction of the primary photon before the collision must be equal to that after the collision, and that the component normal to this direction after the collision must be zero. Hence one has further from (10.2.7) and the second of the equations (10.3.15)

$$\frac{h\nu}{c} = \frac{h\nu'}{c}\cos\theta + \frac{m^0 v}{\sqrt{1 - v^2/c^2}}\cos\varphi, \tag{10.3.19}$$

and

$$0 = \frac{h\nu'}{c}\sin\theta - \frac{m^0 v}{\sqrt{1 - v^2/c^2}}\sin\varphi. \tag{10.3.20}$$

Elimination of φ from the equations (10.3.19) and (10.3.20) results in

$$(h\nu)^2 + (h\nu')^2 - 2h^2\nu\nu'\cos\theta = \frac{m^0 c^2 v^2}{1 - v^2/c^2},$$

and (10.3.18) can be written in the form

$$(h\nu)^2 + (h\nu')^2 - 2h^2\nu\nu' + 2hm^0c^2(\nu - \nu') = \frac{m^2c^2v^2}{1 - v^2/c^2}.$$

Subtraction of these equations eliminates v and leads to an equation between ν, ν', and θ alone, namely

$$m^0c^2(\nu - \nu') = h\nu\nu'(1 - \cos\theta).$$

This equation can be re-written in terms of the wavelengths $\lambda = c/\nu$ and $\lambda' = c/\nu'$ of the primary and the scattered radiation:

$$\lambda' - \lambda = \frac{2h\sin^2(\theta/2)}{m^0 c}. \tag{10.3.21}$$

Introducing the universal constant Λ of the dimension of a length

$$\Lambda \equiv \frac{h}{m^0 c} = 0{\cdot}024\ \text{Å}, \tag{10.3.22}$$

called the "*Compton wavelength*", one obtains finally for the increase $\Delta\lambda$ of the wavelength of the radiation in the process of scattering through an angle θ:

$$\Delta\lambda \equiv \lambda' - \lambda = 2\Lambda\sin^2\theta/2. \tag{10.3.23}$$

This formula shows that the increase of wavelength depends only on the scattering angle (and not on the wavelength of the primary radiation), and that $\Delta\lambda$ has its maximum value for "back-scattering" ($\theta = \pi$) when $\Delta\lambda = 2\Lambda$, that further for scattering under 90° ($\theta = \pi/2$) $\Delta\lambda = \Lambda$. Owing to the smallness of Λ the effect is only observable for short-wave radiation (X-rays or γ-rays), where measurements completely confirmed the correctness of the relation (10.3.23).

10.4. Relativistic electrodynamics

In order to derive the laws of relativistic electrodynamics we start from the *Lorentz–Maxwell equations* for the electromagnetic field of moving charged particles which were introduced in section 9.3. It was stated there that these equations were not invariant with respect to the Galileo transformations, but it can be shown that they are invariant with respect to the Lorentz transformations and are therefore valid, not only when the Lorentzian immovable ether is used as frame of reference, but also for all frames of reference which are admissible in the special theory of relativity. The Lorentz transformations were in fact introduced by Lorentz for the purpose of developing a theory of electromagnetic phenomena in moving bodies which would account for the absence of electromagnetic ether-drift effects. Einstein's first paper, in which he proclaimed the principles of the special theory of relativity was also devoted to the same purpose.

We start this section with the proof of the invariance of the Lorentz–Maxwell equations (9.3.26)

$$\operatorname{div} \mathbf{E} = 4\pi\varrho, \tag{10.4.1}$$

$$\operatorname{div} \mathbf{H} = 0, \tag{10.4.2}$$

$$\operatorname{curl} \mathbf{H} = \frac{4\pi}{c}(\varrho\mathbf{v}) + \frac{1}{c}\dot{\mathbf{E}}, \tag{10.4.3}$$

$$\operatorname{curl} \mathbf{E} = -\frac{1}{c}\dot{\mathbf{H}}, \tag{10.4.4}$$

with respect to the Lorentz transformations (10.1.10), and the derivation of the transformation equations for the electromagnetic field quantities. We shall do this by first simply stating the transformation equations for the electric and magnetic field strengths and by then verifying the invariance of the equations (10.4.1) to (10.4.4) when the Lorentz transformations and the electromagnetic field transformations are applied.

If, as usual, the two systems of coordinates S and S' are supposed to move relative to each other with speed V in the direction of the common x-axes, it is maintained that the components E'_x, E'_y, E'_z, H'_x, H'_y, H'_z of the electromagnetic field vectors, measured in S', are connected with the components E_x, E_y, E_z, H_x, H_y, H_z, measured in S by the equations

$$\left.\begin{aligned} E'_x &= E_x, & H'_x &= H_x, \\ E'_y &= \beta(E_y - VH_z/c), & H'_y &= \beta(H_y + VE_z/c), \\ E'_z &= \beta(E_z + VH_y/c), & H'_z &= \beta(H_z - VE_y/c). \end{aligned}\right\} \tag{10.4.5}$$

It follows from the inverse Lorentz transformations (10.1.11) that

$$\left.\begin{aligned}\frac{\partial}{\partial x'} = \beta\left(\frac{\partial}{\partial x} + \frac{V}{c^2}\frac{\partial}{\partial t}\right), \quad \frac{\partial}{\partial y'} = \frac{\partial}{\partial y}, \quad \frac{\partial}{\partial z'} = \frac{\partial}{\partial z},\\ \frac{\partial}{\partial t'} = \beta\left(\frac{\partial}{\partial t} + V\frac{\partial}{\partial x}\right).\end{aligned}\right\} \tag{10.4.6}$$

From (10.4.5) and (10.4.6) one obtains now

$$\frac{\partial \mathrm{H}'_x}{\partial x'} + \frac{\partial \mathrm{H}'_y}{\partial y'} + \frac{\partial \mathrm{H}'_z}{\partial z'}$$

$$= \beta\left(\frac{\partial \mathrm{H}_x}{\partial x} + \frac{\partial \mathrm{H}_y}{\partial y} + \frac{\partial \mathrm{H}_z}{\partial z}\right) + \frac{\beta V}{c^2}\left(\frac{\partial \mathrm{H}_x}{\partial t} + c\frac{\partial \mathrm{E}_z}{\partial y} - c\frac{\partial \mathrm{E}_y}{\partial z}\right).$$

The first term in brackets on the right-hand side of this equation is zero on account of (10.4.2), and the second term in brackets also vanishes by reason of the x-component of (10.4.4). Thus one has

$$\operatorname{div}' \mathbf{H}' \equiv \frac{\partial \mathrm{H}'_x}{\partial x'} + \frac{\partial \mathrm{H}'_y}{\partial y'} + \frac{\partial \mathrm{H}'_z}{\partial z'} = 0, \tag{10.4.7}$$

which proves the invariance of (10.4.2).

Similarly one obtains

$$\frac{\partial \mathrm{E}'_x}{\partial x'} + \frac{\partial \mathrm{E}'_y}{\partial y'} + \frac{\partial \mathrm{E}'_z}{\partial z'} = \beta\left(\frac{\partial \mathrm{E}_x}{\partial x} + \frac{\partial \mathrm{E}_y}{\partial y} + \frac{\partial \mathrm{E}_z}{\partial z}\right)$$

$$+ \frac{\beta V}{c^2}\left(\frac{\partial \mathrm{E}_x}{\partial t} - c\frac{\partial \mathrm{H}_z}{\partial y} + c\frac{\partial \mathrm{H}_y}{\partial z}\right),$$

from which, with the help of (10.4.1) and the x-component of (10.4.3), follows

$$\frac{\partial \mathrm{E}'_x}{\partial x'} + \frac{\partial \mathrm{E}'_y}{\partial y'} + \frac{\partial \mathrm{E}'_z}{\partial z'} = 4\pi\varrho\beta - 4\pi\varrho\beta\frac{Vv_x}{c^2}.$$

If one now introduces a quantity

$$\varrho' = \varrho\beta\,(1 - Vv_x/c^2), \tag{10.4.8}$$

the above equation takes the form

$$\operatorname{div}' \mathbf{E}' \equiv \frac{\partial \mathrm{E}'_x}{\partial x'} + \frac{\partial \mathrm{E}'_y}{\partial y'} + \frac{\partial \mathrm{E}'_z}{\partial z'} = 4\pi\varrho'. \tag{10.4.9}$$

Thus the field equation (10.4.1) is invariant if ϱ' is interpreted as charge density in the system S'. Equation (10.4.8) is therefore the transformation equation for charge density.

Next we calculate $\text{curl}'_x\, \mathbf{E}'$:

$$\frac{\partial \mathsf{E}'_z}{\partial y'} - \frac{\partial \mathsf{E}'_y}{\partial z'} = \beta \left(\frac{\partial \mathsf{E}_z}{\partial y} - \frac{\partial \mathsf{E}_y}{\partial z} \right) + \beta \frac{V}{c} \left(\frac{\partial \mathsf{H}_y}{\partial y} + \frac{\partial \mathsf{H}_z}{\partial z} \right)$$

$$= -\frac{\beta}{c} \left(\frac{\partial \mathsf{H}_x}{\partial t} + V \frac{\partial \mathsf{H}_x}{\partial x} \right) = -\frac{1}{c} \frac{\partial \mathsf{H}'_x}{\partial t'}.$$

This shows that the x-component of (10.4.4) is invariant, and the same can be proved for the y- and z-components as well. Thus we have

$$\text{curl}'\, \mathbf{E}' = -\frac{1}{c} \frac{\partial \mathbf{H}'}{\partial t'}, \tag{10.4.10}$$

indicating that the field equation (10.4.4) is invariant.

Finally, we calculate $\text{curl}'_x\, \mathbf{H}'$:

$$\frac{\partial \mathsf{H}'_z}{\partial y'} - \frac{\partial \mathsf{H}'_y}{\partial z'} = \beta \left(\frac{\partial \mathsf{H}_z}{\partial y} - \frac{\partial \mathsf{H}_y}{\partial z} \right) - \frac{\beta V}{c} \left(\frac{\partial \mathsf{E}_y}{\partial y} + \frac{\partial \mathsf{E}_z}{\partial z} \right)$$

$$= \frac{4\pi\varrho\beta}{c} (v_x - V) + \frac{\beta}{c} \left(\frac{\partial \mathsf{E}_x}{\partial t} + V \frac{\partial \mathsf{E}_x}{\partial x} \right).$$

Further, we have

$$\frac{\partial \mathsf{E}'_x}{\partial t'} = \beta \left(\frac{\partial \mathsf{E}_x}{\partial t} + V \frac{\partial \mathsf{E}_x}{\partial x} \right),$$

and from (10.4.8) and the first of the equations (10.1.22)

$$\varrho' v'_x = \varrho\beta (v_x - V).$$

Thus we get

$$\frac{\partial \mathsf{H}'_z}{\partial y'} - \frac{\partial \mathsf{H}'_y}{\partial z'} = \frac{4\pi}{c} \varrho' v'_x + \frac{1}{c} \frac{\partial \mathsf{E}'_x}{\partial t'}$$

and similar equations for the other two components of curl$'$ $\mathbf{H}$. Hence the vector equation

$$\text{curl}'\, \mathbf{H}' = \frac{4\pi}{c} \varrho' \mathbf{v}' + \frac{1}{c} \frac{\partial \mathbf{E}'}{\partial t'} \tag{10.4.11}$$

holds, which proves that the field equation (10.4.3) is also invariant. This concludes the proof of the invariance of the Lorentz–Maxwell equations with respect to the Lorentz transformations.

It remains to be shown that the equations (10.4.5) satisfy the principle of relativity. For this purpose we solve these equations for E_x, E_y, E_z, H_x, H_y, H_z, in order to get the inverse transformations and obtain

$$\left.\begin{aligned} \mathsf{E}_x &= \mathsf{E}'_x, & \mathsf{H}_x &= \mathsf{H}'_x, \\ \mathsf{E}_y &= \beta\left(\mathsf{E}'_y + \frac{V}{c}\mathsf{H}'_z\right), & \mathsf{H}_y &= \beta\left(\mathsf{H}'_y - \frac{V}{c}\mathsf{E}'_z\right), \\ \mathsf{E}_z &= \beta\left(\mathsf{E}'_z - \frac{V}{c}\mathsf{H}'_y\right), & \mathsf{H}_z &= \beta\left(\mathsf{H}'_z + \frac{V}{c}\mathsf{E}'_y\right). \end{aligned}\right\} \quad (10.4.12)$$

These equations are formally identical with the equations (10.4.5), except that V is replaced by $-V$, from which follows that the two frames of reference S and S' are completely equivalent.

The same holds for the transformation equation (10.4.8) for charge density, for with the help of the first of the velocity transformation equations (10.1.24) it can be written in the inverse form

$$\varrho = \varrho'\beta\,(1 + Vv_x/c^2). \quad (10.4.13)$$

We shall now discuss some of the consequences of the transformation equations for the electromagnetic field. Let us first consider a volume element $\varDelta V$, stationary in S and containing electric charge of density ϱ. Its total electric charge is then

$$Q = \varrho \varDelta V. \quad (10.4.14)$$

Because of $v = 0$ we have, from (10.4.8),

$$\varrho' = \varrho\beta. \quad (10.4.15)$$

The size $\varDelta V'$ of the volume element, as measured by an observer in S', is smaller by a factor $\sqrt{1 - V^2/c^2}$ than $\varDelta V$, because its dimensions in the x-direction are reduced by this factor, while the dimensions normal to this direction are unaffected. Hence

$$\varDelta V' = \frac{1}{\beta}\varDelta V, \quad (10.4.16)$$

and from (10.4.14), (10.4.15), and (10.4.16) follows

$$Q' = \varrho'\,\varDelta V' = \varrho\,\varDelta V = Q, \quad (10.4.17)$$

that is the electric charge is invariant with respect to a Lorentz transformation. Thus the *law of conservation of electric charge* holds in relativistic electrodynamics, irrespective of the frame of reference used for the charge measurement.

The same result can, of course, also be derived from the inverse equation

(10.4.13). For if $\Delta V'$ is now supposed to be at rest in S', that is if $\mathbf{v}' = 0$, then $\varrho = \varrho'\beta$ and $\Delta V = \frac{1}{\beta}\Delta V'$, so that again $\varrho\,\Delta V = \varrho'\,\Delta V'$.

The law of conservation of electric charge further demands that the continuity equation (9.3.27) for the coordinate system S

$$\frac{\partial \varrho}{\partial t} + \operatorname{div}(\varrho \mathbf{v}) = 0 \tag{10.4.18}$$

must also be satisfied for the system S', or that equation (10.4.18) be invariant with respect to the Lorentz transformations. This can also be proved direct. For it follows from (10.4.6), (10.4.8), and the velocity transformation equations (10.1.24) that

$$\begin{aligned}
\frac{\partial \varrho'}{\partial t'} &= \beta\left(\frac{\partial \varrho'}{\partial t} + V\frac{\partial \varrho'}{\partial x}\right)\\
&= \beta^2\left[\left(\frac{\partial \varrho}{\partial t} + V\frac{\partial \varrho}{\partial x}\right) - \frac{V}{c^2}\left(\frac{\partial(\varrho v_x)}{\partial t} + V\frac{\partial(\varrho v_x)}{\partial x}\right)\right],\\
\frac{\partial(\varrho' v_x')}{\partial x'} &= \beta\left(\frac{\partial(\varrho' v_x')}{\partial x} + \frac{V}{c^2}\frac{\partial(\varrho' v_x')}{\partial t}\right)\\
&= \beta^2\left[-V\left(\frac{\partial \varrho}{\partial x} + \frac{V}{c^2}\frac{\partial \varrho}{\partial t}\right) + \left(\frac{\partial(\varrho v_x)}{\partial x} + \frac{V}{c^2}\frac{\partial(\varrho v_x)}{\partial t}\right)\right],\\
\frac{\partial \varrho' v_y'}{\partial y'} &= \frac{\partial(\varrho v_y)}{\partial y}, \quad \frac{\partial(\varrho' v_z')}{\partial z'} = \frac{\partial(\varrho v_z)}{\partial z},
\end{aligned}$$

so that, indeed,

$$\frac{\partial \varrho'}{\partial t'} + \operatorname{div}'(\varrho'\mathbf{v}') = \frac{\partial \varrho}{\partial t} + \operatorname{div}(\varrho\mathbf{v}) = 0. \tag{10.4.19}$$

Let us now consider a system of charged bodies which are at rest in the frame of reference S'. These then generate an electric field $\mathbf{E}'$ but no magnetic field, hence $\mathbf{H}' = 0$. Consequently we obtain from the transformation equations (10.4.5) the relations

$$\mathsf{H}_x = 0, \ \mathsf{H}_y = -V\mathsf{E}_z/c, \ \mathsf{H}_z = V\mathsf{E}_y/c, \tag{10.4.20}$$

which may be combined into the vector equation

$$\mathbf{H} = \frac{1}{c}(\mathbf{V} \wedge \mathbf{E}) \quad (\mathbf{H}' = 0). \tag{10.4.21}$$

This equation is evidently valid, irrespective of the special choice of the coordinate system S.

The formulae (10.4.20) and (10.4.21) indicate that an observer in S, with respect to whom the mentioned field generating bodies move with velocity **V**, will observe not only an electric field **E** but also a magnetic field **H** at right angles to **V** and **E**. In particular the expressions apply to the field surrounding a charged particle that moves with respect to the observer with velocity **V**.

A formula of the same appearance as (10.4.21) had previously been derived from classical considerations, based on the equivalence of moving charges and "current elements", for the electromagnetic field surrounding charged bodies which move through the *ether* with velocity **v** (9.3.33). The present formula, however, follows straightforward from the relativistic field transformation equations, and **V** is the *relative* velocity of the charged body with respect to the *observer*, without any reference to a hypothetical ether.

From the inverse transformations (10.4.12) one further obtains for $\mathbf{H}' = 0$:

$$\mathsf{E}_x = \mathsf{E}'_x, \quad \mathsf{E}_y = \beta \mathsf{E}'_y, \quad \mathsf{E}_z = \beta \mathsf{E}'_z. \tag{10.4.22}$$

These equations indicate that the electric field of charges, moving with respect to an observer, is different from that of the same charges at rest with respect to the same observer, the component of the field strength *in* the direction of movement remaining *unchanged* and the components *normal* to this direction being *increased* by a factor $1/\sqrt{1 - V^2/c^2}$. The latter change is negligibly small for slow moving charged particles, but becomes very significant for "relativistic" particles whose velocities are comparable with c, because it affects their ionizing power.

As a further example for the application of the transformation equations for the electromagnetic field, we consider a situation where $\mathbf{E} = 0$, that is where there is a purely magnetic field **H** present in the system S. We then obtain from (10.4.5)

$$\mathsf{E}'_x = 0, \quad \mathsf{E}'_y = -\beta V \mathsf{H}_z/c, \quad \mathsf{E}'_z = \beta V \mathsf{H}_y/c. \tag{10.4.23}$$

These equations can be combined into the vector equation

$$\mathbf{E}' = \frac{\beta}{c}(\mathbf{V} \wedge \mathbf{H}) \quad (\mathbf{E} = 0), \tag{10.4.24}$$

which holds for any direction of **V**.

The meaning of the relation (10.4.24) is that a body, moving with speed V in a *magnetic* field **H**, experiences an additional *electric* field $\mathbf{E}'$ at right angles to **V** and **H**. Its equivalence to the classical relationship (9.3.23) is evident, since the field called $\mathbf{E}_b$ there is identical with $\mathbf{E}'$ here, and the quantity **B** is to be replaced by **H** in the present case; but the factor β, which is a typical feature of relativistic electrodynamics, is missing in the classical formula. Furthermore, in the latter **v** is the velocity with respect to the ether, whereas in the expression (10.4.24) **V** is the velocity with respect to the observer.

If, on the other hand, we assume $\mathbf{E}' = 0$, that is to say that there is only a magnetic field $\mathbf{H}'$ in the reference system S', one obtains from (10.4.5)

$$\mathbf{E} = -\frac{1}{c}(\mathbf{V} \wedge \mathbf{H}) \quad (\mathbf{E}' = 0), \tag{10.4.25}$$

which indicates that the movement of a magnetized body relative to an observer "generates" an electric field **E**. This is the phenomenon of "unipolar induction" which is thus seen to be a direct consequence of relativistic electrodynamical principles but cannot be deduced from the Maxwell equations. For macroscopic bodies one has to replace the field strength **H** by **B**, and it can be shown that the tentative expression (9.3.21) for the unipolar induction effect follows from (10.4.25).

From the inverse transformations (10.4.12) one has further for $\mathbf{E}' = 0$:

$$\mathsf{H}_x = \mathsf{H}'_x, \quad \mathsf{H}_y = \beta \mathsf{H}'_y, \quad \mathsf{H}_z = \beta \mathsf{H}'_z. \tag{10.4.26}$$

These equations are completely analogous to the equations (10.4.22) and indicate a similar distortion of the magnetic field of a moving magnetic body.

We finally consider the Lorentz expression (9.3.28) for the force **F**, exerted on a particle of charge e, that moves with velocity **v** in an electromagnetic field **E**, **H**, as measured by an observer in S:

$$\mathbf{F} = e\left[\mathbf{E} + \frac{1}{c}(\mathbf{v} \wedge \mathbf{H})\right]. \tag{10.4.27}$$

If the particle is *at rest* in S, then **F** will be the *Newtonian* force

$$\mathbf{F} = e\mathbf{E} \quad (\mathbf{v} = 0). \tag{10.4.28}$$

For an observer in S' the particle moves with velocity $-\mathbf{V}$ in the x-direction, and (10.4.27) then reads

$$\mathbf{F}' = e\left[\mathbf{E}' - \frac{1}{c}(\mathbf{V} \wedge \mathbf{H}')\right] \tag{10.4.29}$$

or, in coordinate form,

$$F'_x = e\mathsf{E}'_x, \quad F'_y = e(\mathsf{E}'_y + V\mathsf{H}'_z/c), \quad F'_z = e(\mathsf{E}'_z - V\,\mathsf{H}'_y/c).$$

With the help of the transformations (10.4.12) one then obtains

$$F'_x = e\mathsf{E}_x, \quad F'_y = e\mathsf{E}_y/\beta, \quad F'_z = e\mathsf{E}_z/\beta. \tag{10.4.30}$$

A comparison of (10.4.28) and (10.4.30) yields the transformation equations for the electromagnetic force on the particle

$$F'_x = F_x, F'_y = F_y\sqrt{1 - V^2/c^2}, \quad F'_z = F_z\sqrt{1 - V^2/c^2}. \tag{10.4.31}$$

One notices at once that these equations are identical with the general force transformation equations (10.2.37), derived from relativistic mechanics. This is, of course, a necessary condition for experiments, like the Trouton–Noble experiment (see section 9.3), to have a negative outcome. For in order that the balance between the electromagnetic forces, acting on the charged condenser, and the elastic forces of the suspension be maintained, irrespective of the frame of reference used, the transformation law for these two types of forces must be the same.

We conclude this section by showing that the Lorentz–Maxwell equations can be written in four-dimensional notation, which reveals at once that they are invariant with respect to the Lorentz transformations.

For this purpose we first introduce a "four-vector", representing current density and charge density, as follows:

$$j_1 = \varrho v_x,\ j_2 = \varrho v_y,\ j_3 = \varrho v_z,\ j_4 = i\varrho c \tag{10.4.32}$$

(analogous to the "momentum–energy vector") (10.2.19). Making use of the four-dimensional coordinate system (10.1.29),

$$x_1 = x,\ x_2 = y,\ x_3 = z,\ x_4 = ict, \tag{10.4.33}$$

we can at once show that the equation

$$\frac{\partial j_1}{\partial x_1} + \frac{\partial j_2}{\partial x_2} + \frac{\partial j_3}{\partial x_3} + \frac{\partial j_4}{\partial x_4} = 0 \tag{10.4.34}$$

is equivalent to the continuity equation (10.4.18), for it follows from (10.4.32) and (10.4.33) that

$$\frac{\partial j_1}{\partial x_1} + \frac{\partial j_2}{\partial x_2} + \frac{\partial j_3}{\partial x_3} = \operatorname{div}(\varrho \mathbf{v})$$

and

$$\frac{\partial j_4}{\partial x_4} = \frac{\partial \varrho}{\partial t}.$$

The expression on the left-hand side of (10.4.34) is the four-dimensional divergence of the vector field $\mathbf{j}$ and must therefore be invariant with respect to a rotation in Minkowski space, that is a Lorentz transformation.

We further introduce a four-dimensional "antisymmetric tensor", called the "*electromagnetic field tensor*", by the definition

$$G_{kl} = \begin{pmatrix} 0 & \mathsf{H}_z & -\mathsf{H}_y & -i\mathsf{E}_x \\ -\mathsf{H}_z & 0 & \mathsf{H}_x & -i\mathsf{E}_y \\ \mathsf{H}_y & -\mathsf{H}_x & 0 & -i\mathsf{E}_z \\ i\mathsf{E}_x & i\mathsf{E}_y & i\mathsf{E}_z & 0 \end{pmatrix} \tag{10.4.35}$$

which represents both field vectors **E** and **H**. It can now easily be verified that the eight equations

$$\sum_{l=1}^{4} \frac{\partial G_{kl}}{\partial x_l} = \frac{4\pi j_k}{c} \quad (k = 1, 2, 3, 4) \tag{10.4.36}$$

and

$$\frac{\partial G_{kl}}{\partial x_m} + \frac{\partial G_{lm}}{\partial x_k} + \frac{\partial G_{mk}}{\partial x_l} = 0 \tag{10.4.37}$$

$$(k, l, m = 2, 3, 4; 3, 4, 1; 4, 1, 2; 1, 2, 3)$$

are equivalent to the two scalar equations (10.4.1), (10.4.2), and the two vector equations (10.4.3) and (10.4.4).

For example, for $k = 1$ one has from (10.4.32), (10.4.33), (10.4.35), (10.4.36)

$$\sum_{l=1}^{4} \frac{\partial G_{1l}}{\partial x_l} = \frac{\partial H_z}{\partial y} - \frac{\partial H_y}{\partial z} - \frac{1}{c}\frac{\partial E_x}{\partial t} = \frac{4\pi j_1}{c} = \frac{4\pi}{c}(\varrho v_x),$$

which is the x-component of (10.4.3), and for $k = 4$

$$\frac{1}{i}\sum_{l=1}^{4} \frac{\partial G_{4l}}{\partial x_l} = \frac{\partial E_x}{\partial x} + \frac{\partial E_y}{\partial y} + \frac{\partial E_z}{\partial z} = \frac{4\pi}{ic} j_4 = 4\pi\varrho,$$

which is the equation (10.4.1). Further, for $k = 2$, $l = 3$, $m = 4$ one has, from (10.4.37),

$$i\left(\frac{\partial G_{23}}{\partial x_4} + \frac{\partial G_{34}}{\partial x_2} + \frac{\partial G_{42}}{\partial x_3}\right) = \frac{1}{c}\frac{\partial H_x}{\partial t} + \frac{\partial E_z}{\partial y} - \frac{\partial E_y}{\partial z} = 0,$$

which is the x-component of (10.4.4), and for $k = 1$, $l = 2$, $m = 3$,

$$\frac{\partial G_{12}}{\partial x_3} + \frac{\partial G_{23}}{\partial x_1} + \frac{\partial G_{31}}{\partial x_2} = \frac{\partial H_z}{\partial z} + \frac{\partial H_x}{\partial x} + \frac{\partial H_y}{\partial y} = 0,$$

which is eqn. (10.4.2).

Problems and exercises

P. 10.1. An observer moves with velocity V in the positive x-direction with respect to a sphere of radius R. What shape will he attribute to the sphere from measurements with instruments in the moving frame of reference S'?

P. 10.2. Write the Lorentz transformations in terms of the coordinates x_1, x_2, x_3, x_4 in Minkowski space, and show that they are orthogonal. Comment on the significance of this result.

P. 10.3. According to its definition (10.1.30) the "world distance" D between two events can be real, zero, or imaginary. Explain in physical terms the meaning of $D = 0$ and show that it corresponds to a cone in a 4-dimensional space with coordinates x, y, z, ct. Where do the points corresponding to real and to imaginary values of D lie with respect to this cone?

P. 10.4. A particle of rest mass m^0, moving with velocity v in the positive x-direction, impinges on a resting particle of the same rest mass. After the impact the two particles move with velocities v_1 and v_2 in directions making angles θ_1 and θ_2 with the x-axis. Use the laws of conservation of energy and momentum of relativistic mechanics for deriving equations from which v_1, v_2 and θ_2 can be calculated for any given θ_1.

P. 10.5. Two spheres, made of an inelastic material, of rest masses m^0 move in opposite direction with speeds v until they collide head-on and stop. Their kinetic energy is then transformed into heat. Show that the heat content Q increases the rest masses to an amount $m^{0\prime} = m^0 + Q/c^2$.

P. 10.6. A particle of rest mass m^0 is acted upon by a constant force F in the x-direction. If it starts from rest at $t = 0$ derive an expression for the displacement x as a function of t. Compare this result with that valid in classical mechanics.

P. 10.7. What impression about the distribution of stars in the sky would an observer obtain if he were moving with a velocity comparable with that of light relative to the system of the fixed stars, as a result of the relativistic aberration of light?

P. 10.8. Two molecular beams, whose particles are emitting monochromatic radiation of frequency ν', are moving in opposite directions with speeds V. An observer, looking towards a point on the beams so that his line of vision makes an angle φ with the beam direction, observes two different frequencies ν_1 and ν_2, coming from molecules moving in the two opposite directions. Calculate the relative frequency shift $\Delta\nu/\nu' = (\nu_1 - \nu_2)/\nu'$.

P. 10.9. Describe the process of the collision between a photon and an electron in a coordinate system in which the total momentum of the two particles before the collision is zero. Show that, if the collision results in a reversal of the original photon direction, the frequency of the photon is changed in accordance with Compton's formula if measured in a frame of reference in which the electron is at rest before the collision.

P. 10.10. An electron of charge e moves with velocity V along the x-axis. Derive expressions for the components of the electric and magnetic field strengths $\mathbf{E}$ and $\mathbf{H}$ of the electromagnetic field surrounding the electron, as measured by a resting observer, at any point with coordinates x, y, z at a time t when the x-coordinate of the electron is equal to X. Also calculate the magnitudes of $\mathbf{E}$ and $\mathbf{H}$ and comment on the results obtained.

P. 10.11. Prove that the expressions $(\mathbf{EH})$ and $(\mathrm{E}^2 - \mathrm{H}^2)$ are invariant with respect to the Lorentz transformations.

P. 10.12. Complete the verification of eqns. (10.4.36) and (10.4.37), carried out in section 10.4, by showing that the equations hold for any combination of subscripts.

CHAPTER 11

Connections Between the Special Theory of Relativity and Quantum Theory

THE treatment of quantum theory given in Part I of this book is entirely non-relativistic, being based on classical mechanics. On the other hand, the special theory of relativity, as treated in Chapter 10, has no quantum features except that in section 10.3 the laws of relativistic kinematics and mechanics were applied to photons. In the present chapter we shall deal with the most important connections between relativity and quantum theory, namely, on the one hand, the ideas which led to the development of a relativistic quantum mechanics, and, on the other hand, with the principle underlying the so-called quantum field theory, which is a development of the idea of the dualistic nature of light waves and photons (see section 3.2) extended to general electromagnetic and the so-called nuclear fields.

In view of the great mathematical and conceptual difficulties of these theories and their still rather fluid state, we must restrict ourselves to an elementary introduction into these subjects.

11.1. Elements of relativistic quantum mechanics

We begin this section with a brief review of the argument which led De Broglie to the concept of the wave nature of material particles with finite rest mass, which is based on the theory of relativity, as mentioned in section 2.2. His starting point was to assign a frequency ν^0 of oscillation to a stationary particle of rest mass m^0, in analogy to the Einstein relation for photons, by making use of the "principle of equivalence of mass and energy" (10.2.13) and putting

$$E_0 = m^0 c^2 = h\nu^0. \tag{11.1.1}$$

This oscillation is represented by a "wave function"

$$\varphi_0 = A_0\, e^{2\pi i \nu^0 t'} \tag{11.1.2}$$

in a system S' in which the particle is at rest. In order to obtain the wave function φ by which an observer in S has to represent the particle, which is moving with respect to him with speed V in the positive x-direction, one has to subject the exponent in (11.1.2) to the Lorentz transformation (10.1.10) for time. Thus

$$\varphi = A\, e^{2\pi i\nu^0\beta(t - xV/c^2)}. \tag{11.1.3}$$

Introducing the quantities

$$\nu = \frac{\nu^0}{\sqrt{1 - V^2/c^2}}, \quad \mathbf{k} = \nu\mathbf{V}/c^2, \tag{11.1.4}$$

we can write (11.1.3) in the form

$$\varphi = A\, e^{2\pi i(\nu t - kx)} \tag{11.1.5}$$

which represents a running wave in the direction of the particle velocity $\mathbf{V}$ with frequency ν and wave number k. This is the "De Broglie wave" associated with the particle.

From (10.2.14), (11.1.1), and (11.1.4) one has now

$$E = \frac{E_0}{\sqrt{1 - V^2/c^2}} = \frac{h\nu^0}{\sqrt{1 - V^2/c^2}} = h\nu. \tag{11.1.6}$$

This shows that the relationship between energy and frequency is relativistically invariant which is, of course, a necessary prerequisite for the logical consistency of the theory.

Further, from (10.2.8), (10.2.15), (11.1.4), and (11.1.6),

$$\mathbf{p} = m\mathbf{V} = \frac{E\mathbf{V}}{c^2} = \frac{h\nu\mathbf{V}}{c^2} = h\mathbf{k}, \tag{11.1.7}$$

one obtains the basic De Broglie relationship between the momentum of a particle and the wave vector of the associated De Broglie wave which was introduced in (2.2.18) without proof.

The fact that the concept of De Broglie waves has its origin in relativistic considerations but that the quantum-mechanical wave equation, which is based on this concept, is, nevertheless, derived from the non-relativistic Hamiltonian function, clearly points to an inconsistency in the fundamental principles. Attempts were therefore made quite early in the development of quantum mechanics to formulate a wave equation which would conform to the principles of the theory of relativity. It is immediately evident that the time-dependent wave equations (6.1.1) or (6.1.4) cannot be relativistically invariant because they are linear in the derivative with respect to time but of second order in the derivatives with respect to the coordinates.

In 1926 it occurred to Schrödinger (1926 b), O. Klein, W. Gordon and V. Fock, independently of each other, to use as a basis for the formulation of a relativistically invariant wave equation the energy–momentum equation (10.2.16) of the special theory of relativity

$$p^2 - \frac{E^2}{c^2} + (m^0 c)^2 = 0, \tag{11.1.8}$$

and make this into an operator equation by replacing **p** and E by the operators (3.1.7) and (6.1.2) respectively. This procedure leads at once to the following wave equation for a particle in the absence of an external field

$$\nabla^2 \varphi - \frac{1}{c^2}\frac{\partial^2 \varphi}{\partial t^2} - \frac{m^{02} c^2}{\hbar^2}\varphi = 0 \tag{11.1.9}$$

with a time-dependent wave function $\varphi\,(x, y, z, t)$. This equation is generally known under the name "*Klein–Gordon equation*".

It is seen that the Klein–Gordon equation is of the second order in all four variables x, y, z, t, and in order to show that it is indeed relativistically invariant, that is to say invariant with respect to the Lorentz transformations, one can make use of the four-dimensional coordinate system (10.1.29) and write (11.1.9) in the form

$$\left(\Box^2 - \frac{1}{\Lambda^2}\right)\varphi = 0, \tag{11.1.10}$$

where $\Box$ denotes the four-dimensional operator

$$\Box \equiv \frac{\partial^2}{\partial x_1^2} + \frac{\partial^2}{\partial x_2^2} + \frac{\partial^2}{\partial x_3^2} + \frac{\partial^2}{\partial x_4^2} \tag{11.1.11}$$

and Λ is the Compton wavelength (10.3.22) of the particle of rest mass m^0. Thus, provided φ is a scalar, the Klein–Gordon equation is indeed relativistically invariant.

The expressions (6.2.1) for the charge density and (6.2.13) for the current density, corresponding to a particle of charge e, cannot both be correct in relativistic quantum mechanics because they cannot be combined to form a four-vector (10.4.32). However, if the expression (6.2.1) is changed to

$$\varrho = \frac{\hbar e i}{2m^0 c^2}\left(\varphi^* \frac{\partial \varphi}{\partial t} - \varphi\frac{\partial \varphi^*}{\partial t}\right), \tag{11.1.12}$$

and the expression for the current density vector (6.2.13) is retained, namely

$$\mathbf{j} = \frac{\hbar e}{2m^0 i}(\varphi^* \nabla\varphi - \varphi\,\nabla\varphi^*), \tag{11.1.13}$$

one can write (11.1.12) in the form

$$j_4 \equiv i\varrho c = \frac{\hbar e}{2m^0 i}\left(\varphi^* \frac{\partial\varphi}{\partial x_4} - \varphi\frac{\partial\varphi^*}{\partial x_4}\right),$$

and this can be combined with (11.1.13) to the four-vector

$$j_k = \frac{\hbar e}{2m^0 i}\left(\varphi^* \frac{\partial \varphi}{\partial x_k} - \varphi \frac{\partial \varphi^*}{\partial x_k}\right) \quad (k = 1, 2, 3, 4). \tag{11.1.14}$$

Using the notation (11.1.11) one obtains from (11.1.14)

$$\sum_k \frac{\partial j_k}{\partial x_k} = \frac{\hbar e}{2m^0 i}(\varphi^* \square \varphi - \varphi \square \varphi^*), \tag{11.1.15}$$

and if use is made of the fact that φ satisfies the wave equation (11.1.10), and evidently also its conjugate complex φ^*, one sees immediately that the right-hand side of (11.1.15) vanishes. Thus the continuity equation (10.4.34) is satisfied, which is a necessary condition for the compatibility of the relations (11.1.12) and (11.1.13).

For a stationary quantum state with energy E one can again assume φ to have the form (6.1.5)

$$\varphi(x, y, z, t) = e^{-iEt/\hbar}\psi(x, y, z) \tag{11.1.16}$$

and accordingly

$$\varphi^* = e^{iEt/\hbar}\psi^*. \tag{11.1.17}$$

Substituting for φ and φ^* from these equations into the expression (11.1.12) for ϱ one obtains, with the help of (10.2.15),

$$\varrho = \frac{eE}{m^0c^2}\varphi\varphi^* = e\frac{m}{m^0}\varphi\varphi^* = e\frac{m}{m^0}\psi\psi^* > 0, \tag{11.1.18}$$

which differs from the non-relativistic expression (6.2.1) only by the factor m/m^0. Thus for energies which are small compared with the rest energy of the particle, the relationship (6.2.1) holds approximately and the statistical interpretation of $|\varphi|^2 = |\psi|^2$ as a probability density is still valid.

Under non-stationary conditions, however, $\partial\varphi/\partial t$ is not determined by φ, and ϱ need not always be positive. Thus the Klein–Gordon equation is not in general consistent with the usual statistical interpretation of quantum mechanics.

This difficulty would not arise if the wave equation were *linear* in the derivatives with respect to time and space. For then $\partial\varphi/\partial t$ would be determined by φ itself, as is the case in non-relativistic quantum mechanics. The simplest way for obtaining such a linear wave equation would be to split up the operator $\left(\square^2 - \frac{1}{\Lambda^2}\right)$ on the left-hand side of equation (11.1.10) into a product of the two operators

$$\sum_k \mathfrak{a}_k \frac{\partial}{\partial x_\kappa} + \frac{1}{\Lambda} \quad \text{and} \quad \sum_k \mathfrak{a}_k \frac{\partial}{\partial x_k} - \frac{1}{\Lambda}, \tag{11.1.19}$$

where the $\mathfrak{a}_k$ are constant quantities.

If the condition

$$\sum_{k,j=1}^{4} \mathfrak{a}_k \mathfrak{a}_j \frac{\partial^2}{\partial x_k \, \partial x_j} - \frac{1}{\Lambda^2} = \Box^2 - \frac{1}{\Lambda^2}$$

is to hold, these quantities must satisfy the relations

$$\left. \begin{aligned} \mathfrak{a}_k \mathfrak{a}_j + \mathfrak{a}_j \mathfrak{a}_k &= 0 \quad \text{for} \quad k \gtrless j \\ \mathfrak{a}_k \mathfrak{a}_k &= 1. \end{aligned} \right\} \tag{11.1.20}$$

These conditions can evidently not be fulfilled by ordinary numbers, but they may be imposed on matrices which then will be of the "anticommutative" type (see section 7.1).

As there are four variables x_k, the simplest choice for the $\mathfrak{a}_k$ is to represent them by finite matrices with four rows and four columns. The wave function, on which the operators (11.1.19) are to operate, must then also be represented by matrices, and here again the simplest assumption is to replace the scalar φ of the Klein–Gordon equation by a matrix $\mathfrak{F}$ with one single column (column vector) of four function $f_j(x_1, x_2, x_3, x_4)$.

As the second of the operators (11.1.19) can be obtained from the first by simply changing the signs of the $\mathfrak{a}_k$ we need only consider the latter. The wave equation thus takes the form

$$\sum_k \mathfrak{a}_k \frac{\partial \mathfrak{F}}{\partial x_k} + \frac{1}{\Lambda} \mathfrak{F} = 0, \tag{11.1.21}$$

where the products are defined by the rules for the multiplication of matrices (4.3.2). This equation was first proposed by Dirac in 1928 and is known as the "*Dirac wave equation*". It forms the basis for all subsequent developments of relativistic quantum mechanics.

The quantities $\mathfrak{a}_k$ can be represented by matrices in many different ways. A suitable choice for the following considerations is:

$$\left. \begin{aligned} \mathfrak{a}_1 &= \begin{pmatrix} 0 & 0 & 0 & i \\ 0 & 0 & i & 0 \\ 0 & -i & 0 & 0 \\ -i & 0 & 0 & 0 \end{pmatrix}, \quad \mathfrak{a}_2 = \begin{pmatrix} 0 & 0 & 0 & 1 \\ 0 & 0 & -1 & 0 \\ 0 & -1 & 0 & 0 \\ 1 & 0 & 0 & 0 \end{pmatrix} \\ \mathfrak{a}_3 &= \begin{pmatrix} 0 & 0 & i & 0 \\ 0 & 0 & 0 & -i \\ -i & 0 & 0 & 0 \\ 0 & i & 0 & 0 \end{pmatrix}, \quad \mathfrak{a}_4 = \begin{pmatrix} 1 & 0 & 0 & 0 \\ 0 & 1 & 0 & 0 \\ 0 & 0 & -1 & 0 \\ 0 & 0 & 0 & -1 \end{pmatrix}. \end{aligned} \right\} \tag{11.1.22}$$

It is easily verified that the matrices (11.1.22) satisfy the conditions (11.1.20).

For example:

$$(\mathfrak{a}_1\mathfrak{a}_2) = \begin{pmatrix} i & 0 & 0 & 0 \\ 0 & -i & 0 & 0 \\ 0 & 0 & i & 0 \\ 0 & 0 & 0 & -i \end{pmatrix}, \quad (\mathfrak{a}_2\mathfrak{a}_1) = \begin{pmatrix} -i & 0 & 0 & 0 \\ 0 & i & 0 & 0 \\ 0 & 0 & -i & 0 \\ 0 & 0 & 0 & i \end{pmatrix} = -(\mathfrak{a}_1\mathfrak{a}_2),$$

$$(\mathfrak{a}_1\mathfrak{a}_2) = \begin{pmatrix} 1 & 0 & 0 & 0 \\ 0 & 1 & 0 & 0 \\ 0 & 0 & 1 & 0 \\ 0 & 0 & 0 & 1 \end{pmatrix} = \mathfrak{J}.$$

It is to be noticed that the 2×2 sub-matrices appearing in $\mathfrak{a}_1$, $\mathfrak{a}_2$, $\mathfrak{a}_3$ are obtained by multiplying the Pauli matrices (7.3.29) by $2i/\hbar$ and $-2i/\hbar$ respectively. This suggests that relativistic quantum mechanics may be able to account for the spin of elementary particles. This is, indeed, the case; we shall return to this point later on in this section.

In order that (11.1.21) be relativistically invariant, the four components of $\mathfrak{F}$ have to transform in a certain way if the x_k are subjected to a Lorentz transformation. $\mathfrak{F}$ is then said to have the character of a "*spinor*", the term indicating the above-mentioned connection with spin. We cannot go into the details of the "spinor calculus" within the framework of this book.

We shall now show that the Dirac equation (11.1.21) for a particle in field-free space is satisfied by a wave function of the character of a plane De Broglie wave. Making use of the four-vector $\mathbf{J}$ (10.2.19) and the space–time coordinates (10.1.29), we can then put

$$\mathfrak{F} = \mathfrak{g}e^{\pm \frac{i}{\hbar}\sum_k J_k x_k}, \tag{11.1.23}$$

where $\mathfrak{g}$ is a one-column matrix, independent of the x_k. In the following we choose the positive sign of the exponent, in keeping with the convention (11.1.16).

From (11.1.23) then follows

$$\frac{\partial \mathfrak{F}}{\partial x_k} = \frac{i}{\hbar}\mathfrak{g}J_k\, e^{\frac{i}{\hbar}\sum_k J_k x_k}. \tag{11.1.24}$$

From (11.1.22) one has further, according to the rules of matrix multiplication,

$$\mathfrak{a}_1\mathfrak{g} = i\begin{vmatrix} g_4 \\ g_3 \\ -g_2 \\ -g_1 \end{vmatrix}, \quad \mathfrak{a}_2\mathfrak{g} = \begin{vmatrix} g_4 \\ -g_3 \\ -g_2 \\ g_1 \end{vmatrix}, \quad \mathfrak{a}_3\mathfrak{g} = i\begin{vmatrix} g_3 \\ -g_4 \\ -g_1 \\ g_2 \end{vmatrix}, \quad \mathfrak{a}_4\mathfrak{g} = \begin{vmatrix} g_1 \\ g_2 \\ -g_3 \\ -g_4 \end{vmatrix}. \tag{11.1.25}$$

Substituting from (11.1.23), (11.1.24), and (11.1.25) into (11.1.21), we find that that equation is identically satisfied if the quantities g_1, g_2, g_3, g_4 are solutions of the four homogeneous algebraic equations

$$\left.\begin{aligned}
\left(\frac{\hbar}{\Lambda} + iJ_4\right) g_1 - J_3 g_3 - (J_1 - iJ_2)\, g_4 &= 0,\\
\left(\frac{\hbar}{\Lambda} + iJ_4\right) g_2 - (J_1 + iJ_2)\, g_3 + J_3 g_4 &= 0,\\
J_3 g_1 + (J_1 - iJ_2)\, g_2 + \left(\frac{\hbar}{\Lambda} - iJ_4\right) g_3 &= 0,\\
(J_1 + iJ_2)\, g_1 - J_2 g_2 + \left(\frac{\hbar}{\Lambda} - iJ_4\right) g_4 &= 0.
\end{aligned}\right\} \qquad (11.1.26)$$

By making use of the relation (11.1.8) it is easily verified that the equations (11.1.26) are compatible with each other and that there are two independent solutions, namely,

$$\left.\begin{aligned}
&g_1 = \gamma J_3, \quad g_2 = \gamma (J_1 + iJ_2), \quad g_3 = \gamma \left(\frac{\hbar}{\Lambda} + iJ_4\right)\\
&\quad = \gamma (m^0 c - E/c), \quad g_4 = 0;\\
&g_1 = \gamma (J_1 - iJ_2), \quad g_2 = -\gamma J_3, \quad g_3 = 0,\\
&g_4 = \gamma (m^0 c - E/c),
\end{aligned}\right\} \qquad (11.1.27)$$

where γ is an arbitrary factor.

The fact that the amplitude $\mathfrak{g}$ of a progressing De Broglie wave has three non-vanishing components, like a vector, in Dirac's relativistic quantum mechanics, in contrast to the scalar De Broglie waves of non-relativistic quantum theory, is of great significance. For it indicates that the De Broglie waves are *polarized*, or that the particles, represented by such waves, must have an *axis* associated with them which can be orientated at various angles towards the direction of their velocity. This axis can only be an axis of intrinsic angular momentum of the particle, and thus the existence of spin is an essential feature of relativistic quantum mechanics.

In a system of coordinates, in which the particle is at rest, the linear momentum is zero and the energy is equal to the rest energy, or $J_1 = J_2 = J_3 = 0$, $\frac{\hbar}{\Lambda} + iJ_4 = m^0 c - E/c = 0$. Hence from the third and fourth of the equations (11.1.26) one has $g_3 = g_4 = 0$, and thus the first and second of these equations are identically satisfied, irrespective of the values of g_1 and g_2. Therefore in this case there is a twofold infinity of solutions which may be constructed as linear combinations of the two special solutions

$$\mathfrak{g}^{(1)} = \begin{vmatrix} 1 \\ 0 \\ 0 \\ 0 \end{vmatrix}, \quad \mathfrak{g}^{(2)} = \begin{vmatrix} 0 \\ 1 \\ 0 \\ 0 \end{vmatrix}, \qquad (11.1.28)$$

with arbitrary coefficients. This indicates that the spin axis can have an arbitrary direction in space.

Had we chosen the negative sign in the exponent of (11.1.23) this would have been equivalent to a change of sign of Λ in the Dirac equation (11.1.21) or to have chosen the second instead of the first of the two operators (11.1.19). It is easy to see that in that case the corresponding solution of the equations (11.1.26) for the resting particle would be

$$\mathfrak{g}^{(3)} = \begin{vmatrix} 0 \\ 0 \\ 1 \\ 0 \end{vmatrix}, \quad \mathfrak{g}^{(4)} = \begin{vmatrix} 0 \\ 0 \\ 0 \\ 1 \end{vmatrix}. \tag{11.1.29}$$

The significance of the solutions (11.1.28) and (11.1.29) becomes apparent if one considers the eigenvalue problem of the matrix operator

$$\mathfrak{S} = \begin{vmatrix} s & 0 & 0 & 0 \\ 0 & -s & 0 & 0 \\ 0 & 0 & s & 0 \\ 0 & 0 & 0 & -s \end{vmatrix}, \tag{11.1.30}$$

that is, according to (7.1.9), the solution of the equation

$$\mathfrak{S}\mathfrak{g} = \sigma\mathfrak{g}. \tag{11.1.31}$$

Substituting for $\mathfrak{g}$ in turn the four quantities $\mathfrak{g}^{(J)}$ one sees at once that they are indeed solutions of (11.1.31) and that the corresponding eigenvalues are

$$\sigma^{(1)} = s, \quad \sigma^{(2)} = -s, \quad \sigma^{(3)} = s, \quad \sigma^{(4)} = -s. \tag{11.1.32}$$

Since the submatrices in (11.1.30) are equal to the Pauli spin matrix $\mathfrak{s}_z$ (7.3.29) (with s instead of $\hbar/2$), this means that for the solutions $\mathfrak{g}^{(1)}$ and $\mathfrak{g}^{(3)}$ the spin of magnitude s is directed in the positive z-direction and for $\mathfrak{g}^{(2)}$ and $\mathfrak{g}^{(4)}$ in the negative z-direction. Moreover, it can be proved (which we shall not do here), that, indeed, $s = \hbar/2$ for all "permanent" elementary particles.

The Dirac equation (11.1.21) for a free particle can be modified to describe the behaviour of a particle in an electromagnetic field which may be used instead of the non-relativistic Schrödinger equation in cases where the particle velocity is not negligibly small compared with that of light. This is, for example, of particular importance if one wants to account for the "relativistic fine structure" of spectral lines and the scattering of fast-moving particles in the electric field of a nucleus. Dirac's theory has indeed been able to account for these and other phenomena in agreement with observation. But, as indicated in the introduction to this chapter, these matters cannot be dealt with on the level of this textbook.

11.2. Photon and meson fields

The duality between light waves and photons was already commented upon in sections 2.2 and 3.2. According to this duality, a plane electromagnetic wave of frequency ν and wave vector $\mathbf{k}$ is equivalent to a photon of energy $E = h\nu$ and momentum $\mathbf{p} = h\mathbf{k}$. As electromagnetic waves are only special cases of general electromagnetic fields one is logically led to the conclusion that such fields should be equivalent to assemblies of photons. If this is really the case, it ought to be possible to derive the field equations for electromagnetic fields from suitable equations describing the behaviour of photons. That this is indeed possible may be demonstrated by the following simple consideration.

Let us assume that the Klein–Gordon equation (11.1.9) applies to photons. As photons have rest mass zero this equation reduces to

$$\frac{\partial^2 \varphi}{\partial t^2} = c^2 \nabla^2 \varphi, \tag{11.2.1}$$

which, according to (1.2.2), is formally identical with the optical wave equation. If φ is independent of time, (11.2.1) reduces to the Laplace equation

$$\frac{\partial^2 \varphi}{\partial x^2} + \frac{\partial^2 \varphi}{\partial y^2} + \frac{\partial^2 \varphi}{\partial z^2} = 0 \tag{11.2.2}$$

of the classical theory of potentials. The scalar function φ in (11.2.1) may therefore be identified with the scalar potential of the electromagnetic field in the absence of space charges ϱ. The condition $\varrho = 0$ does not, however, exclude the possible presence of "point charges". If in particular φ is a function only of the distance r from such a point charge the equation (11.2.2) simplifies to

$$\frac{d^2 \varphi}{dr^2} + \frac{2}{r}\frac{d\varphi}{dr} = 0. \tag{11.2.3}$$

It can be verified readily that the solution of this equation under the condition that φ vanishes at infinity is

$$\varphi(r) = e/r, \tag{11.2.4}$$

where the constant e is the finite electric charge at $r = 0$.

Thus the theory is consistent with the existence of infinitely small charged elementary particles. These, however, would have an infinitely large electric "self energy", so that (11.2.1) cannot be the correct field equation for the electromagnetic field. Apart from this the insufficiency of the theory is also shown by the fact that the electromagnetic field is a *vector* field which demands that, apart from the scalar potential φ, there should be a *vector potential* $\mathbf{A}$ that also satisfies the wave equation (11.2.1) [see (1.2.2)] in the absence of space currents $\mathbf{i} = \varrho\mathbf{v}$. Since the quantum mechanical wave function in the Klein–Gordon equation is scalar, this equation cannot be taken to be the correct "equation of motion" for photons. It also cannot explain the fact that photons have spins with angular momentum $\hbar$. Finally, photons are not "permanent" particles as they can be emitted and absorbed by matter or "created" and "annihilated".

For all these reasons a proper quantum theory of the electromagnetic field has to take as its basis a wave equation for the photons which is similar in type to the Dirac equation rather than the Klein–Gordon equation and in which also certain operators for creation and annihilation of photons are incorporated. It is not possible to deal with this problem within the framework of the present book. We may, however, give a rough picture of the dualistic relationship between the continuous electromagnetic field and the photons, which emerges from the theory, as follows. The electromagnetic field surrounding an isolated charged particle is equivalent to a cloud of photons which are perpetually created and annihilated, and the electromagnetic interaction between pairs of particles is equivalent to a perpetual interchange of photons between them. The latter process is somewhat similar to the exchange of electrons between molecules which is responsible for the "exchange forces" acting between them, or the formation of chemical bonds, as explained in section 8.2.

A similar situation is encountered in the theory of *nuclear forces*. In order to explain why the "nucleons" which constitute atomic nuclei (protons and neutrons) are held together in close proximity, in spite of the electric *repulsion* between the protons, it is necessary to assume that a specific type of *attractive* force, not envisaged in classical theory, is acting between nucleons which extends only over very short distances of the order of magnitude of the nuclear dimensions. Yukawa in 1935 conceived the idea of attributing these forces to a "*nuclear field*" to which a new type of non-permanent particles are associated in a similar dualistic way as the photons are associated to the electromagnetic field.

He took as his starting point again the Klein–Gordon equation (11.1.10) but assumed that the particles concerned had a finite rest mass μ so that the field equation for the potential of the nuclear field had the form

$$\Box^2\varphi = \frac{1}{\beta^2}\,\varphi \tag{11.2.5}$$

with

$$\beta = \hbar/\mu c. \tag{11.2.6}$$

This equation is called the "*Yukawa equation*".

For a static nuclear field, where φ is independent of time, the Yukawa equation simplifies to

$$\nabla^2\varphi = \frac{1}{\beta^2}\varphi. \tag{11.2.7}$$

If φ again depends on the distance r from a centre only, the solution of (11.2.7) is

$$\varphi(r) = -\frac{q}{r}e^{-r/\beta}, \tag{11.2.8}$$

where q is a constant which plays a role similar to the electric charge e for the electric potential and the negative sign indicates that the force exerted on a similar particle is attracting.

One notices that, because of the presence of the exponential factor, the Yukawa potential decreases with increasing distance from the centre much more rapidly than the electric potential (11.2.4) and therefore represents a "short-range force" in contrast to the "long-range" electromagnetic forces. The approximate range is of the order of magnitude of β, and if this is to be of the order 10^{-13} cm one finds from (11.2.6) that μ must be of the order of a few hundred electronic masses. The particles associated with the nuclear fields are therefore called "*mesons*" because their masses are between those of the electrons (and positrons) and those of the nucleons.

The discovery that particles with masses of the predicted order of magnitude are indeed emitted and absorbed in nuclear processes, just as photons are emitted and absorbed in processes occurring in the outer shells of the atoms, very strongly supports the basic idea of the meson field theory of nuclear forces. These particles are found to have spin zero so that φ could be supposed to be a scalar potential. However, there are several reasons why the primitive Yukawa theory, as sketched out above, cannot be correct.

Firstly, the potential (11.2.8) would imply an infinite self-energy of a nucleon, even if it were electrically neutral like the neutron. Secondly, there exists a number of types of mesons which may be either uncharged or carry a positive or a negative elementary charge e; they form two groups, one with mesons of approximately 140 electronic masses, called "*π-mesons*" (or "*pions*") and another with approximately 500 electronic masses, called "*K-mesons*" (or "*Kaons*"). Thirdly, all these particles are non-permanent, like the photons, and can be created or annihilated and transformed into other particles in a bewildering variety of ways. Finally, the interaction forces between nucleons, as observed by means of suitable experiments, are not in agreement with Yukawa's law (11.2.8).

Great progress has since then be made in the development of the meson theory of nuclear fields, on lines similar to those followed in the photon theory of the electromagnetic field, and of a theory which would account for the mutual interactions of these fields or the various associated particles.

However, the theory is still fluid and in many ways unsatisfactory, and we have still no clear explanation for the observed mass ratios of these particles. The rough picture of the dualistic relationship between nuclear field and mesons is similar to that previously sketched for the electromagnetic field: an isolated nucleon must be imagined to be surrounded by a cloud of mesons which are perpetually emitted and reabsorbed so as to form the static nuclear field of the nucleon, and the interaction between pairs of nucleons can be interpreted as an exchange force due to the perpetual exchange of mesons between them. A more detailed study of these problems would be quite outside the scope of this textbook.

Problems and exercises

P.11.1. Verify that the matrices (11.1.22) satisfy the relations (11.1.20) for all values of the indices.

P.11.2. Prove that the equations (11.1.26) are compatible with one another, and that the expressions (11.1.27) are a pair of independent solutions of these equations.

P.11.3. Derive the analogue to the equations (11.1.26) for the case of the ***negative*** sign of the exponent in the expression (11.1.23), and hence prove the validity of the relation (11.1.29) for the matrices $\mathfrak{g}^{(3)}$ and $\mathfrak{g}^{(4)}$.

P.11.4. Derive the general solution of the differential equation (11.2.3) and show that it reduces to the expressions (11.2.4) under the condition $\varphi = 0$ at $r = \infty$.

P.11.5. Prove that Laplace's equation is generally satisfied if there is a finite number of point-charges e_k at the points with position vectors $\mathbf{r}_k$.

P.11.6. Derive the general solution of the differential equation (11.2.7) in the case that φ is a function of r only, and show that it reduces to (11.2.8) if φ is to be finite everywhere except at $r = 0$.

CHAPTER 12

Elements of the General Theory of Relativity

THE special theory of relativity, to which Chapter 10 was devoted, is based on Einstein's "restricted principle of relativity", according to which the laws of physics are valid with respect to all *inertial* frames of reference, that is all systems of coordinates which move with constant velocity relative to the system of fixed stars as a whole (section 9.1). In 1915 Einstein extended the theory to what is now known as the "*general theory of relativity*". This theory is based on the postulate that the laws of physics must have the property of being valid with respect to *any* frame of reference; this is the "*general principle of relativity*".

In order to be able to formulate the laws of the general relativity theory in mathematical form Einstein found it necessary to replace the Euclidean geometry of classical physics by the "non-Euclidean geometry" developed by Gauss and Riemann, and to make use of the "generalized calculus of tensors" due to Ricci and Levi-Civita. As the readers of this textbook are not supposed to be acquainted with these complicated mathematical subjects, we shall restrict ourselves to an explanation of the fundamental principles of the theory in this chapter as far as it is possible to do without the mentioned mathematical techniques, and to a discussion of the various physical implications of the theory which are capable of experimental verification.

12.1. The basic principles of the general theory of relativity

At first sight it seems impossible that the laws of dynamics should be valid for *all* frames of reference. For it was shown in section 9.1 that an observer in a closed laboratory could, by means of mechanical experiments, find out whether the laboratory system was an inertial frame of reference or not, because of the appearance of "inertial forces" acting on his test bodies in the latter case which are absent only in the former case.

Newton was of the opinion that the inertial forces on a body were caused by its *absolute* acceleration, and he illustrated this by his famous "bucket experiment". Suppose that a bucket filled with water is put into a rotational movement about a vertical axis. To begin with the water is at rest with respect to the bucket walls and its surface is horizontal which shows that no inertial forces are acting on it. For some time the water surface will still remain horizontal, although now water and bucket are rotating relative to one another; but gradually, due to friction, the water will attain the same angular velocity as the bucket and its surface will become concave. This reveals that now the water is acted upon by inertial (centrifugal) forces although it is at rest relative to the bucket. If now the rotation of the bucket is stopped suddenly, the water surface will still remain concave for some time while the water rotates relative to the bucket until finally the original situation is restored. From this Newton concluded that the inertial forces could not be produced by the *relative* motion of the water with respect to the bucket walls and must therefore be due to its *absolute* rotation in space.

This argument was criticized by E. Mach as long ago as 1883 in his famous book on the history of the principles of mechanics, on the grounds that the possible influence of the distant masses in the universe on the mechanical behaviour of material bodies cannot be discounted. The observed centrifugal forces on the water may be caused by the rotation of the water relative to those masses, and the fact that the relative rotation of the water with respect to the bucket walls apparently does not produce any inertial forces on the former may be due to the enormous preponderance of the distant masses over the mass of the bucket. He therefore concluded that it remained to be seen whether such forces would not after all become observable if the "bucket" had walls of many miles of thickness.

It has already been emphasized in section 9.1 that the inertial forces are proportional to the masses of the bodies on which they act, irrespective of the material of these bodies. This is, of course, due to the fact that the "*inertial masses*" m^i are defined by the Newtonian law of motion (1.1.1)

$$\mathbf{F} = m^i \mathbf{a}. \tag{12.1.1}$$

On the other hand, according to Newton's law of gravitation, the force exerted by a gravitational field of field strength $\mathbf{g}$ is given by

$$\mathbf{F} = m^g \mathbf{g}, \tag{12.1.2}$$

where the quantity m^g, which is independent of the field, is usually called the "*gravitational mass*" of the body.

Combining (12.1.1) and (12.1.2) one obtains for the acceleration $\mathbf{a}^g$ of the body by gravity the expression

$$\mathbf{a}^g = \frac{m^g}{m^i}\,\mathbf{g}. \tag{12.1.3}$$

It was first established by Galileo in the sixteenth century that a^g at some fixed location was the same for all bodies, which implies that m^g must be proportional to m^i, the constant of proportionality only depending on the unit used for measuring g. If m^g and m^i are supposed to have the same physical dimension then g has the dimension of acceleration, and if then a^g and g are measured in the same units one has $m^g = m^i$: *the gravitational mass of a body is equal to its inertial mass.*

Observations on free-falling bodies can establish the validity of this important relationship only with a very restricted accuracy. It is easy to see that it implies that pendulums with bobs, made of materials of different density but of the same shape and size, ought to have the same oscillation frequency. Experiments of this kind were first carried out by Newton in order to provide further evidence for his law of gravitation, and later more refined experiments showed that the relationship held with an accuracy greater than 0·01 per cent. The investigations of R. v. Eötvös and collaborators in the second decade of this century made it possible to establish the law of the equality of gravitational and inertial mass with enormously increased accuracy. These experiments consisted in comparing the combined force of the gravitational attraction by the earth and the centrifugal force of the earth's rotation on equal masses of different materials by means of the "gravitational balance". The accuracy achieved was at least 10^{-6}, and a recent repetition of these experiments resulted in an accuracy of 10^{-10}.

While contemplating the extension of the principle of relativity to general frames of reference Einstein realized that the precise equality of two completely differently defined quantities, the inertial and the gravitational mass, must mean that the phenomena of inertia and gravitation are only two aspects of one and the same fundamental physical law. This is also in keeping with the above-mentioned ideas of Mach, according to which the phenomenon of inertia has its origin in the distant masses of the universe.

An observer in a closed system is indeed not able to conclude from the outcome of mechanical experiments, carried out within the system, whether the forces exerted on the test bodies used are due to an acceleration of the system with respect to the fixed stars or to a gravitational field. In particular, an observer in a laboratory which performs a uniformly accelerated movement with respect to the fixed stars in a direction towards the ceiling of that laboratory will experience the same mechanical effects as if the laboratory was an inertial frame of reference but under the influence of a homogeneous gravitational field directed towards the floor of the laboratory.

These considerations led Einstein to the formulation of the so-called "*principle of equivalence*" which demands that it be impossible to distinguish between inertial and gravitational forces and contends that they are in fact one and the same. We may therefore say that an acceleration of the frame of reference is equivalent to the "generation" of a gravitational field in a direction opposite to the acceleration (a phenomenon which is very forcefully encountered by astronauts in the initial stages of their flight), and that a gravitational field can be made to "disappear" if the frame of reference moves with the acceleration g (producing, for example, the phenomenon of weightlessness in a space ship as soon as its engines have stopped).

The principle of equivalence is a necessary supplement to the general principle of relativity, for it makes it possible to formulate the laws of mechanics *with the inclusion of gravitation* in such a way that they are invariant with respect to *any* transformation of the coordinate system. The general theory of relativity therefore *incorporates* a theory of gravitation in contrast to classical theory where the Newtonian laws of dynamics and the Newtonian law of gravitation appear as *independent* fundamental laws.

But the principle of the constancy of the velocity of light which, as was shown in Chapter 10, is an essential feature of the special theory of relativity, can no longer be retained in the general theory of relativity without some modification. For a ray of light which moves in a straight line in empty space, as seen by an observer in an inertial frame of reference, will in general follow a curved path in a non-inertial system. According to the principle of equivalence this means that a light ray ought to be deflected in a gravitational field. The same conclusion can be drawn on the basis of the photon theory. For according to the principle of equivalence of mass and energy the energy $h\nu$ of a photon must constitute a relativistic inertial mass $h\nu/c^2$, and this ought to provide the photon with an equal gravitational mass as well; thus the photon should be deflected in a gravitational field. We shall discuss this particular consequence of the general relativity theory in more detail in section 12.2.

It was already indicated in the introduction to this chapter that, in order to formulate the laws of physics in conformity with the principles of the general theory of relativity, it was necessary to modify even the very laws of Euclidean geometry. This can be understood by the following simple consideration. Suppose that an observer wishes to establish experimentally that the ratio between the circumference and the diameter of a circular disk is equal to π, a relationship which follows from the Euclidean principles. He can do this by using short measuring rods whose lengths are very small compared with the circumference of the disk. If the disk of radius R is at rest in an inertial system in a field-free laboratory, the lengths of the measuring rods are evidently independent of their location and orientation and the observer will indeed find the predictions from Euclidean geometry confirmed within the accuracy of his method of measurement. But if the disk with the observer on it rotates about its symmetry axis with constant angular velocity ω (like a roundabout) he is no longer entitled to make the above-mentioned assumptions about his measuring rods. In order to perform the required measurements he can then compare distances Δs between corresponding points on the periphery or a diameter of his rotating and another stationary disk. He will find that all distances Δs on the *periphery* of the stationary disk are shorter than those on the rotating disk by a factor $\sqrt{1 - R^2\omega^2/c^2}$ on account of the

Lorentz contraction (see section 10.1). But the distance along the corresponding *diameters* of the two disks will be found to be the same because the Δs move at right angles to the relative velocity. The observer will therefore conclude that the ratio between the circumference and the diameter of his disk is *greater* than π, a finding which is clearly incompatible with the basic axioms of Euclidean geometry.

Geometries based on axioms different from those of Euclid have been conceived by mathematicians as a purely intellectual exercise long before the advent of the theory of relativity. A special type of non-Euclidean geometry developed by Gauss can be envisaged without great difficulty as a generalization of the two-dimensional geometry on a curved surface. On a plane surface a straight line is defined as a line which has the property that the distance between any two of its points, measured along it, is a minimum. There are, of course, in general no straight lines on a curved surface, but there exist lines which have the same minimum property. They are called "*geodetic lines*" or, in short, "*geodetics*", and they play the same part in the geometry of curved surfaces as the straight lines in plane geometry.

A two-dimensional geometry, based on the concept of geodetics instead of straight lines, can be built up in close analogy to and as a generalization of plane geometry, and it turns out that some of the theorems of such a geometry differ essentially from the familiar theorems of the Euclidean plane geometry. In particular the sum of the angles of a triangle is in general different from π, and the ratio of the circumference of a circle to its diameter is also different from π. A simple example for a geometry of this type is the geometry on a spherical surface, where the geodetics are the greatest circles.

A generalization of the Gaussian principles is due to Riemann who extended them to three and more dimensions. It is, of course, not possible to visualize a "curved" space as we can a curved surface. But analytically there is no essential difference between the two-dimensional Gaussian and the multi-dimensional Riemannian geometries. Whereas, according to the rules of the Euclidean geometry, the length of a line element in ordinary "plane" three-dimensional space is given by the relationship

$$ds^2 = dx^2 + dy^2 + dz^2,$$

in a "curved" space the corresponding relationship has on its right-hand side a linear combination between the products and squares of the differentials dx, dy, dz, the coefficients of which represent the particular geometrical properties of this space, including its "curvature".

From considerations like the one concerning the measurement of lengths on a rotating disc it appears proper to use Riemannian geometry in a curved space instead of Euclidean geometry in all cases where inertial forces are encountered. Now according to the equivalence principle such forces are equivalent to gravitational forces. We may therefore say that the presence of a gravitational field affects the geometrical properties of space by making its geometry non-Euclidean.

We shall now show that, apart from the modifications of the classical rules of space measurement, which have to be applied in order to comply with the general principle of relativity, we must also apply similar modifications to the classical rules for time measurement. Within an inertial system the rate of clocks does not depend on their situation in space, and hence any number of clocks, which are stationary in that system, can be precisely synchronized with each other. But one sees at once that this is not possible in a system S' which rotates with constant angular velocity ω relative to an inertial system S. For a point at a distance R from the rotation axis moves at any moment with a speed $v = R\omega$ with respect to a point on the axis. Thus, according to the clock retardation theorem of the special theory of relativity (10.1.14) and (10.1.15), a clock at the former point will for an observer in S show a slower rate than a clock at the latter point. Thus an observer in S' will notice that the rate of his clocks depends on their locations within his frame of reference in a manner which is apparently closely connected with the observed centrifugal forces. For the ratio of the time intervals $(\Delta t)_2$, measured by the clock at the distance R from the centre, and $(\Delta t)_1$, measured by the clock at the centre, is

$$\frac{(\Delta t)_2}{(\Delta t)_1} = \sqrt{1 - \frac{R^2\omega^2}{c^2}} = \sqrt{1 - \frac{g_c R}{c^2}}, \tag{12.1.4}$$

where g_c is the centrifugal acceleration.

According to the principle of equivalence the observer is justified in interpreting g_c as the intensity of a gravitational field. We may therefore introduce a dimensionless gravitational potential γ by the definition

$$\Gamma_c \equiv \frac{g_c}{c^2} = -\frac{\partial \gamma}{\partial R} \quad (\gamma = 0 \quad \text{for} \quad R = 0) \tag{12.1.5}$$

which leads to

$$\gamma = -\frac{R^2\omega^2}{2c^2}, \tag{12.1.6}$$

so that (12.1.4) can be written in the form

$$\frac{(\Delta t)_2}{(\Delta t)_1} = \sqrt{1 + 2\gamma}. \tag{12.1.7}$$

If, as will normally be the case, γ is very small we obtain for the relative retardation δt of the clock at R with respect to a clock at the centre of rotation the simple relation

$$\delta t \equiv \frac{(\Delta t)_2}{(\Delta t)_1} - 1 \simeq \gamma. \tag{12.1.8}$$

That the relations (12.1.7) and (12.1.8) are generally valid for *any* gravitational fields can be made plausible by proving that they hold in a *homogeneous* gravitational field. Suppose that two identical clocks are both at the same location in such a field at $t = 0$ and that one of them, C_1, is kept there artificially while the other, C_2, is allowed to fall freely in the direction of the field. C_1 is therefore under the influence of the gravitational field of strength g but C_2 is not. When C_2 has fallen through a distance l it has acquired a velocity $v = \sqrt{2gl}$. The clock C_1 will then have this relative velocity with respect to the clock C_2 and, according to the retardation theorem, its rate will show a relative slow-down

$$\frac{(\Delta t)_2}{(\Delta t)_1} = \sqrt{1 - \frac{2gl}{c^2}} = \sqrt{1 - 2\Gamma l}. \qquad (12.1.9)$$

The gravitational potential γ is, in the case of a homogeneous field, by the definition (12.1.5) equal to

$$\gamma = -l\Gamma. \qquad (12.1.10)$$

Thus, substituting from (12.1.10) into (12.1.9) we obtain again the relation (12.1.7) or, for small γ, the relation (12.1.8) for the relative retardation of the clock which is subject to the gravitational field, relative to the clock in the field free space.

In connection with the subject of the influence of inertial or gravitational forces on the rate of clocks it is appropriate to make a few remarks concerning the so-called "*clock paradox*" which is from time to time claimed as revealing an internal logical inconsistency in the theory of relativity. Suppose that of two precisely synchronized clocks one, C_1, remains stationary at a point O in an inertial frame of reference S while the other, C_2, goes on a journey and finally returns to O. An observer in S will then find that, according to (10.1.14), the clock C_1 is in advance of C_2. But an observer in a system S' relative to whom the clock C_2 remains stationary, will then, in virtue of the relation (10.1.15), expect to find C_2 in advance of C_1, which is a clear contradiction.

The fallacy of this argument lies in the fact that the complete equivalence of the systems S and S', on which the relationships (10.1.14) and (10.1.15) are based, only holds in the special theory of relativity when both systems of reference are inertial. But this is obviously not the case here, for if C_2 is to return to C_1 the former must move along a curved path, and hence must suffer acceleration at least temporarily, so that S' cannot be an inertial frame of reference. From the point of view of the observer in S the clock C_1 is stationary while the clock C_2 is accelerated and thus under the influence of inertial forces which, in view of the foregoing discussion, slow down its movement. From the point of view of the observer in S' the clock C_2 is under the influence of a gravitational field while the clock C_1 performs an unconstrained movement in that gravitational field. Hence again C_2 must be retarded compared with C_1.

In section 10.1 we introduced the concept of a four-dimensional Minkowski space in which the coordinates x_1, x_2, x_3, x_4 represent point events. In the special theory of relativity it is assumed that Euclidean geometry holds in this space, or that the relation (10.1.31) for the line element $d\sigma$ of a world line

$$d\sigma^2 = (dx_1)^2 + (dx_2)^2 + (dx_3)^2 + (dx_4)^2 \tag{12.1.11}$$

is satisfied. If one wishes to extend the concept of the Minkowski space to the general theory of relativity one has, on account of the foregoing discussion concerning space and time measurement, to assume that Euclidean geometry does no longer apply to this space if the coordinate system used is non-inertial or, in other words, if gravitational forces are encountered. Einstein therefore extended Riemann's geometry to the four-dimensional Minkowski space which necessitated the replacement of the relation (12.1.11) for the line element $d\sigma$ by the more general relation

$$d\sigma^2 = \sum_{i,j=1}^{4} g_{ij}\, dx_i\, dx_j \quad (g_{ij} = g_{ji}). \tag{12.1.12}$$

The six coefficients g_{ij} in this equation, which in general will be functions of the coordinates, form what is called a "symmetric tensor of the second rank". This tensor is called the "*fundamental metric tensor*".

From the preceding discussion on the influence of gravitational (or inertial) forces on space and time measurements it follows that measurements in two adjacent points in space at two very shortly following times are not affected by such forces. $d\sigma$ is therefore not only invariant with respect to the Lorentz transformations of the special theory of relativity (section 10.1), but generally invariant against any transformations of the coordinates x_1, x_2, x_3, x_4. It follows from this that the "proper time" element $d\tau$, which is related to $d\sigma$ by the relation (10.1.32), namely

$$d\tau^2 = -d\sigma^2/c^2, \tag{12.1.13}$$

is an invariant measure of time.

One can always choose a system of coordinates x_1, x_2, x_3, x_4 in such a way that the g_{ij} are all equal to unity in any particular world point. In the immediate neighbourhood of this point the gravitational field will then be "transformed away". But apart from this infinitesimal region the g_{ij} will then in general be different from one. It is the dependence of the components of the metric tensor on the coordinates which determines the geometric properties of the space–time continuum, in particular its "curvature". On the other hand, as we have seen, it is the gravitational field which is responsible for the deviation of this geometry from the Euclidean. We are thus led to the conclusion that the metric tensor is the mathematical representation of the gravitational field in the framework of the general theory of relativity.

If the gravitational field is sufficiently weak one can put

$$g_{ij} = 1 + \gamma_{ij}, \qquad (12.1.14)$$

where the quantities γ_{ij} are small compared with unity. The dimensionless quantities γ_{ij} therefore form a measure of the deviation from the Euclidean character of the space–time continuum due to the gravitational field. This suggests to interpret the tensor γ_{ij} as a generalization of the gravitational potential of this field. We shall justify this idea later on more convincingly.

In the special theory of relativity the world lines of mass points on which no forces are acting are straight lines with equations

$$\frac{d^2x_k}{d\tau^2} = 0 \quad (k = 1, 2, 3, 4), \qquad (12.1.15)$$

for the last of these equations, according to the definition (10.1.27) of $d\tau$ means that the speed of the particle is constant, and the first three equations then mean that the velocity vector $\mathbf{v}$ is also constant.

It was mentioned before that in a non-Euclidean geometry the geodetic lines play the part of the straight lines of the Euclidean geometry. In the general theory of relativity it is therefore assumed that a particle, which is under the influence of gravitational (or inertial) forces only, will move in such a way that its world line is a geodetic. It can be shown that the differential equations for a world line will then have the form

$$\frac{d^2x_k}{d\sigma^2} + \sum_{i,j=1}^{4} \Gamma^k_{ij} \frac{dx_i}{d\sigma} \frac{dx_j}{d\sigma} = 0 \quad (k = 1, 2, 3, 4), \qquad (12.1.16)$$

and that these equations are invariant with respect to any transformation of the coordinates .The quantities Γ^k_{ij} are function of the g_{ij} and their derivatives with respect to the coordinates. They form a tensor of the third rank which may be interpreted as a generalization of the gravitational field strength

$$\Gamma = -\operatorname{grad} \gamma \qquad (12.1.17)$$

of the Newtonian theory.

If all the components Γ^k_{ij} of the gravitational field are zero, (12.1.16) reduces to (12.1.15), as it should. If the gravitational field is not zero but sufficiently weak, we can make use of the approximation (12.1.14). It can then be shown that the Γ^k_{ij} are connected with the γ_{ij} by the tensor relations

$$\Gamma^k_{ij} = \frac{1}{2}\left(\frac{\partial \gamma_{ij}}{\partial x_k} - \frac{\partial \gamma_{ik}}{\partial x_j} - \frac{\partial \gamma_{jk}}{\partial x_i}\right) \quad (i, j, k = 1, 2, 3, 4) \qquad (12.1.18)$$

which are generalizations of the vector equation (12.1.17).

If it is further assumed that the velocities of the particles are small compared with c and that the γ_{ij} vary only slowly in time the quantities $\frac{dx_1}{d\sigma}, \frac{dx_2}{d\sigma}, \frac{dx_3}{d\sigma}$ become all small compared with $\frac{dx_4}{d\sigma} \simeq 1$, and the equations (12.1.16) reduce, in virtue of (12.1.13), to

$$\frac{d^2 x_k}{d\tau^2} = c^2 \Gamma^k_{44} \quad (k = 1, 2, 3) \tag{12.1.19}$$

and the expressions (12.1.18) reduce to

$$\Gamma^k_{44} = \frac{1}{2} \frac{\partial \gamma_{44}}{\partial x_k} \quad (k = 1, 2, 3). \tag{12.1.20}$$

The latter relations are equivalent to the vector equation (12.1.17) if one identifies $-\gamma_{44}/2$ with the dimensionless gravitational potential γ and $c^2\Gamma^k_{44}$ with the components of the gravitational force per unit mass. Equation (12.1.19) is thus seen to be the Newtonian equation of motion of a particle under the influence of the gravitational field.

These results justify *a posteriori* our previous assumption concerning the character of the quantities g_{ij} and Γ^j_{ik}. It is also seen that the quantities γ_{ij} are small compared with unity as long as the velocities of material particles under the influence of a gravitational field are small compared with c. As this is always the case for the movement of celestial bodies, we are justified in considering the gravitational fields in the universe as "weak" in the previously used sense. Nevertheless, (12.1.19) is not *precisely* valid in the general theory of relativity, and hence very small deviations of the orbits of the planets from those predicted by the classical celestial mechanics based on Newton's equations must be expected. We shall return to this point in section 12.2.

In classical physics the Newtonian gravitational potential φ_g satisfies Poisson's differential equation

$$\nabla^2 \varphi_g \equiv \frac{\partial^2 \varphi_g}{\partial x^2} + \frac{\partial^2 \varphi_g}{\partial y^2} + \frac{\partial^2 \varphi_g}{\partial z^2} = 4\pi G \varrho_g, \tag{12.1.21}$$

where G is the universal gravitational constant and ϱ_g the density of gravitational mass. The solution of this equation can be written in the form of an integral

$$\varphi_g(x, y, z) = -G \iiint \frac{\varrho_g(\xi, \eta, \zeta)}{r} \, d\xi \, d\eta \, d\zeta, \tag{12.1.22}$$

where r is the distance between the points with coordinates x, y, z and ξ, η, ζ, and the triple integral is to be extended over all space. Newton's law of gravitational attraction between two mass points follows from (12.1.22) as a special case, for the right-hand member then reduces to $-Gm/r$ and thus we have for the gravitational potential at a distance r from m

$$\varphi_g(r) = -\frac{Gm}{r}. \quad (12.1.23)$$

It therefore follows from (12.1.2) for the force exerted by the mass m on another mass m'

$$F(r) = -m' \frac{d\varphi_g}{dr} = -G\frac{mm'}{r^2}, \quad (12.1.24)$$

the negative sign indicating that the force is directed towards m.

The equation (12.1.21) is only valid in an inertial frame of reference. In order to obtain the law of gravitation in the general theory of relativity one must try to replace it by a relationship of the same type which is relativistically invariant against any transformation from one frame of reference to another. There is no unique way towards the solution of this problem, but Einstein was led by the following guiding principles.

The scalar quantity ϱ_g on the right-hand side of Poisson's equation (12.1.21) has, according to the principle of equivalence of gravitational and inertial mass and the principle of equivalence of mass and energy, the character of an *energy density*. In analogy to the replacement of the Newtonian scalar gravitational potential by the tensor g_{ij} it seems plausible to replace ϱ_g by a *tensor* T_{ij} of the second rank which represents energy and momentum density in space, similarly as in the special theory of relativity energy and momentum of a particle are represented by the momentum–energy vector $\mathbf{J}$ (10.2.19). Accordingly, the quantity $\nabla^2\varphi_g$ on the left-hand side of equation (12.1.21) must also be replaced by a differential tensor R_{ij} of the second rank in the derivatives of the gravitational potentials g_{ij} with respect to the coordinates x_1, x_2, x_3, x_4 in such a way that it resembles $\nabla^2\varphi_g$ most closely and transforms in the same way as T_{ij}. Einstein indeed succeeded to set up a gravitational field equation in this way, in accordance with the principle of relativity; it has the form

$$R_{ij} = \frac{K}{c^2} T_{ig} \quad (i, j = 1, 2, 3, 4), \quad (12.1.25)$$

where K is a universal constant which must be closely related to Newton's gravitational constant G. It can indeed be seen easily that the dimension of K is that of G/c^2.

A necessary condition which these equations have to satisfy is that in a sufficiently weak field they reduce to the Poisson equation (12.1.21) or that the gravitational potentials are given by expressions of the form (12.1.22). It can indeed be shown that under these circumstances, and if the field is practically stationary, the tensor γ_{ij} reduces to

$$\left.\begin{aligned} \gamma_{11} = \gamma_{22} = \gamma_{33} = -\gamma_{44} = 2\gamma, \\ \gamma_{ij} = 0 \quad \text{for} \quad i \gtrless j, \end{aligned}\right\} \quad (12.1.26)$$

where $\gamma = \varphi_g/c^2$ satisfies the equation (12.1.22), provided that the numerical value of the constant K in the field equations (12.1.25) is suitably chosen. The Newtonian classical law of gravitation (12.1.24) is thus seen to be a first approximation to a more general and complicated law.

It will now be shown that the previously discussed behaviour of measuring rods and clocks in gravitational fields can also be deduced from the relations (12.1.12), (12.1.14), and (12.1.26). Let us assume that an observer, situated within a small region of space, uses an inertial frame of reference S so that the g_{ij} there are all equal to 1. It then follows that within this region

$$(d\sigma)^2 = ds^2 - c^2\,dt^2. \tag{12.1.27}$$

For another observer in another frame of reference S', which in general will not be inertial, according to (12.1.26) the equation

$$d\sigma'^2 = (1 + 2\gamma)\,ds'^2 - (1 - 2\gamma)\,c^2\,dt'^2 \tag{12.1.28}$$

will be valid instead. It follows from the invariance of σ^2 that the right-hand sides of (12.1.27) and (12.1.28) must be equal or that

$$\left.\begin{aligned} \frac{ds'}{ds} &= (1 + 2\gamma)^{-1/2} \simeq 1 - \gamma, \\ \frac{dt'}{dt} &= (1 - 2\gamma)^{-1/2} \simeq 1 + \gamma. \end{aligned}\right\} \tag{12.1.29}$$

The latter equation is identical with (12.1.7) for small γ, indicating that the rate of a clock at a point with gravitational potential γ is *retarded* in comparison with a clock at a point where $\gamma = 0$. The former relation indicates that all length dimensions are *increased* by the presence of a gravitational field by a relative amount γ. This is in conformity with our previous conclusion concerning the rotating disk, and it shows that, even in a first approximation, the deviations from Euclidean geometry have to be taken into account.

If one attempts to obtain solutions of the Einstein field equations to a higher approximation in some special cases one finds that the equations (12.1.25) have to be replaced by more complicated relations for the components g_{ij} of the gravitational potential tensor and that the Newtonian law (12.1.22) is no longer strictly valid. In particular, owing to the fact that the field equations hold for the four-dimensional Minkowski space and therefore contain the second derivatives of the potentials not only with respect to the space coordinates but also with respect to time, these equations have really the character of wave equations. As a result it is found that the changes of the gravitational field due to the movement of the field-generating masses do not occur instantaneously but propagate through space in the form of "gravitational waves" with the speed of light c.

As already emphasized in this section, the close relationship between gravitation and "space curvature" is one of the most characteristic features of the general theory of relativity. Some authors go so far as to maintain that the space curvature is the *cause* of the phenomenon of gravitation. This point of view is, however, not consistent with Mach's principle which is one of the pillars on which the general theory of relativity rests, and according to which the phenomena of gravitation and inertia are simply two aspects of the interaction between matter. Thus it seems more appropriate to regard the geometry of the space–time continuum as a convenient mathematical representation of the laws for these interactions.

This is of particular importance if one wishes to draw a picture of the structure of the universe as a whole and its changes in time, which is the subject of "*cosmology*". Newton had already speculated whether one should imagine the number of stars to be finite and occupying a finite part of an infinite "empty" space, or whether one should assume the number of stars to be infinite and filling the infinite space with approximately equal density. According to the first notion the universe would in fact be finite and bounded because the idea of an infinite empty space is physically meaningless. The second possibility is excluded by physical reasons because an infinite number of radiating stars would cause the intensity of the gravitational field and the density of radiation to be infinitely large everywhere.

If one assumes that the geometry of the space–time continuum is Euclidean at sufficiently great distances from gravitating bodies, then the four-dimensional universe could be imagined as a four-dimensional analogue of a two-dimensional flat, but slightly buckled, infinite surface. But if one admits the possibility of the geometry of the space–time continuum as a whole to be non-Euclidean than one can imagine the above-mentioned two-dimensional *infinite* and unbounded surface to be replaced by an unbounded but *closed* surface, for example like a slightly buckled spherical surface with a finite radius and area. In this case the four-dimensional universe would still be unbounded in space and time but would have a finite volume and could be occupied by a finite number of stars with finite density so that the previously mentioned difficulties of the Newtonian concepts would disappear.

From dimensional considerations it is evident that the "radius" R_u of this closed universe would have to be of the order of magnitude

$$R_u = GM_u/c^2, \tag{12.1.30}$$

where M_u would be the total mass in the universe. If one assumes this mass to be distributed with approximately uniform density ϱ_u, then (12.1.30) can be re-written in the form

$$R_u = c\left(\frac{4\pi}{3}\,G\varrho_u\right)^{-1/2}. \tag{12.1.31}$$

From astronomical observations the value of ϱ_u is estimated to be of the order of 10^{-30} g/cm³ and $\dfrac{4\pi G}{3c^2} \simeq 3\times 10^{-28}$. Thus R_u would be of the order of magnitude 6×10^{28} cm or about 50,000 million light years.

12.2. Some consequences of the theory

In this section we shall discuss some consequences of the general theory of relativity to the extent to which they can be deduced from the contents of the preceding section. There are comparatively very few of these consequences which can be put to the test by observation and experiment, unlike the situation in the special theory of relativity. The reason for this is that, as can be seen from (12.1.29), the effects to be expected are of the order of magnitude $\delta\gamma = \delta\varphi_g/c^2$, where $\delta\varphi_g$ is the difference between the Newtonian potential at two points P_1 and P_2 of observation.

It follows from the definition of φ_g that $m^0\delta\varphi_g = \delta U$ is the difference between the gravitational potential energies of a particle of rest mass m^0 in the two positions at P_1 and P_2. Thus according to (10.2.12)

$$\delta\gamma = \frac{\delta\varphi_g}{c^2} = \frac{\delta U}{m^0 c^2} = \frac{\delta m}{m^0}, \tag{12.2.1}$$

where δm is the change of the rest mass of the particle when it is moved from P_1 to P_2. Owing to the comparative weakness of the *gravitational* fields the quantity (12.2.1) is always extremely small.

On the other hand, the potential energy differences of a particle with an electric charge e in an *electric* field with potential φ_e, for example in the case of the electric fields in the interior of atomic nuclei or in particle accelerators can become comparable with, or even large, compared with its rest energy; consequently δm can become comparable or even large compared with the rest mass m^0, as was discussed in section 10.2. This is responsible for the fact that relativistic effects, predicted by the special theory of relativity, can under certain circumstances be very conspicuous.

We first consider a mechanical consequence of the general theory of relativity concerning the movement of the planets which was first deduced by Einstein (1915) and K. Schwarzschild (1916). The classical (Keplerian) laws for the movement of the planets are obtained by combining Newton's equations of motion for a material particle with his gravitational law for the interaction between particles. It follows in particular that, if only the gravitational force between a planet and the sun is taken into account, the planet performs a closed elliptical orbit about the common mass centre of sun and planet. But we have seen in the preceding section that Newton's equations of motion in the form (12.1.19) are only a first approximation to the differential equations (12.1.16), and that the gravitational field equation (12.1.22), from which Newton's law of attraction between mass points (12.1.24) follows, is also only a first approximation to Einstein's field equations (12.1.25). It is therefore to be expected that in a second approximation the orbit of a planet will be found to deviate slightly from an ellipse. The orbit will in fact not be strictly closed, but it may be described as an ellipse which performs a precessional rotation in its own plane with a periodic time T' which is very long compared with the classical periodic time T of the orbital movement, so that $\varepsilon = T/T'$ is a small fraction.

It is not possible to give the complete derivation of the formula for ε on the narrow basis of section 12.1, but the general shape of this formula can be deduced directly from the above argument that ε may be expected to be of the order of magnitude (12.2.1). In the present case $\delta\varphi_g$ is essentially the difference between the potentials of the sun's gravitational field at infinite distance and at the distance r of the planet from the sun (under the assumption that the orbit is almost circular); thus from (12.1.23)

$$\delta\varphi_g \simeq GM/r, \tag{12.2.2}$$

where M is the mass of the sun. The equality of the centrifugal and the gravitational forces on the planet, moving with orbital velocity v, further demands that

$$mv^2/r = GMm/r^2. \tag{12.2.3}$$

From (12.2.2) and (12.2.3) one has

$$\delta\varphi_g \simeq v^2 = 4\pi^2 r^2/T^2,$$

and thus

$$\varepsilon \simeq \delta\varphi_g/c^2 \simeq 4\pi^2 r^2/T^2 c^2. \tag{12.2.4}$$

Finally, according to the definition of ε the precession angle α (in radians) per revolution of the planet is

$$\alpha = 2\pi T/T' = 2\pi\varepsilon \simeq 8\pi^3 r^2/T^2 c^2. \tag{12.2.5}$$

The precise formula, as derived from the theory, for a planet with semi-major axis a of its quasi-elliptic orbit and numerical excentricity e, is

$$\alpha = \frac{24\pi^3 a^2}{(1-e^2)\, c^2 T^2}, \tag{12.2.6}$$

which has the same form as (12.2.5).

Formula (12.2.2) shows that the greatest effect is to be expected for the planets nearest to the sun. Astronomers have indeed found that the planet Mercury exhibited a movement of its perihelion which could not be explained by perturbations from the other planets and which was estimated to amount to between 41·6 and 43·5 sec of arc per century. The value calculated from the relativistic formula (12.2.6) is 43 sec of arc, which is seen to be in complete agreement with observation.

We now turn to an optical consequence of the general theory of relativity, namely the deflection of light beams in gravitational fields. It was already mentioned in section 12.1 that the occurrence of such a phenomenon can be predicted from the equivalence theorem, and that on account of it the postulate of the constancy of the velocity of light is no longer valid in the general theory of relativity.

For the mathematical treatment of this phenomenon in weak gravitational fields we can make use of the equations (12.1.27) and (12.1.28). In a region where the gravitational forces are negligibly small and therefore the velocity of light is exactly equal to c formula (12.1.27) demands that $d\sigma^2$ along the world line of a light ray must be zero. It then follows from (10.1.28) because of the invariance of $d\sigma$ ($d\sigma = d\sigma'$) that

$$c' \equiv \frac{ds'}{dt'} = c\sqrt{\frac{1-2\gamma}{1+2\gamma}} \simeq c(1-2\gamma) = c\left(1 - \frac{2\varphi_g}{c^2}\right). \tag{12.2.7}$$

This relation expresses the dependence of the light velocity on the gravitational potential φ_g.

If the gravitational field is generated by a body of mass M the potential φ_g is given by (12.1.23) and hence (12.2.7) takes the form

$$c' \simeq c\left(1 + \frac{2GM}{rc^2}\right), \tag{12.2.8}$$

where r is the distance from M. If now a light ray travels in the direction of r one obtains

$$\frac{dc'}{dt} = -\frac{2GM}{r^2c}\frac{dr}{dt} \simeq -\frac{2GM}{r^2}. \tag{12.2.9}$$

We can interpret this in the language of the photon theory of light as meaning that a photon at a distance r from M is accelerated *towards* M by an amount $g = 2GM/r^2$, which may be attributed to a gravitational attraction of the photon by M following a law similar to Newton's gravitational law (12.1.24), as already indicated in section 12.1.

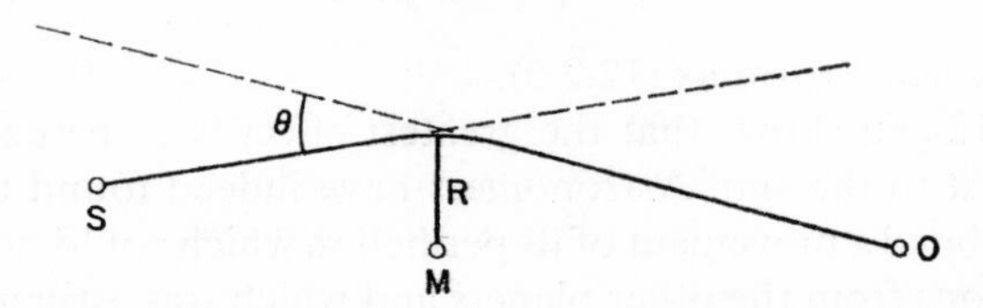

FIG. 26. Deflection of a light ray from a star by the gravitational field of the sun.

Let us now assume that a light ray from a distant star S (Fig. 26) on its way to a terrestrial observer O passes a gravitational centre of mass M at a short distance R. According to what has just been found it will become slight-y bent *towards* M so that the observer will see the star image displaced *away* rom M by a small angle θ.

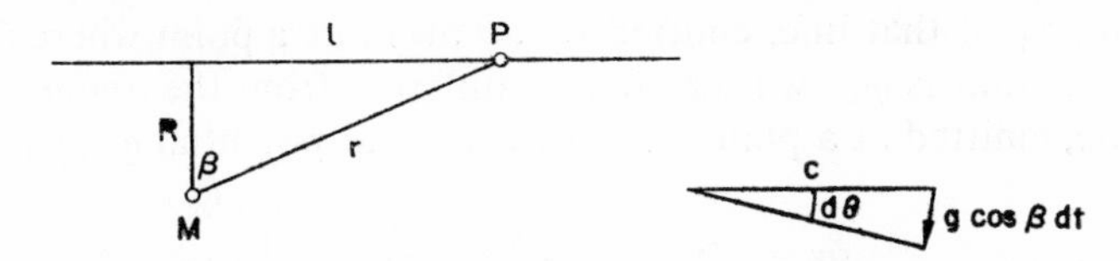

FIG. 27. Diagram illustrating the derivation of the formula for the gravitational deflection of light.

In order to calculate θ we notice (Fig. 27) that the angular deviation $d\theta$ of the ray within a time dt near a point P is equal to

$$d\theta = \frac{g \cos\beta \, dt}{c} = \frac{2GM \cos\beta \, dt}{cr^2} = \frac{2GM \cos^3 \beta \, dt}{cR^2}. \tag{12.2.10}$$

Now from $l = R \tan\beta$ follows

$$dl = \frac{R}{\cos^2\beta} \, d\beta = c \, dt$$

and hence

$$dt = \frac{R \, d\beta}{c \cos^2 \beta}. \tag{12.2.11}$$

Thus from (12.2.10) and (12.2.11)

$$d\theta = \frac{2GM \cos\beta}{c^2 R} \, d\beta \tag{12.2.12}$$

and, finally,

$$\theta = \frac{2GM}{c^2 R} \int_{-\pi/2}^{+\pi/2} \cos\beta \, d\beta = \frac{4GM}{c^2 R}. \tag{12.2.13}$$

If formula (12.2.13) is applied to a light ray passing near the surface of the sun then M is the mass and R the radius of the sun, and the numerical value of θ is 1·75 sec of arc. Einstein accordingly predicted that the images of stars seen very near to the sun's limb during a total eclipse should be displaced radially away from the sun by this amount relative to their normal positions. This prediction was first confirmed by the observations of two British astronomical expeditions in 1919 (Dyson, Eddington and Davidson, 1920). Later, more accurate observations seem to indicate that the actual displacements are about 20 per cent greater than those predicted from the general theory of relativity. It is possible that this discrepancy may be due to the superposition of some other effect upon the relativistic one.

Another optical phenomenon predicted by the general theory of relativity is an immediate consequence of the relation (12.1.8) concerning the influence of gravitational fields on clocks. For an atom, emitting a certain spectral line of frequency ν can be regarded as a clock, and it is thus to be expected that the frequency ν_1 of that line, emitted by an atom at a point where the gravitational potential is φ_1, will be slightly different from the frequency ν_2 of the same line, emitted at a point with gravitational potential φ_2, so that

$$\delta\nu = \frac{\nu_2 - \nu_1}{\nu_1} = \frac{\varphi_2 - \varphi_1}{c^2} = \frac{\delta\varphi}{c^2}. \tag{12.2.14}$$

Accordingly there will be a slight shift $\delta\lambda$ of the wavelength λ

$$\delta\lambda = \frac{\lambda_2 - \lambda_1}{\lambda_1} = \frac{\varphi_1 - \varphi_2}{c^2} = -\frac{\delta\varphi}{c^2}. \tag{12.2.15}$$

Formula (12.2.15) indicates that the wavelength of the line, emitted at the point with *smaller* φ compared with the point of observation, will be *larger* than the wavelength of the locally emitted line, or that the observer will notice a *red-shift* of the spectral line of magnitude $\delta\varphi/c^2$.

The same result can be derived from a quantum theoretical consideration. The expressions (2.2.13) and (5.3.17) for the energy levels of the hydrogen atom contain the electronic mass m as a factor, and consequently the expressions for the frequency of the spectral lines of hydrogen also contain this factor. The same holds for all atomic spectra. Now if an atom is moved from a point with lower gravitational potential to one with higher potential the potential energies of its electrons are increased by $m\delta\varphi$ and hence, according to the equivalence of mass and energy, their rest masses are increased by an amount $\delta m = m\delta\varphi/c^2$. Consequently the frequencies of the spectral lines are increased by

$$\delta\nu = \delta m/m = \delta\varphi/c^2$$

in agreement with formula (12.2.14).

This theorem can immediately be applied to the spectral lines emitted (or absorbed) by atoms on the surface of the sun. According to the formula (12.1.23) we have in this case (because the distance between earth and sun is large compared with the sun's radius R) $\delta\varphi = GM/R$, and hence the relativistic red-shift for the spectral lines of the sun is expected to be

$$\delta\lambda = \frac{GM}{Rc^2} \simeq 2\times 10^{-6}. \tag{12.2.16}$$

This amounts to an absolute red-shift for the Fraunhofer D-line of about 0·01 Å. Observations, in particular those of M.G. Adam, have definitely established the existence of the effect, although its magnitude varies over the disk of the sun and seems to be about 50 per cent larger than predicted by (12.2.16). The effect ought to be considerably larger on the so-called "white dwarf" stars whose radii are much smaller than that of the sun, and measurements of this kind have shown that a red-shift of the expected order of magnitude indeed exists.

In recent years it has become possible to establish the existence of the gravitational red-shift by means of laboratory experiments by making use of the "Mössbauer effect" (see section 10.3). In these experiments a source of soft γ-rays with extremely narrow frequency width was placed on top of an evacuated tube of height l and an absorber of the same frequency at the bottom. A frequency shift due to the action of the earth's gravitational field should then result in a change in the amount of absorption. With $l = 25$ m one has $\delta\varphi/c^2 = lg/c^2 \simeq 2{\cdot}5 \times 10^{-15}$. The relative width of the spectral line used is about 1300×10^{-15} so that the expected $\delta\nu$ is only $\frac{1}{500}$ of the line width. Nevertheless, the accuracy of the experiment is so great that the existence of the effect could be confirmed within 10 per cent of the predicted magnitude.

Finally, the experiments with the rotating disk described in section 10.3, whose results were interpreted there as a verification of the existence of the transverse Doppler effect, can also be interpreted as another confirmation of the gravitational red-shift. For, from the point of view of an observer in a reference system which rotates with the disk, the source and the absorber of the radiation are at rest with respect to one another but subject to an inertial centrifugal field. According to the equivalence theorem this is equivalent to a gravitational field directed away from the centre so that the gravitational potential at the rim is lower by the amount (12.1.6) than that at the centre. As already discussed in section 12.1, this leads to a red-shift, measured by an observer at the centre, exactly equal to the one due to the transverse Doppler effect.

Problems and exercises

P. 12.1. Use the cosine theorem of spherical trigonometry for deriving an expression for the angles α of a unilateral spherical triangle with sides a, and hence show that the sum of the angles must always be larger than π and that the "excess" increases with a.

P. 12.2. Derive a formula for the ratio of the circumference s and the radius r of a circle on a spherical surface of radius R, on the one hand, and the ratio of the area a of that circle and the square of r, on the other hand. Discuss how this result may be used for determining the curvature of the surface from geodetic measurements on the surface alone.

P. 12.3. A system of curvilinear coordinates ξ, η on the surface of a hemisphere of radius r is defined by the network of lines $\xi = \text{const}$, $\eta = \text{const}$ which are the projections of the lines $x = \text{const}$, $y = \text{const}$ of a Cartesian coordinate system in the base plane of the hemisphere upon its surface. Calculate the coefficients in the expression for the line element $ds^2 = g_{11}\xi^2 + g_{22}\eta^2 + 2g_{12}\xi\eta$ as functions of x, y.

P. 12.4. Making use of Gauss's theorem of the theory of potentials verify that the expression (12.1.22) is the solution of the Poisson equation (12.1.21).

P. 12.5. A point source of light at a point A is observed by an observer at a point B, first in a field-free space and then in the presence of a homogeneous gravitational field of field strength $\mathbf{g}$ to which the line AB is inclined at an angle α. In the latter case the observer will notice a deflection of the position of the light source by a small angle $\delta\alpha$. Obtain the direction and magnitude of $\delta\alpha$ from the principle of equivalence of the general theory of relativity.

P. 12.6. If in the experiment of P. 12.5 the light emitted by the source is monochromatic with frequency ν the observer will find that the gravitational field has produced a frequency shift $\delta\nu$. Calculate $\delta\nu$ in terms of g and α by means of formula (12.2.14).

PART III

Statistical Mechanics

CHAPTER 13

The Theory of Ensembles

THE foundations of statistical mechanics, as a generalization of the kinetic theory of gases, were laid during the last three decades of the last century, mainly by Boltzmann, Maxwell, and Gibbs, with the purpose of deducing the macroscopic laws for the physical behaviour of material bodies, in particular their thermal properties, from the microscopic classical laws governing the movement of their constituent particles. These latter were supposed to be the strictly *causal* laws of classical dynamics which were combined with *statistical* considerations concerning the behaviour of very large numbers of these particles. This resulted in what is now usually called "*classical statistical mechanics*".

It was already pointed out in the introduction that the combination of causal and chance aspects in one and the same theory led to considerable fundamental difficulties. These were only overcome when the laws of classical mechanics were replaced by those of quantum mechanics which, as explained in detail in Part I, are inherently statistical in character and allow strict prediction to be made not for individual systems but only for *ensembles* of identical systems. This led to the development of "*quantum statistical mechanics*".

In this chapter the fundamental principles of the statistical mechanics of ensembles will be presented, and general laws, applicable to arbitrary systems, will be derived. It will further be shown how the basic laws of *thermodynamics* may be deduced from these laws.

13.1. Short review of the classical statistical mechanics of ensembles

In order to enable the reader to get a better understanding of the difficulties inherent in the fundamental principles of classical statistical mechanics, and to appreciate more fully the great improvement brought about by quantum statistical mechanics, we shall in this section give a short review of the basic notions of the classical theory.

The subjects of the theory are arbitrary mechanical systems of N degrees of freedom. These systems may either be single, elementary particles, or atoms or molecules, or macroscopic bodies, made up of atoms and molecules interacting with each other, known under the name of "*assemblies*" of particles.† The "state" of any such system at any particular time t is mathematically described by the values of the N generalized coordinates $q_1, \ldots, q_N$ and the corresponding "generalized momenta" $p_1, \ldots, p_N$, as described in section 1.1. In what is to follow we shall denote these $2N$ quantities, in arbitrary order, by $x_1, \ldots, x_{2N}$.

It is customary and convenient to represent these quantities geometrically as (Cartesian) coordinates in a $2N$-dimensional abstract space, called the "*phase-space*". The state of the system under consideration at any time t is then represented by a point—its "*phase point*"—in phase space with position vector $\mathbf{r}\,(x_1, \ldots, x_{2N})$. The state will in general change in the course of time, that is $\mathbf{r}$ will be a function of time, and hence the phase point of the system will describe a trajectory in phase space, called its "*phase curve*".

The shape of the phase curve depends not only on the *internal* mechanism of the system but also on external influences. Thus if repeated experiments are made on a given system by letting it start from one and the same initial state, the subsequent development will differ to some degree from one experiment to another, because the system can be only incompletely shielded from the external influences which are beyond the control of the experimenter. Hence the corresponding phase curves, which all start from the same point in phase space, will diverge more and more as time goes on.

Instead of making repeated experiments with one system one can imagine a large number of exact replicas of the system and let them all start from the same initial state. Thus their phase points, all represented in the same phase space, will start from the same point but will, after a time t be distributed over the phase space according to a certain "*distribution function*" $\varrho\,(\mathbf{r}, t)$ which expresses the relative number density of the phase points in the various regions of the phase space as a function of time, that is the number of phase points per unit volume in the neighbourhood of the point $\mathbf{r}$, divided by the total number of systems.

We can now again make repeated experiments with our multitude of identical systems, starting always from the same point in phase space. Because of the unavoidable external disturbances the observer will find the distribution function to be different in each experiment. But he will also find that these differences become smaller and smaller the larger the number of systems is, provided that the disturbances are of a "random" character, or to use a modern term, are "*stochastic*" in nature. It is therefore plausible to assume that the distribution would become completely determined when the number of systems was infinitely large. It is very important to realize that this so-

† The statistical mechanics of assemblies is treated in detail in Chapter 14.

called "*law of great numbers*" can neither be proved mathematically from the probability calculus, nor from the laws of classical physics, and that it is indeed alien to the idea of causality which dominates classical physics. In the framework of classical statistical mechanics it must be regarded as an independent empirically established law of nature.

Adopting the nomenclature of Gibbs we shall call an imaginary *infinite* multitude of identical systems an "*ensemble*" (see also section 6.1). (In German literature an ensemble is usually called "*virtuelle Gesamtheit*".) It is a fundamental theorem of classical statistical mechanics that an ensemble, in which all the systems start from one and the same initial state, will have a precisely defined continuous distribution function $\varrho(\mathbf{r}, t)$ for all subsequent times. According to the definition of ϱ as a relative number density it measures at the same time the "probability density" (see also section 3.2) for any member of the ensemble to be found at time t in a certain state represented by $\mathbf{r}$, or more precisely, $\varrho\, dV$ is the probability of finding the phase point of the system in the $2N$-dimensional volume element dV of the phase space, surrounding the point with position vector $\mathbf{r}$ at time t.

More generally, if an ensemble has an initial distribution $\varrho(\mathbf{r}, 0)$ at time $t = 0$ its distribution function $\varrho(\mathbf{r}, t)$ at any subsequent time is supposed to be completely determined. In the theory of *stochastic processes*, which deals with such problems, it is assumed that ϱ satisfies a certain partial differential equation whose solution under given "boundary conditions" and for the given initial condition $\varrho(\mathbf{r}, 0)$ yields the distribution function $\varrho(\mathbf{r}, t)$ for all time. We shall not deal with classical stochastic theory in this textbook.

An ensemble for which the distribution function is *independent of time* is said to be "*stationary*" or in "*statistical equilibrium*". It is believed that any ensemble will eventually automatically reach statistical equilibrium after a sufficiently long time has elapsed, and that the corresponding distribution function can for any given ensemble in principle be obtained as the time-independent solution of the above-mentioned stochastic differential equation applying to the ensemble. We shall see, however, that very general theorems about stationary distribution functions may be derived from much simpler considerations (section 13.3.) Statistical mechanics in the usual sense of this term is restricted to ensembles in statistical equilibrium, and its principal task is the determination of stationary distribution functions.

In the applications of statistical mechanics one is mainly interested in the distribution functions of certain *observable parameters* which determine the *macroscopic* properties of material bodies. Any such quantity A must be a function of the *microscopic* parameters $x_1, \ldots, x_{2N}$ of the system. To any fixed value a of A belong all points in phase space which lie on a $(2N - 1)$-dimensional "hypersurface" with the equation $A(\mathbf{r}) = a$. Thus the distribution function or probability density function $f(a, t)$ follows from the condition that the probability $f(a, t)\, \delta a$ of finding the phase point of the system in the thin "shell" in phase space between the hypersurfaces $A = a$ and $A = a + \delta a$

is

$$f(a, t)\,\delta a = \int_{\omega} \cdots \int \varrho\,(\mathbf{r}, t)\,dV, \tag{13.1.1}$$

where the integral is extended over the volume ω of the shell. In a stationary ensemble the distribution functions $f(a)$ are, of course, independent of time.

In particular, if A is the energy E of the system then $A(\mathbf{r})$ is its Hamiltonian function H, and $f(E, t)$ is called the "*energy distribution function*". The surfaces $H = E$ are called "*energy surfaces*" and the shell between $H = E$ and $H = E + \delta E$ is called the "*energy shell*". Again, in a stationary ensemble the energy distribution function $f(E)$ is independent of time.

Once the distribution function for an observable A is known one can calculate the "*ensemble average*" $\langle a \rangle$ of A over all members of the ensemble and the mean square deviation from the average, or the "*variance*" of A

$$\sigma_a^2 \equiv \langle (a - \langle a \rangle)^2 \rangle = \langle a^2 \rangle - (\langle a \rangle)^2 \tag{13.1.2}$$

from the equations

$$\left.\begin{aligned} \langle a(t) \rangle &= \int a f(a, t)\,da, \\ \sigma_a^2(t) &= \int a^2 f(a, t)\,da - (\langle a(t) \rangle)^2. \end{aligned}\right\} \tag{13.1.3}$$

In a stationary ensemble the quantities $\langle a \rangle$ and σ_a^2 are constants. It is plausible to identify $\langle a \rangle$ with the strictly fixed value of the parameter A for the system in equilibrium according to *macroscopic*, phenomenological theory. The quantity σ_a^2 then measures the deviation to be expected from this value, due to the *microscopic* mechanism of the system.

Let us now consider a system whose energy E is kept constant. Its classical Hamiltonian function H then satisfies the canonical Hamiltonian equations (1.1.12)

$$\dot{q}_i = \frac{\partial H}{\partial p_i}, \quad \dot{p}_i = -\frac{\partial H}{\partial q_i} \quad (i = 1, 2, \ldots, N), \tag{13.1.4}$$

from which follows

$$\frac{\partial \dot{q}_i}{\partial q_i} + \frac{\partial \dot{p}_i}{\partial p_i} = 0 \quad (i = 1, 2, \ldots, N). \tag{13.1.5}$$

In the present notation these equations take the form

$$\sum_{j=1}^{2N} \frac{\partial \dot{x}_j}{\partial x_j} = 0. \tag{13.1.6}$$

The relative number density ϱ of the phase points of an ensemble of such systems in phase space must satisfy a "continuity equation" which is the $2N$-dimensional generalization of the three-dimensional continuity equation of hydrodynamics (which has the same form as the continuity equation (6.2.14) of electrodynamics). Thus the relation

$$\frac{\partial \varrho}{\partial t} + \sum_{j=1}^{2N} \frac{\partial (\varrho \dot{x}_j)}{\partial x_j} = \frac{\partial \varrho}{\partial t} + \varrho \sum_{j=1}^{2N} \frac{\partial \dot{x}_j}{\partial x_j} + \sum_{j=1}^{2N} \frac{\partial \varrho}{\partial x_j} \dot{x}_j = 0 \qquad (13.1.7)$$

must hold.

From (13.1.6) and (13.1.7) one obtains

$$\frac{d\varrho}{dt} \equiv \frac{\delta \varrho}{\partial t} + \sum_{j=1}^{2N} \frac{\partial \varrho}{\partial x_j} \frac{dx_j}{dt} = \frac{\partial \varrho}{\partial t} + \sum_{j=1}^{2N} \frac{\partial \varrho}{\partial x_j} \dot{x}_j = 0. \qquad (13.1.8)$$

This is "*Liouville's theorem*". It means that the density ϱ in the neighbourhood of the phase point of any of the members of the ensemble remains constant while it is moving along its phase curve which lies entirely on the hypersurface $H = E$.

We shall now relax the latter condition and assume that the energy of the members of a *stationary* ensemble is restricted to values between E and $E + \delta E$. In this case all the phase curves must be contained in the corresponding energy shell. It is then possible to derive an important theorem concerning the distribution function ϱ of the ensemble from Liouville's theorem if an additional hypothesis, the so-called "*quasi-ergodic hypothesis*" (due to Boltzmann and Ehrenfest) is introduced. This maintains that the phase curve of any member of a stationary ensemble will, in the course of time, come arbitrarily near to any point within the accessible region of phase space. This assertion cannot be proved on the basis of classical dynamics and must, like the law of great numbers, be regarded as an independent law of nature within the framework of classical statistical mechanics.

If the ergodic hypothesis is accepted it follows at once from Liouville's theorem that ϱ must have a *constant* value throughout the whole energy shell, provided the ensemble is stationary. Vice versa it follows from the equation (13.1.8) that, if ϱ is independent of the coordinates within the energy shell, one must have: $\partial\varrho/\partial t = 0$, that is the ensemble must be stationary.

If one and the same system is repeatedly started from the same state at $t = t_0$, corresponding to a point with position vector $\mathbf{r}_0$ in phase space, the functions $\mathbf{r}(t)$ for $t > t_0$ will, because of the randomness of the above-mentioned external influences, in general be different from each other, and the corresponding phase curves will not coincide. The sequence of states through which the system proceeds in an individual experiment is therefore to a certain extent subject to chance. Such a sequence is called a "*stochastic time series*" ("*Zeitgesamtheit*" in German literature), and instead of considering an ensemble of systems one can apply statistical considerations to the time series of an individual system. It is indeed possible to develop statistical mechanics on the basis of the statistics of stochastic time series alone, as was first attempted by Einstein in 1902, independently of Gibbs.

In particular one can define a "*time distribution function*" $\theta(\mathbf{r})$ for a stochastic time series in the following way: let τ be the time, within a total time T of observation, which the phase point of the system spends within a certain volume δV of the phase space, surrounding a position $\mathbf{r}$. The relative "*time of sojourn*" τ/T will then measure the probability of finding the system, at any randomly chosen moment within T, in a state corresponding to a point within δV. The time distribution function is then defined by

$$\theta(\mathbf{r}) = \lim_{T\to\infty,\delta V\to 0} (\tau/T\delta V). \tag{13.1.9}$$

A time series is said to be *stationary* if $\theta(\mathbf{r})$ is independent of t_0, that is if $\theta(\mathbf{r})$ is invariant with respect to a time translation. It follows that an ensemble of systems whose time series are stationary must also be stationary. But the stationarity of an ensemble does not necessarily imply that the time series of its members be stationary, unless the ergodic hypothesis holds, according to which each of the time series must cover all states the system is capable of. Under these circumstances $\theta(\mathbf{r})$ becomes identical with $\varrho(\mathbf{r})$ and the time series is then called "*ergodic*".

The theorem that the time distribution function of the ergodic time series of a system is identical with the distribution function of a stationary ensemble of the same systems enables one to calculate the *time averages* of the macroscopic parameters of the system and the magnitude of the *fluctuations in time* of these parameters about their averages observed on an individual system, by means of the formulae (13.1.3). It further follows from this theorem that for an *isolated* system, whose phase point is restricted to an energy shell, θ is a constant, and hence the time of sojourn of the system in a volume V of the energy shell is simply proportional to that volume.

13.2. Basic notions of quantum statistical mechanics

Whereas in classical mechanics the state of a system of N degrees of freedom can be precisely defined by the set of values $q_1, \ldots, q_N$ of the generalized coordinates and the set of values $p_1, \ldots, p_N$ of the generalized momenta, Heisenberg's uncertainty principle of quantum mechanics (3.2.13) maintains that the combined values of any pair q_i, p_i of coordinates and momenta can only be determined with an uncertainty of the order of Planck's constant h:

$$\Delta p_i \Delta q_i \simeq h \quad (i = 1, 2, \ldots, N). \tag{13.2.1}$$

Hence the state of the system can only be defined with the overall uncertainty

$$\Delta x_1, \Delta x_2, \ldots, \Delta x_{2N} \simeq h^N \tag{13.2.2}$$

of the quantities $x_1, \ldots, x_{2N}$. This is equivalent to saying that the phase space of the system is divided into elementary "cells" of volume h^N, which may be numbered in arbitrary sequence, each cell representing a quantum state with a set of N quantum numbers. The state of the system is then completely defined at any particular time t if it is known in which of these cells its phase point is situated at that time.

The concept of a discontinuous cell structure of the phase space with cells of finite size was first introduced into statistical mechanics by Planck on the basis of the ideas of the older quantum theory. This can be best understood when one considers a system with one degree of freedom whose phase space is a plane with coordinates q, p. The classical phase curves of such a system in its stationary quantum states are closed lines which, according to Sommerfeld's quantum conditions (2.2.14), include areas that are integer multiples of h. Thus we can imagine the cell boundaries to be defined by the phase curves for two neighbouring quantum states; the cells will then all have areas h. For a simple rotator the cells will be rectangular in shape, and for a harmonic oscillator they will be bordered by concentric ellipses, as shown by Fig. 28.

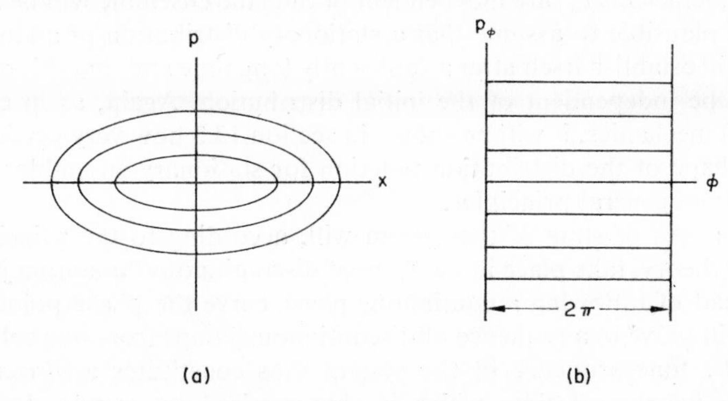

FIG. 28. Cell structure of the two-dimensional phase-space of (a) a harmonic oscillator, (b) a simple rotator.

An ensemble of identical systems will again be represented by the distribution function of the relative numbers of phase points in the various cells which now, however, is a discontinuous function $\varrho_k(t)$ of the cell index k at time t. From an obvious generalization of the considerations in section 6.1 we may identify the quantities ϱ_k with the squares of the magnitudes of the expansion coefficients c_k of the wave function $\varphi\,(q_1, \ldots, q_N; t)$ of the system according to a formula of the type (6.1.16)

$$\varphi\,(q_1, \ldots, q_N, t) = \sum_k c_k(t)\,\varphi_k, \tag{13.2.3}$$

where the φ_k are the wave functions belonging to the eigenvalues E_k of the system in its stationary states which form an orthonormal set.

If one now assumes that at $t = 0$ all the members of the ensemble are in the same quantum state $k = 1$ so that all the coefficients except c_1 are zero, one can, at least in principle, solve the wave equation under this initial condition for all subsequent t. The expansion of φ then yields the coefficients c_{1j}, and from these one obtains the density distribution of the ensemble at time t: $\varrho_j(t) = |c_{1j}(t)|^2$. Once the matrix of the coefficients c_{kj} for all combinations of indices k, j has been evaluated in this way the general problem of constructing the distribution function $\varrho_k(t)$ at time t, when the distribution function $\varrho_k(0)$ at time $t = 0$ is given, is solved.

We have shown in section 13.1 that in classical statistical theory a stochastic differential equation has to be set up which, although connected with the Hamiltonian mechanical equations of motion of the system, cannot be derived entirely from them without the introduction of notions of randomness, alien to classical dynamics. In quantum statistical mechanics the "stochastic" differential equation is identical with the fundamental quantum mechanical wave equation itself. This demonstrates clearly the superiority of the fundamental concepts of quantum statistics over classical statistical theory.

If the coefficients c_{kj} are independent of time the ensemble will be stationary. It is plausible to assume that a stationary distribution of a closed ensemble will establish itself after a sufficiently long time and that this distribution will be independent of the initial distribution. Again, as in classical statistical mechanics, it will be shown in section 13.3 how very general laws for the shape of the distribution function for stationary ensembles can be derived from general principles.

The changes of state of the system will, according to the principles of quantum theory, take place in the form of discontinuous "quantum jumps", and instead of following a continuous phase curve the phase point of the system will move in a sequence of discontinuous jumps from one cell to another. The time sequence of the system thus constitutes a *discontinuous stochastic function* of time which is characterized by certain "*transition probabilities*". From what has been said in section 6.1 it appears that the quantities $|c_{kj}(t)|^2$ can in fact also be interpreted as the transition probabilties for the movement of the phase point of an *individual* member of the ensemble from its position in the kth cell at $t = 0$ to the jth cell in time t. The matrix of the c_{kj} is therefore called the "*transition probability matrix*".

If the external as well as the internal disturbances which give rise to the transitions are of a random character, the transition probabilities are, for short periods of time, proportional to t and we can thus, as already explained in section 6.1, introduce time independent "*transition probability rates*" w_{kj} such that $w_{kj}\,dt$ measures the probability for the phase point to move from cell k to cell j in time dt.

It will now be shown that for conservative systems the matrix of the w_{kj} is symmetrical, that is that for all pairs of values of k and j

$$w_{kj} = w_{jk}. \tag{13.2.4}$$

This is called the "*principle of microscopic reversibility*", due to Onsager (1931).

In classical mechanics all changes occurring in conservative systems are strictly reversible. This follows immediately from the Hamiltonian equations (1.1.12); for if the sign of t in these equations is reversed and simultaneously the signs of the momenta p_i, they remain invariant because the Hamiltonian function H is of the second order in the p_i. It follows that if at a certain moment all velocities are reversed the system will move precisely backward on the same trajectory.

Similarly it was shown by the argument at the end of section 6.3 that in quantum mechanics conservative systems are also strictly reversible. From this fact the validity of the principle of microscopic reversibility, as expressed by (13.2.4), follows at once.

The relative number of transitions from cell k to cell j during a time interval dt in an ensemble of identical conservative systems is, in virtue of the definition of the transition probability rates, equal to $\varrho_k w_{kj}\, dt$, and the number of transitions in the opposite direction is equal to $\varrho_j w_{jk} dt$. Thus it follows for the change $d\varrho_k$ of ϱ_k in time dt,

$$d\varrho_k = -\sum_j \varrho_k w_{kj}\, dt + \sum_j \varrho_j w_{jk}\, dt. \tag{13.2.5}$$

Combining the equations (13.2.4) and (13.2.5) we obtain

$$\frac{d\varrho_k}{dt} = \sum_j (\varrho_j - \varrho_k)\, w_{kj}. \tag{13.2.6}$$

It is seen immediately from (13.2.6) that $d\varrho_k/dt = 0$ if all the ϱ_k are equal, or that the ensemble is stationary if the density of the phase points is constant within the energy shell. This corresponds exactly to the result obtained in the preceding section by means of Liouville's theorem in classical statistical mechanics. Vice versa it can also be proved that for an ensemble to be stationary the equality of all the ϱ_k is a necessary condition. For if the ϱ_k are not all equal at any particular time then at least one, say ϱ_r, must have the largest value. It follows that all terms on the right-hand side of equation (13.2.6) must be negative for $k > r$, because the quantities w_{kj} are essentially positive, and hence $d\varrho_r/dt$ cannot vanish, and the ensemble cannot be stationary unless there is no largest value of the ϱ_k. Similarly, if follows that the ensemble cannot be stationary unless there is no smallest value of the ϱ_k. Thus in order that the ensemble be stationary all the ϱ_k must be equal to one another.

Whereas, as we have shown in section 13.1, it was necessary to make use of the ergodic hypothesis in order to prove the theorem of the constancy of the phase point density within the energy shell of a stationary system in classical statistical mechanics, it appears now that no such hypothesis needs to be introduced into quantum statistical mechanics; the theorem is seen to be a consequence of the fundamental principles of quantum mechanics.

We now turn to the problem of calculating the ensemble averages of observable quantities in quantum statistics. It was explained in section 7.1 that in quantum mechanics an "observable "A in a certain system is defined by a differential operator $\mathscr{A}$ in the generalized coordinates of the system, with discrete eigenvalues a_n, which are the solutions of differential equations of the type (7.1.9). The (in general time-dependent) expectation value $\bar{A}$ of A for repeated measurements under indentical conditions is given by the relation (7.1.19)

$$\bar{A}(t) = (\varphi \cdot \mathscr{A}\varphi), \tag{13.2.7}$$

where the right-hand member is the inner product of the functions φ and $\mathscr{A}\varphi$, according to the definition (4.1.31), and φ is the wave function of the system. Instead of making repeated measurements on an individual system we can, of course, also make measurements on an ensemble of identical systems, so that the expectation value $\bar{A}(t)$ also represents the ensemble average $\langle A(t) \rangle$ of the observable A.

According to the expansion theorem (4.2.8) the expression (13.2.3) for $\varphi(q_1, \ldots, q_N; t)$ remains valid for any particular time if the φ_k are replaced by an arbitrary orthonormal set of independent functions $\psi_k(q_1, \ldots, q_N)$ and the coefficients c_k are certain functions of time. We can then define quantities

$$\varrho_{kj} = c_j^* c_k \tag{13.2.8}$$

forming a matrix $\mathfrak{R}$, called the "*density matrix*" of the system. It follows at once from the definition (13.2.8) that

$$\varrho_{jk} = \varrho_{kj}^*, \tag{13.2.9}$$

which means that the density matrix is Hermitian and therefore, as pointed out in section 7.1, represents an observable in the quantum mechanical sense. The diagonal elements of the density matrix are

$$\varrho_{kk} = (c_k)^2, \tag{13.2.10}$$

and it follows from the orthonormality of the functions ψ_k that the sum of the elements ϱ_{kk} (usually denoted as the "*trace*" of the matrix) is equal to unity:

$$\sum_k \varrho_{kk} \equiv \mathrm{Tr}\,\mathfrak{R} = 1, \tag{13.2.11}$$

independently of the choice of the functions ψ_k.

From (13.2.3) and (13.2.7) we now obtain for the ensemble average $\langle A \rangle$ of the observable A:

$$\langle A \rangle = \sum_{kj} ([c_k\psi_k] \cdot [\mathscr{A} c_j\psi_j]) = \sum_{kj} c_k^* c_j (\psi_k \cdot \mathscr{A}\psi_j). \tag{13.2.12}$$

Now according to the definition (7.1.13) the matrix $\mathfrak{A}$ that represents the operator $\mathscr{A}$ is defined by the following set of relations for its elements:

$$a_{kj} = (\psi_k \cdot \mathscr{A}\psi_j).$$

Making use of these relations and the relation (13.2.8) we obtain further from (13.2.12) and the rule (4.3.2) for the multiplication of matrices

$$\langle A \rangle = \sum_{kj} \varrho_{jk} a_{kj} = \mathrm{Tr}\,(\mathfrak{R}\mathfrak{A}). \tag{13.2.13}$$

If in particular the functions ψ_k are the eigenfunctions of the operator $\mathscr{A}$ according to (7.1.20) the matrix $\mathfrak{A}$ is diagonal and its elements are the eigenvalues a_k of $\mathscr{A}$. Thus, from (13.2.10) and (13.2.13),

$$\langle A(t) \rangle = \sum_n \varrho_{nn}(t)\, a_n = \sum_n |c_n|^2 a_n. \tag{13.2.14}$$

This relation is identical with (7.1.27) and shows, in combination with (13.2.11), that under the above specified circumstances the diagonal elements ϱ_{nn} of the density matrix are the probabilities of finding a specified member of the ensemble in a state with eigenvalue a_k of the observable A at time t, or in other words, they are the relative numbers of phase points to be found in a region in phase space representing the condition $a = a_n$. The quantity ϱ_{nn} is thus the quantum mechanical analogue of the classical quantity (13.1.1), and equation (13.2.14) is the quantum analogue of the first of the classical relations (13.1.3). If for example $\mathscr{A}$ is the Hamiltonian operator $\mathscr{H}$, then the a_n are the energy levels E_n, and ϱ_{nn} are the relative numbers of phase points in the "energy shells" E_n of the phase space.

13.3. Special stationary ensembles

In the preceding sections we have already considered one particular case of a stationary ensemble which is defined by the condition that the energies of the members of the ensemble should be confined to values between E_0 and $E_0 + \delta E$, which means that the corresponding phase points are restricted to an energy shell between the hypersurfaces $E = E_0$ and $E = E_0 + \delta E$. We have shown that the relative phase point density ϱ is then constant within the shell. Following the nomenclature of Gibbs we shall call such an ensemble a "*microcanonical ensemble*" (for reasons which will be made clear later in this section).

We wish to calculate the microcanonical ensemble average $\langle a \rangle_m$ of an observable quantity A under "classical" conditions. It follows then from (13.1.1) that in this case

$$f(a)\,\delta a = \varrho\,\delta\omega, \tag{13.3.1}$$

where now $\delta\omega$ is the $2N$-dimensional volume of that part of the energy shell which corresponds to the range δa of a. Further, from the first of the relations (13.1.3) and from the condition $\varrho\omega = 1$,

$$\langle a \rangle_m = \frac{1}{\omega} \int \cdots_{2N} \int a\, d\omega. \tag{13.3.2}$$

Now the "thickness" of the shell is $\delta n = \frac{dn}{dE}\,\delta E$ so that the volume of the shell belonging to a $(2N-1)$-dimensional surface element ds of the shell is

$$ds\,\delta n = ds\,\frac{dn}{dE}\,\delta E.$$

Hence, finally, from (13.3.2)

$$\langle a \rangle_m = \frac{\delta E}{\omega} \int \underset{2N-1}{\cdots} \int a(\mathbf{r})\,\frac{dn}{dE}\,ds, \tag{13.3.3}$$

where the integral extends over the whole surface of the energy shell and dE/dn is the magnitude of the gradient of E in the direction of the normal to the shell surface.

Let us now in particular consider the phase function

$$a_i = x_i\,\frac{\partial H}{\partial x_i}, \tag{13.3.4}$$

where $H(x_1, \ldots, x_{2N})$ is the Hamiltonian function of the system. Substituting from (13.3.4) into (13.3.3) we obtain for the ensemble average of a_i

$$\langle a_i \rangle_m = \frac{\delta E}{\omega} \int \underset{2N-1}{\cdots} \int x_i\,\frac{\partial H}{\partial x_i}\,\frac{1}{|\nabla H|}\,ds. \tag{13.3.5}$$

$\frac{\partial H}{\partial x_i}\frac{1}{|\nabla H|}$ is evidently the cosine of the angle between the direction of the x_i-axis and the normal to the shell surface at any particular point of that surface. Thus $\frac{\partial H}{\partial x_i}\frac{ds}{|\nabla H|}$ is the projection of the surface element ds upon the "hyperplane" that is perpendicular to the x_i-axis. Hence the integral in (13.3.5) is simply equal to the total volume Ω surrounded by the energy shell, and we obtain the relation

$$\langle a_i \rangle_m = \delta E\,\frac{\Omega}{\omega}, \tag{13.3.6}$$

which shows that the microcanonical ensemble averages of the quantities defined by (13.3.4) are all equal to each other, depending only on the value of E_0.

If x_i is one of the generalized momenta p_i of the system, it follows, with the help of (1.1.10) and (1.1.12), that

$$\langle a_i \rangle_m = \left\langle p_i\,\frac{\partial H}{\partial p_i} \right\rangle_m = \langle p_i \dot{q}_i \rangle_m = \left\langle \dot{q}_i\,\frac{\partial K}{\partial \dot{q}_i} \right\rangle_m, \tag{13.3.7}$$

K being the total kinetic energy of the system. As K is a homogeneous function of the second order in the generalized velocities $\dot{q}_i$ one has

$$K = \frac{1}{2} \sum_i \dot{q}_i \frac{\partial K}{\partial \dot{q}_i} = \sum_i K_i, \tag{13.3.8}$$

where K_i is that part of the kinetic energy of the system which belongs to the ith generalized coordinate. Thus from (13.3.6), (13.3.7), and (13.3.8)

$$\langle K_i \rangle_m = \delta E \frac{\Omega}{2\omega}. \tag{13.3.9}$$

This relation expresses the fact that all the K_i are equal and depend on E_0 and δE only. This is called the "*equipartition theorem*" (due to Boltzmann) which is valid for an arbitrary choice of the generalized coordinates used, provided that "classical" conditions prevail, that is that the Hamiltonian function is a continuous function of the generalized coordinates and momenta. It will be shown at the end of this section that the equipartition theorem does *not* hold in quantum statistics where this condition is not satisfied.

According to the argument in the last paragraph of section 13.1. we can identify the ensemble averages $\langle K_i \rangle$ with the *time* averages $\overline{K}_i$ for individual members of the ensemble over long periods of observation, and the equipartition theorem then maintains that these time averages should be equal for all degrees of freedom. If, for example, the system consists of a "colloidal solution" of small, but microscopically visible particles, suspended in a liquid, then according to the equipartition theorem the mean kinetic energies of each of the six components of the translational and the rotational movement of all the particles should have the same value, irrespective of the size of the particles, namely that of the mean kinetic energy per degree of freedom of the molecules of the surrounding liquid. The resulting irregular translational and rotational movement of colloidal particles is known as "*Brownian movement*". The theory of this phenomenon was, on the basis of the equipartition theorem, developed independently by Einstein and Smoluchowski (1905–6).

If x_i is one of the generalized coordinates q_i of the system, one has, instead of (13.3.7), on account of (1.1.11) and (1.1.12),

$$\langle a_i \rangle_m = \left\langle q_i \frac{\partial H}{\partial q_i} \right\rangle_m = \left\langle q_i \frac{\partial U}{\partial q_i} \right\rangle_m = -\langle q_i F_i \rangle_m, \tag{13.3.10}$$

where $U(q_1, \ldots, q_N)$ is the potential energy function of the system and F_i are the components of the generalized forces.

In the special case when the F_i are "elastic" forces, that is when U is a homogeneous function of the second order in the q_i one has, in analogy to (13.3.8)

$$U = \frac{1}{2} \sum_i q_i \frac{\partial H}{\partial q_i} = \sum_i U_i, \tag{13.3.11}$$

and hence

$$\langle U_i \rangle_m = \delta E \frac{\Omega}{2\omega}. \tag{13.3.12}$$

This means that the ensemble averages of the potential energies belonging to the ith degree of freedom are equal to the ensemble averages of the kinetic energies belonging to the same degree of freedom, so that the total mean energies E_i per degree of freedom are all equal to one another. This is an extension of the equipartition theorem which, for example, is valid for systems of the character of elastic solid bodies. We shall make use of this theorem later in sections 15.2 and 15.3.

Generally it follows from (13.3.6), (13.3.8), (13.3.9), and (13.3.10) that

$$\left\langle \sum_i (q_i F_i) \right\rangle_m = -2\langle K \rangle_m. \tag{13.3.13}$$

The quantity on the left-hand side of this equation is a generalization of the so-called "*virial*", which was originally introduced into the kinetic theory of gases by Clausius, and the relation (13.3.13) is called the "*virial theorem*", according to which the virial of a system is equal and opposite in sign to twice the mean kinetic energy of the system.

We now drop the condition that the energy of each member of the ensemble be restricted to a small energy range δE and assume that it can freely exchange energy with its surroundings. At any moment we can then subdivide the ensemble into sub-ensembles, each belonging to a different energy range between E_i and $E_i + \delta E$. In phase space such an ensemble will be represented by a density distribution $\varrho(\mathbf{r})$ of phase points in which the phase points of the various sub-ensembles occupy energy shells which enclose each other like onion shells. If such an ensemble is to be stationary the distribution functions ϱ_i for the sub-ensembles must all be independent of time. Each of the sub-ensembles must therefore be microcanonical, and it follows from what has been said before that ϱ_i is a function of the energy E_i alone: $\varrho_i = \varrho(E_i)$. An ensemble which satisfies this condition is, after Gibbs, called a "*canonical ensemble*" and the function $\varrho(E)$ is called the "*canonical distribution function*". The fact that a canonical ensemble is made up of microcanonical sub-ensembles is the reason for the choice of the descriptive title for the latter.

It will now be shown that the energy distribution function of a canonical ensemble is a *universal* function which is independent of the particular mechanism of the system. For this purpose we consider a system whose energy is confined to the energy region between E and $E + \delta E$, and which is composed of n weakly interacting identical sub-systems. Let $\varrho^{(l)}(E^{(l)})$ be the energy distribution function for the lth sub-system, $E^{(l)}$ being its energy, so that $\varrho^{(l)}(E^{(l)})\, dv^{(l)}$ is the probability of finding that system in a state corresponding to an infinitesimal range $dv^{(l)}$ of its set of coordinates $x^{(l)}$. Thus the probability of finding the whole system in a state corresponding to an infinitesimal range dv of the whole set of coordinates $x_1 \cdots x_{2N}$ is

$$\prod_l \varrho^{(l)}(E^{(l)})\, dv^{(l)},$$

because the systems are practically independent.

On the other hand, as the whole system is a member of a microcanonical ensemble, this probability must also be equal to $\varrho(E)\, dv$ so that the relation

$$\prod_l \varrho^{(l)}(E^{(l)})\, dv^{(l)} = \varrho(E)\, dv \tag{13.3.14}$$

must hold. But since

$$\prod_l dv^{(l)} = dv \tag{13.3.15}$$

and

$$\sum_l E^{(l)} = E, \tag{13.3.16}$$

we have further

$$\prod_l \varrho^{(l)}(E^{(l)}) = \varrho\left(\sum_l E^{(l)}\right). \tag{13.3.17}$$

Taking logarithms on both sides of equation (13.3.17) we obtain the relation

$$\sum_l \ln \varrho^{(l)}(E^{(l)}) = \ln \varrho\left(\sum_l E^{(l)}\right). \tag{13.3.18}$$

This is a functional equation of the type

$$\sum_l f_l(x_l) = f\left(\sum_l x_l\right),$$

which can only be satisfied if the f_l are linear functions of their arguments. Hence the solution of (13.3.18) must have the form

$$\ln \varrho^{(l)}(E^{(l)}) = \alpha_l - \frac{E^{(l)}}{\theta}, \tag{13.3.19}$$

where the α_l and θ are constants. Thus one obtains finally for the canonical distribution function of the lth subsystem the expression

$$\varrho^l(E^{(l)}) = C_l \exp(-E^{(l)}/\theta). \tag{13.3.20}$$

Formula (13.3.20) must hold quite generally for any canonical ensemble because we can always imagine its members to be sub-systems of a larger system, the rest of the latter constituting the "surroundings" with which the system in question can freely exchange energy, according to the above formulated definition of a canonical ensemble. The constant θ, called the "*modulus*" of the ensemble, is then a characteristic of the surroundings whose physical meaning will be discussed in the following section.

Under classical conditions $E = H(\mathbf{r})$ is the continuous Hamiltonian function of the generalized coordinates and momenta of the system, and the canonical distribution function is then expressed by the continuous density function in phase space:

$$\varrho(\mathbf{r}) = C \exp(-H(\mathbf{r})/\theta), \tag{13.3.21}$$

where C is a constant which is determined by the normalizing condition for ϱ

$$\int \cdots \int \varrho(x_1, \ldots, x_{2N})\, dV = 1, \tag{13.3.22}$$

the integration extending over the whole phase space, and the modulus θ is a characteristic of the medium surrounding the system.

From (13.3.21) it follows immediately that the relative number of members of a canonical ensemble, whose energies lie in the narrow range between E and $E + \delta E$, is equal to

$$f(E)\,\delta E = C\omega(E) \exp(-E/\theta) \tag{13.3.23}$$

where, as before, ω is the volume of the relevant energy shell. The function $f(E)$ is called the "*energy distribution function*".

Under quantum conditions the canonical distribution law takes the discontinuous form

$$\varrho_k = C \exp(-E_k/\theta), \tag{13.3.24}$$

where, as in section 13.2, ϱ_k is the density of phase points in the cell with index k (k representing a set of N quantum numbers) and E_k the corresponding energy. The constant C is now determined by the condition

$$\sum_k \varrho_k = 1. \tag{13.3.25}$$

The energy distribution function f_n, that is the relative number of members of the ensemble which belong to the quantum energy level E_n, is obtained by multiplying the right-hand side of (13.3.24) by the number g_n of quantum states which belong to one and the same "degenerate" energy eigenvalue of the system:

$$f(n) = \frac{g_n \exp(-E_n/\theta)}{\sum_n g_n \exp(-E_n/\theta)}. \tag{13.3.26}$$

The numbers g_n are called the "*statistical weights*" of the energy states E_n.

Under classical conditions the canonical ensemble average of any quantity A which is a function of the coordinates $x_1, \ldots, x_{2N}$ is obtained from (13.3.21) according to the definition

$$\langle a \rangle_c = \frac{\int \cdots \int a(\mathbf{r}) \exp(-H(\mathbf{r})/\theta)\, dV}{\int \cdots \int \exp(-H(\mathbf{r})/\theta)\, dV}. \tag{13.3.27}$$

We shall use this formula for calculating the canonical ensemble average of one of the quantities x_i, say x_s, on which the Hamiltonian of the system depends quadratically, that is, when H can be written in the form

$$H(\mathbf{r}) = \alpha x_s^2 + H'(x_1, \ldots, x_{s-1}, x_{s+1}, \ldots, x_{2N}). \tag{13.3.28}$$

Using the abbreviation $\beta = \alpha/\theta$ we obtain from (13.3.27) and (13.3.28)

$$\langle x_s^2 \rangle_c = \frac{\int_{-\infty}^{+\infty} x_s^2 \exp(-\beta x_s^2)\, dx_s}{\int_{-\infty}^{+\infty} \exp(-\beta x_s^2)\, dx_s}.$$

The integral I in the denominator has the form of a Laplace integral and is easily evaluated to

$$I = \sqrt{\pi/\beta}$$

so that

$$\langle x_s^2 \rangle_c = -\frac{dI}{d\beta}\frac{1}{I} = -\frac{d(\ln I)}{d\beta} = \frac{1}{2\beta}.$$

Thus

$$\langle x_s^2 \rangle_c = \frac{\theta}{2\alpha}, \tag{13.3.29}$$

and the part $\langle E_s \rangle_c$ of the average total energy of the system belonging to the parameter x_s is equal to

$$\langle E_s \rangle_c = \alpha \langle x_s^2 \rangle_c = \theta/2. \tag{13.3.30}$$

If x_s is one of the generalized *momenta* of the system, then E_s is the kinetic energy of the system belonging to that momentum and the relation (13.3.30) again expresses the classical equipartition theorem which is thus seen to hold for canonical ensembles as well as for microcanonical ensembles. If x_s is a generalized *coordinate* on which the potential energy of the system depends quadratically, one obtains the analogue of the theorem expressed by (13.3.12) for a microcanonical ensemble, and we see that for each degree of freedom of a system which behaves like a harmonic oscillator the total average energy is simply equal to θ.

Under quantum conditions one obtains the canonical ensemble average of an observable which is defined by an operator $\mathscr{A}$ from the following consideration. Let $\mathscr{H}$, as usual, be the Hamiltonian operator. We can now define another operator

$$\mathscr{T} \equiv \exp(-\mathscr{H}/\theta) \equiv \sum_{r=0}^{\infty} \frac{1}{r!}[-\mathscr{H}/\theta]^r, \tag{13.3.31}$$

and represent this operator by a matrix $\mathfrak{T}$ as discussed in section 7.1. We can then replace the canonical distribution law (13.3.21) by a matrix equation of the form

$$\mathfrak{R} = C\mathfrak{T}, \tag{13.3.32}$$

in which $\mathfrak{R}$ is the density matrix in that particular representation and C a normalizing factor. If, as before, $\mathfrak{A}$ is the matrix representing the operator $\mathscr{A}$

in the same representation one obtains the canonical ensemble average $\langle A \rangle_c$ of A from the previously derived relation (13.2.13).

If in particular, the representation chosen is in terms of the eigenfunctions of $\mathscr{A}$ then, as pointed out in section 7.2, the matrix $\mathfrak{A}$ is diagonal with matrix elements a_n which are the eigenvalues of $\mathscr{A}$, and we can evaluate $\langle A \rangle_c$ from (13.2.14), where ϱ_{nn} are the diagonal elements of the matrix $\mathfrak{R}$. In the simplest case, when $\mathscr{A} = \mathscr{H}$, the a_n are the energy eigenvalues E_n, and the matrix $\mathfrak{R}$ becomes also diagonal with elements f_n which are given by (13.3.26). Thus the ensemble average of the energy is obtained from the relation

$$\langle E \rangle_c = \frac{\sum\limits_n E_n g_n \exp(-E_n/\theta)}{\sum\limits_n g_n \exp(-E_n/\theta)}. \tag{13.3.33}$$

This shows that, in contrast to the classical relation (13.3.27), $\langle E \rangle_c$ is not determined by the Hamiltonian function $H(\mathbf{r})$ but depends explicitly only on the eigenvalues E_n, and that the equipartition theorem does not hold in quantum statistical mechanics.

13.4. Derivation of the laws of thermodynamics

The classical laws of thermodynamics are of a phenomenological nature and quite unrelated to those of classical mechanics. But on the basis of the molecular theory of matter it ought to be possible to derive the former from the latter. This meets with a grave difficulty; for whereas, as we have seen, the fundamental laws of classical as well as quantum mechanics demand strict reversibility of all mechanical processes, thermal phenomena are in general not reversible. For example, as was pointed out in section 6.3, the classical diffusion equation (6.3.17) indicates the irreversibility of diffusion processes, and the same holds for the process of heat conduction. More generally, the classical theorem of the increase of entropy expresses most comprehensively the irreversible nature of the laws of thermodynamics.

This difficulty is resolved if, as explained in the foregoing section, the thermodynamic parameters are interpreted as expectation values of observables, or alternatively as ensemble averages of the observables. For then the laws of thermodynamics, referring to these parameters, are no longer strictly causal but are intrinsically statistical in character. Thus under conditions of stationarity the expectation values of the thermodynamic parameters will be independent of time, which corresponds to "thermodynamic equilibrium", and under non-stationary conditions the macroscopic changes will take place in such a way that a state of smaller expectation value will be followed irreversibly by a state of larger expectation value although, as was shown before, the microscopic changes are governed by the principle of microscopic reversibility.

We are now going to derive the three fundamental theorems of classical thermodynamics from the principles laid down in the previous sections of this chapter, following an argument due to Einstein (1903). For this purpose we first have to give the statistical mechanical definition of the quantities "*internal energy*" U, "*heat energy*" Q, and "*work*" W, appearing in the first law of thermodynamics.

We assume that the member systems of the ensemble are under "external constraint" which is described by certain macroscopic external parameters ξ_r. The classical Hamiltonian function H of the system will then not only depend on the internal microscopic parameters $x_1, \ldots, x_{2N}$ but also on the external parameters $\xi_1, \ldots, \zeta_s$. A change δE of the energy content of the system is then partly due to the changes δx_i of the internal parameters x_i and partly to the changes $\delta \xi_r$ of the external parameters. Thus we have

$$\langle \delta E \rangle = \left\langle \sum_{i=1}^{2N} \frac{\partial H}{\partial x_i} \delta x_i \right\rangle + \left\langle \sum_{n=1}^{s} \frac{\partial H}{\partial \xi_r} \delta \xi_r \right\rangle . \tag{13.4.1}$$

The ensemble average $\langle \delta E \rangle$, which is the expectation value of the increase of energy of the system, is evidently to be identified with the increase δU of the internal energy. The first term on the right-hand side of (13.4.1) is then that part of the increase of energy of the system which occurs while the external parameters are kept constant; it is therefore to be identified with the supplied heat energy δQ. The second term, which represents the increase of energy of the system when no heat is supplied to it, must then be the work δW, done on the system by the change of the external parameters. Thus (13.4.1) is the statistical form of the *first law of thermodynamics*

$$\delta U = \delta Q + \delta W. \tag{13.4.2}$$

We now consider in particular a canonical ensemble with the distribution function (13.3.24) which satisfies the condition (13.3.25). We can write these relations in a different form, namely

$$\varrho_k = \exp\left(\frac{\psi - E_k}{\theta}\right), \tag{13.4.3}$$

$$\sum_k \varrho_k = \sum_k \exp\left(\frac{\psi - E_k}{\theta}\right) = 1, \tag{13.4.4}$$

where the quantities ψ and θ are independent of the internal parameters x_i but will in general be functions of the external parameters ξ_r.

Small variations of the external parameters produce corresponding changes $\delta\psi$ and $\delta\theta$ of ψ and θ which, in virtue of (13.4.3) and (13.4.4), must satisfy the relation

$$\sum_k \delta\left(\frac{\psi - E_k}{\theta}\right) \exp\left(\frac{\psi - E_k}{\theta}\right) = 0$$

or

$$\sum_k \delta\left(\frac{\psi - E_k}{\theta}\right) \varrho_k = 0. \tag{13.4.5}$$

The sum in this equation is by definition the canonical ensemble average of the quantity $\delta\left(\frac{\psi - E_k}{\theta}\right)$. Thus we obtain further

$$\left\langle \delta\left(\frac{\psi - E_k}{\theta}\right)\right\rangle_c = \delta\left(\frac{\psi}{\theta}\right) - \delta\left(\frac{1}{\theta}\right)\langle E\rangle_c - \frac{1}{\theta}\left\langle \sum_k \frac{\partial H}{\partial \xi_r}\,\delta\xi_r\right\rangle_c = 0, \tag{13.4.6}$$

or, in view of the above-discussed thermodynamic interpretation of the ensemble averages,

$$\delta\left(\frac{\psi}{\theta}\right) - U\,\delta\left(\frac{1}{\theta}\right) - \frac{1}{\theta}\,\delta W = 0. \tag{13.4.7}$$

Combining this equation with the relation (13.4.2) one has, finally,

$$\frac{\delta Q}{\theta} = \delta\left(\frac{U - \psi}{\theta}\right). \tag{13.4.8}$$

Now the second law of thermodynamics maintains that the quantity $\delta Q/\theta$ is a perfect differential, namely the differential dS of the state function *entropy*, which is connected with the (Helmholtz) *free energy* F, the internal energy U, and the absolute (Kelvin) *temperature* T by the relation

$$S = \frac{U - F}{T}, \tag{13.4.9}$$

from which follows

$$\delta S = \delta\left(\frac{U - F}{T}\right). \tag{13.4.10}$$

The comparison of (13.4.8) and (13.4.10) suggests relating the modulus θ of the canonical ensemble with the temperature T of the surroundings of its member systems, or "heat reservoirs", with which they are in equilibrium by the relation

$$\theta = kT \tag{13.4.11}$$

in which, for dimensional reasons, k must be a universal constant of the dimension of energy/temperature or the dimension of entropy. k is now generally called "*Boltzmann's constant*", although the universal character of this constant was only later recognized by Planck.

If one now divides both sides of (13.4.8) by k one sees that the left-hand member becomes identical with δS, and (13.4.8) and (13.4.10) become identical if one identifies the quantity ψ with the free energy F. We have thus succeeded in deriving the *second law of thermodynamics* from statistical mechanics.

With the help of the relation (13.4.11) we can now write the equipartition law (13.3.30) in the form

$$\langle E_s \rangle_c = kT/2. \tag{13.4.12}$$

In elementary kinetic theory one derives the equation of state for an ideal gas by equating the mean kinetic energy of the translational movement of a gas molecule to $\frac{3}{2}\frac{RT}{\mathcal{N}}$, R being the absolute gas constant and $\mathcal{N}$ the number of molecules per mole (Avogadro's number). As the molecules have three degrees of freedom of translational movement this is equivalent to ascribing to each of these degrees of freedom a mean kinetic energy $\frac{1}{2}\frac{RT}{\mathcal{N}}$. Comparing this result with the relation (13.4.12) we see that Boltzmann's constant is equal to the gas constant per molecule $R/\mathcal{N}$.

On the basis of the results obtained so far we shall now derive a formula for the free energy of a system in terms of the energy distribution function. For we can now write the relationship (13.3.26) in the form

$$\sum_n g_n \exp\left(\frac{F - E_n}{kT}\right) = 1, \tag{13.4.13}$$

from which follows

$$F = -kT \ln Z, \tag{13.4.14}$$

where

$$Z = \sum_n g_n \exp(-E_n/kT). \tag{13.4.15}$$

The quantity Z, which can be evaluated as a function of the temperature when the eigenvalues E_n of the energy of the system for given values of the external parameters and their statistical weights g_n are known, is usually called the "*partition function*" or "*sum over all states*".

Under classical conditions, when H can be assumed to be a continuous function of the microscopic parameters $x_1, \ldots, x_{2N}$ the partition function (13.4.15) assumes the form of an integral over the whole phase space

$$Z = \frac{1}{h^N} \underbrace{\int \cdots \int}_{2N} \exp\left(\frac{-H(x_1, \ldots, x_{2N})}{kT}\right) dx_1 \cdots dx_{2N}. \tag{13.4.16}$$

From (13.4.14) one obtains the free energy of the system once Z has been evaluated as a function of T and the external parameters $\xi_1, \ldots, \xi_s$, and from this the whole thermodynamic behaviour of the system is determined according to the rules of classical thermodynamics. For example,

$$S = -\frac{\partial F}{\partial T}, \quad U = -T^2 \frac{\partial}{\partial T}\left(\frac{F}{T}\right), \quad \lambda_r = -\frac{\partial f}{\partial \xi_r}. \tag{13.4.17}$$

If the quantities ξ_r are so-called "*extensive parameters*", like volume or strain components, the quantities λ_r are the corresponding "*intensive parameters*", like pressure or the negative stress components, and vice versa. We shall make use of these relations in Chapter 15.

There is still another way for obtaining a statistical expression for the entropy S, namely in terms of the density distribution function ϱ_j for the various quantum states of the system, by means of the formula (13.3.24), which we can now write in the form

$$\varrho_j = \frac{\exp(-E_j/kT)}{\sum_j \exp(-E_j/kT)} = \frac{1}{Z}\exp(-E_j/kT).$$

By taking logarithms on both sides of this equation, multiplying by ϱ_j on both sides, and summing over all j one obtains

$$\sum_j \varrho_j \ln \varrho_j = -\frac{1}{kT}\sum_j E_j\varrho_j - \ln Z = -\frac{U}{kT} + \frac{F}{kT},$$

and from this, with the help of the relation (13.4.9)

$$S = -k\sum_j \varrho_j \ln \varrho_j, \tag{13.4.18}$$

where the sum is to be extended over all quantum states j of which the system is capable.

It will now be shown that the *third law of thermodynamics* (also known as Nernst's theorem) can be given a statistical mechanical interpretation on the basis of the relations (13.4.14) and (13.4.15) for the free energy. This law stipulates that the entropy of any system approaches zero when T approaches the absolute zero point of temperature.

If T is very small all the terms in the sum (13.4.15) become negligibly small compared with the first term which represents the lowest energy level E_1. Thus

$$\begin{aligned}\lim_{T\to 0} F &= \lim_{T\to 0}[-kT\ln(g_1 e^{-E_1/kT})] = \lim_{T\to 0}[E_1 - kT\ln g_1]\\ &= \lim_{T\to 0}[U - kT\ln g_1],\end{aligned} \tag{13.4.19}$$

and combining this relation with (13.4.9) one obtains

$$S(T=0) = k\ln g_1. \tag{13.4.20}$$

If it is now assumed that the lowest quantum state of *any* system is always non-degenerate, then $g_1 = 1$ and $S = 0$ for $T = 0$, which is the third law of thermodynamics.

Although the above-made assumption concerning the non-degeneracy of the lowest quantum level is plausible, its general validity cannot be proved.

However, even if it were not to hold strictly and g_1 were a small integer, the right-hand member of equation (13.4.20) would still be of the order of magnitude k only and thus very small compared with the entropies of macroscopic systems at finite temperatures. Hence the third law of thermodynamics would still hold very approximately.

We now return to the expression (13.4.18) for the entropy and apply it to an ensemble whose members have all the same energy, say E_0, in other words a microcanonical ensemble. In this case, according to section 13.3, all the ϱ_j belonging to the energy level E_0 are equal to one another, say equal to ϱ_0, and all the others are zero. Hence, since $\sum_j \varrho_j = 1$, the relation (13.4.18) turns into

$$S = -k \ln \varrho_0. \tag{13.4.21}$$

If G is the degree of degeneracy of the quantum state E_0 then the number of cells in the energy shell of our microcanonical ensemble is also equal to G, and therefore $\varrho_0 G = 1$. Thus the relation (13.4.21) can be written in the form

$$S = k \ln G. \tag{13.4.22}$$

This is the famous "*Boltzmann theorem*" in the formulation of Planck which connects the entropy S of a system of given "macrostate" with the number G of "microstates" by which the macrostate can be realized. Boltzmann was led to this statistical interpretation of entropy by the following consideration.

Consider a certain material system, consisting of a number of independent and non-interacting sub-systems. If p_r are the probabilities of finding the various sub-systems in certain microstates, then the probability p of finding the whole system in a certain microstate is the product of the p_r. On the other hand, it is known that the entropy S of the system is the sum of the entropies S_r of the independent sub-systems. The relationships

$$S = \sum_r S_r \quad \text{and} \quad p = \prod_r p_r$$

then suggest relations of the form

$$S_r = k \ln p_r, \quad S = k \ln p$$

which interpret entropy as the logarithm of a probability. The quantity G in (13.4.22) thus plays the role of a "probability", although it is by definition an integer; it is therefore called "*thermodynamic probability*".

The third law of thermodynamics also follows from Boltzmann's theorem in the form (13.4.22), for this predicts that the entropy should be zero for a state whose thermodynamic probability is equal to unity, that is a state which can only be materialized in one way.

One of the most important consequences of Boltzmann's statistical interpretation of entropy is the explanation of the intrinsic irreversibility of thermal processes which, as already mentioned before, finds its general expression in the *law of increase of entropy* in a closed system. For if such a system is at a certain time in a state of low probability, it is likely to be found in a state of higher probability at a later time, and to be tending to a state of statistical equilibrium of highest probability.

It will now be shown that the law of increase of entropy holds strictly for ensembles of any kind if it is assumed that the formula (13.4.18), which expresses S in terms of the density distribution function ϱ_j is valid not only for stationary but also for non-stationary ensembles. For differentiating this equation on both sides with respect to time results in

$$\frac{dS}{dt} = -k \sum_j \frac{d\varrho_j}{dt} (\ln \varrho_j + 1). \tag{13.4.23}$$

Substituting here for $d\varrho_j/dt$ from the relation (13.2.6) one obtains further

$$\frac{dS}{dt} = -k \sum_{ij} (\varrho_i - \varrho_j)\, w_{ji} (\ln \varrho_j + 1). \tag{13.4.24a}$$

By interchanging the subscripts j and i we can write this also in the alternative form

$$\frac{dS}{dt} = -k \sum_{ij} (\varrho_j - \varrho_i)\, w_{ji} (\ln \varrho_i + 1), \tag{13.4.24b}$$

if the principle of microscopic reversibility (13.2.4) is taken into account. Adding the equations (13.4.24a) and (13.4.24b) finally leads to

$$\frac{dS}{dt} = \frac{k}{2} \sum_{ij} (\varrho_i - \varrho_j)\, w_{ij} (\ln \varrho_i - \ln \varrho_j). \tag{13.4.25}$$

Since all the $w_{ij} \geqslant 0$ by definition and $(\varrho_i - \varrho_j)$ has always the same sign as $(\ln \varrho_i - \ln \varrho_j)$ one sees that each member of the sum in (13.4.25) is positive and hence

$$\frac{dS}{dt} \geqslant 0. \tag{13.4.26}$$

This indicates that the entropy, as defined for an ensemble of isolated identical systems, must always increase until it reaches "*thermodynamic equilibrium*" in which all the ϱ_i are equal; as we have shown before, this is indeed the necessary and sufficient condition for stationarity in a microcanonical ensemble.

It is important to realize, however, that the law of increase of entropy *does not*, and indeed *cannot* hold precisely for *individual* systems when entropy is a statistically defined quantity. Thus one must expect to find that the entropy of an individual system occasionally *decreases*, in particular if it had already reached its maximum equilibrium value, giving rise to irregular *fluctuations* of macroscopically observable parameters about the thermodynamic equilibrium values. We shall deal with this phenomenon in some detail in section 15.4.

Problems and exercises

P. 13.1. The "multivariate" Gaussian distribution function is defined by

$$\varrho\,(x_1 \cdots x_N) = \text{const}\,\exp\Big(-\tfrac{1}{2}\sum_{ij} a_{ij}x_ix_j\Big) \quad (a_{ij} = a_{ji}),$$

where the sum in the exponent is a positive definite quadric in the variables x_i and the coefficients a_{ij} are in general functions of time. Show that the ensemble averages $\langle x_i \rangle$ of the variables are zero and the ensemble averages $\langle x_i x_j \rangle$ are given by the expressions

$$\langle x_i x_j \rangle = A_{ij}/A,$$

where A is the determinant of the coefficients a_{ij}, and A_{ij} are the complement minors of a_{ij}. (Make use of the fact that $Q = \sum_{ij} a_{ij}x_ix_j = \text{const}$ represents an $(N-1)$-dimensional hyperellipsoid in an N-dimensional space.)

P. 13.2. Assume a distribution function $\varrho = \text{const}\,\exp\,[-\frac{1}{2}(\alpha x^2 + 2\beta xy + \gamma y^2)]$ in a two-dimensional phase-space with coordinates x, y where the coefficients α, β, γ may depend on time. Find the probability density function $f(a)$ for an observable a which is a function of x alone, and hence the ensemble average and variance of a, from the e quations (13.1.3). Show that for the special case $a = x$ the results obtained are in agreement with the general results of P. 13.1.

P. 13.3. Derive expressions for the ensemble averages of the magnitude r of the position vector $\mathbf{r}$ of the phase points in an N-dimensional phase space, and of r^2, when the distribution function is given by

$$\varrho = \text{const}\,\exp\Big(-\alpha \sum_{i=1}^{N} x_i^2\Big) = \text{const}\,e^{-\alpha r^2},$$

in which α may be a function of time. Show that the variance of r tends to zero for very large N, by making use of Sterling's approximation formula for the factorial function.

P. 13.4. Let the wave function $\Phi(\alpha)$ of an ensemble of one-dimensional rotators (defined in the interval $-\pi < \alpha < +\pi$ of the rotation angle α) be equal to a constant A in the interval $-\alpha_0 < \alpha < +\alpha_0$ and zero outside this interval. Obtain the distribution function ϱ_n over the cells of Fig. 28b in the two-dimensional phase space of the ensemble, making use of formulae (13.2.3) and (13.2.10). Show that the ensemble average of the angular momentum P is zero and that of the energy E is infinite. Comment on these facts.

P. 13.5. Let the "square" wave function of P. 13.4 be replaced by a triangular wave function represented by the equations

$$\Phi = A\,(1 - 2\alpha/\pi) \quad \text{for} \quad \alpha \geqslant 0 \quad \text{and} \quad \Phi = A\,(1 + 2\alpha/\pi) \quad \text{for} \quad \alpha \leqslant 0.$$

Show that the ensemble average of the energies of the rotators is now finite and calculate its value. Also calculate the variance σ_P^2 of the distribution of the angular momenta and show that the variance σ_E^2 of the distribution of the energies is infinite.

P. 13.6. Assume the wave function of an ensemble of linear harmonic oscillators to have the triangular shape of P. 4.5 which can be expanded in terms of the eigenfunctions $\psi_k(x)$ of the "particle in the box". Obtain the density matrix ϱ_{kj} of the ensemble, and also the energy matrix E_{kj} in the same representation. Show how one may calculate the ensemble average $\langle \mathbf{E} \rangle$ of the energy from these results.

P.13.7. Let the members of a microcanonical ensemble consist of n "particles in a box" of mass m, which are not subject to any forces and can move freely in the x-direction between $-l/2$ and $l/2$ with energy E_0. Determine the shape of the energy shell between E_0 and $E_0 + \delta E$ in the $2n$-dimensional phase space of these systems.

P.13.8. Consider a microcanonical ensemble of systems, each of which consists of a perfectly elastic particle of mass m and energy between E_0 and $E_0 + \delta E$, moving in a vertical direction under the influence of gravity above a perfectly elastic horizontal base at $x = 0$. Derive expressions for the ensemble averages of the positions x and the magnitudes of the momenta p.

P.13.9. Consider the same ensemble of systems as in P.13.8, but assuming that its distribution is canonical with modulus θ. Derive expressions for the canonical ensemble averages of x, p, the potential energy U, the kinetic energy K of the systems, and their total energy E.

P.13.10. Derive the relations (13.4.17) from the definition of entropy $dS = \delta Q/T$ and the relations (13.4.2) and (13.4.9) (a) for a gas under pressure p when the only extensive parameter is the volume V and (b) for a solid elastic rod, subjected to a tension τ when ξ is the length of the rod.

P.13.11. A vessel of volume V is divided into two compartments of equal size by a thin sliding partition wall. To begin with one compartment contains a liquid solution of n non-interacting particles, the other the solvent only. When the partition wall is removed diffusion of the solvent particles will commence and continue until the concentration in the whole vessel has become uniform. Use the relations (13.4.14) and (13.4.16) to prove that the entropy S of the system has thereby increased by an amount $nk \ln 2$.

P.13.12. Use the relationship (13.4.18) for calculating the ensemble average of the entropy of a one-dimensional rotator at some particular time when the wave function of the ensemble has the triangular form of P.13.5.

P.13.13. Give an alternative solution of the problem put in P.13.11 on the basis of Boltzmann's theorem (section 13.4).

CHAPTER 14

Statistical Mechanics of Assemblies

THE theory of ensembles, to which Chapter 13 was devoted, is applicable to arbitrary systems of N degrees of freedom which are either classically characterized by a Hamiltonian function $H(q_1, \ldots, q_N; p_1, \ldots, p_N)$ or quantum theoretically by a Hamiltonian operator $\mathscr{H}(q_1, \ldots, q_N; t)$. In this chapter we consider in particular such systems which consist of a (usually large) number ν of sub-systems of the same type. The latter can, for example, be elementary particles or atoms or molecules or complexes of molecules. For simplicity we shall call them "particles" and denote systems composed of identical particles as "*assemblies*" of particles. Thus whereas an ensemble is an *imagined infinite* multitude of replicas of a given system (which are, of course, independent of one another), an assembly is a *finite* multitude of *real* identical particles in *space* which may interact with one another by forces (therefore called "*Raumgesamtheit*" in German literature).

In what is to follow we shall deal with the basic principles of the statistical mechanics of assemblies and use them in particular for developing the general thermodynamic properties of such systems.

14.1. Basic principles

The state of an assembly of N degrees of freedom can, as with any other system, be represented by a phase point in a $2N$-dimensional phase space, divided into cells of volume h^N. If the assembly consists of ν "particles", each having s degrees of freedom, one has $N = s\nu$.

The state of any of the particles can similarly be represented geometrically by a phase point in another, $2s$-dimensional phase space, divided into cells of volume h^s. We shall call the latter the "μ-phase space" and the former the "Γ-phase space" (the letters μ and Γ referring to "gas" and "molecule" respectively). The μ-phase space is evidently a subspace of the Γ-phase space.

All the ν particles of the assembly can, of course, be represented by points in one and the same μ-phase space. The state of the whole system is then completely determined if the position of each of the phase points of the ν particles in the cells of the μ-phase space is known, provided that there exist some means, at least in principle, to distinguish the particles from one another. If, however, only the *numbers* ν_i of particle phase points in the cells with index i are given, one speaks of a "*distribution*" of the assembly. Evidently the ν_i must satisfy the condition

$$\sum_i \nu_i = \nu. \tag{14.1.1}$$

(The distribution of an assembly must not be confused with the distribution of an ensemble, as defined in sections 13.1 and 13.2.)

The quantities $w_i = \nu_i/\nu$ may also be interpreted as the probabilities for finding the phase point of any of the particles in the cells with index i in the μ-phase space at some particular time.

The number G_f of ways in which a certain distribution of particles can be realized or, according to the definition in section 13.4, the "thermodynamic probability" of that state, is given by the well-known combinatorial rule

$$G_f = \frac{\nu!}{\prod_i \nu_i!}, \tag{14.1.2}$$

subject to the condition (14.1.1). Hence a certain distribution of the assembly is represented by G_f phase-points in the Γ-phase space of the system which form a symmetrical configuration. Using the nomenclature introduced by Ehrenfest we shall call this configuration a "*star*".

If the particles of the assembly are truly "identical", that is if there is no way of distinguishing them from one another, the quantum mechanical state of the assembly remains unaltered by any interchange of particles, as explained in section 8.1. Hence all the G_f different configurations are described by one and the same N-dimensional wave function Ψ, and the state of the assembly and its changes in time are completely represented by the movement of any one of the G_f points of the star in Γ-phase space.

If the differences between neighbouring energy levels of the particles are very small, that is if their energy spectrum is quasi-continuous (for example if they are "free" molecules of a gas in a container) one can, apart from the above-defined "*fine-grained distribution*" also define a "*coarse-grained distribution*" by subdividing the μ-phase space into domains Δ_j, each containing a large number z_j of cells with energies near an average E_j. Let ζ_j be the number of phase points in the domain Δ_j. The ζ_j must, of course, satisfy the condition

$$\sum_j \zeta_0 = \nu. \tag{14.1.3}$$

Again the ratios $P_j = \zeta_j/\nu$ measure the probabilities of finding the phase point of any of the particles in the domain Δ_j at some particular time.

The thermodynamic probability of a coarse-grained distribution is given by the combinatorial formula

$$G_c = \frac{\nu!}{\prod_j \zeta_j!} \prod_j z_j^{\zeta_j}, \tag{14.1.4}$$

subject to the condition (14.1.3), provided that the particles are individually *distinguishable*. Each of these distributions is represented by a *finite region* of volume $\prod_j z_j h^N$ in Γ-phase space, and all these regions again form a symmetric "star" configuration. The coarse-grained distribution becomes, of course, a fine-grained distribution if all the z_j are equal to unity; the ζ_j then become identical with the ν_j. It is indeed seen that under these circumstances (14.1.4) goes over into (14.1.2).

If the particles are *indistinguishable* a coarse-grained distribution is characterized by the numbers $z_j^{(r)}$ of cells within the domain Δ_j which contain just r phase points. Evidently the $z_j^{(r)}$ must satisfy the conditions

$$\sum_r z_j^{(r)} = z_j, \quad \sum_r r z_j^{(r)} = \zeta_j. \tag{14.1.5}$$

As all fine-grained distributions belonging to a given set $z_j^{(r)}$ are identical, the number of ways in which this distribution can be materialized is simply equal to the number of "complexions" of the cells within the domain, that is

$$G_c = \prod_j \frac{z_j!}{z_j^{(1)}! z_j^{(2)}! \ldots}, \tag{14.1.6}$$

under the conditions (14.1.5); this again constitutes a star configuration in Γ-space.

If the distribution function of an assembly is known, one can calculate the value of any macroscopic parameter A, characterizing the physical state of a homogeneous material body, on the basis of the microscopic mechanism of its constituent particles. We may then assume that

$$A = \nu a, \tag{14.1.7}$$

where a is a function of the coordinates $x_1, \ldots, x_{2s}$ in the μ-phase space of the particles.

If a_i is the value of a belonging to the cell with index i in μ-phase space the average of a over the whole assembly is

$$\bar{a} = \frac{1}{\nu} \sum_i a_i \nu_i, \tag{14.1.8}$$

and hence the mean value $\bar{A}$ of A, which may be identified with the measured value of A for sufficiently large ν (as is normally the case) is from (14.1.7) and (14.1.8)

$$\bar{A} = \sum_i a_i \nu_i. \tag{14.1.9}$$

Similarly, if a_j is the average value of a in the coarse-grained domain Δ_j, one has

$$\bar{A} = \sum_j a_j \zeta_j. \tag{14.1.10}$$

Under classical conditions one may assume $w_j = \zeta_j/\nu$ to become a continuous probability function $w(x_1 \cdots x_{2s})$ and the sum in (14.1.10) to go over into an integral over the whole μ-phase space; hence

$$\bar{A} = \nu \underbrace{\int \cdots \int}_{2s} a(x_1, \ldots, x_{2s})\, w(x_1, \ldots, x_{2s})\, dx_1 \cdots dx_{2s}. \tag{14.1.11}$$

14.2. Stationary and most probable distributions

In keeping with the nomenclature used for ensembles we shall call an assembly *stationary* if the distribution function is independent of time.

It is easy to derive a general expression for the stationary distribution function of an assembly of *weakly interacting* particles with fixed total energy E_0. For one can then use the same argument which was previously employed in section 13.3 for the derivation of the distribution function of a stationary canonical ensemble when the "sub-systems" there are identified with the particles of the assembly here. If the number ν of particles in the assembly is very large, E_0 is also very large and the energies available to the individual particles are practically unlimited. The stationary energy distribution law for the assembly follows then directly from the relation (13.3.20).

In the present notation we thus obtain for a fine-grained distribution the formula

$$\nu_n = C g_n \exp(-E_n/kT), \tag{14.2.1}$$

where the E_n are the energy eigenvalues of the particles, the g_n their statistical weights, and T the temperature of the body, as defined in the usual way. C is a constant which is determined by the condition (14.1.1). For $g_n = 1$ in particular one obtains from (14.2.1)

$$w_n = \frac{\nu_n}{\nu} = \frac{\exp(-E_n/kT)}{\sum_n \exp(-E_n/kT)}, \tag{14.2.2}$$

a formula of which we have already made use in section 2.2 [eqn. (2.2.2)].

The analogue to the distribution law (14.2.1) for coarse-grained distributions is

$$\zeta_j = C' z_j \exp(-E_j/kT), \tag{14.2.3}$$

where E_j is the mean energy of the cells in the domain Δ_j of the μ-phase space of the particles, z_j, as before, the number of cells in that domain, and the constant C' is determined by the condition (14.1.3).

Finally, under classical conditions one has for the continuous *probability density function* w:

$$w(x_1, \ldots, x_{2s}) = C'' \exp[-H(x_1, \ldots, x_{2s}/kT)], \tag{14.2.4}$$

where $H(x_1, \ldots, x_{2s})$ is the Hamiltonian function for the single particles and C'' a constant which is determined by the normalizing condition

$$\underbrace{\int \cdots \int}_{2s} w(x_1, \ldots, x_{2s})\, dx_1 \cdots dx_{2s} = 1. \tag{14.2.5}$$

The formula (14.2.4) was first given by Boltzmann. All three relations (14.2.1), (14.2.3), and (14.2.4) are special forms of what is now generally known under the name "*Boltzmann distribution law*" for stationary assemblies.

It appears from the above argument that if an assembly of a large but finite number of weakly interacting particles is to be in a stationary state it must obey the Boltzmann energy distribution law. But, on the other hand, one can see readily that in fact such an assembly, whose energy is confined to a narrow region between E_0 and $E_0 + \delta E$, cannot be strictly stationary. For if the stochastic time series of the system is ergodic (see section 13.1) its phase curve, in the language of classical statistical mechanics, must in the course of time come arbitrarily near to any point of the energy shell between E_0 and $E_0 + \delta E$ of the Γ-phase space of the system, and the "times of sojourn" of the phase point in any of the cells of the shell are equal. Hence the phase point cannot remain indefinitely in the "star" region which corresponds to the distributions (14.2.1) to (14.2.5) respectively. However, if the volume of this region is very large compared with the volume of the rest of the energy shell, the distribution will be almost stationary, that is, it will not change in the course of time except for small fluctuations. We shall show presently that this is indeed the case if the number ν of particles of the assembly is very large.

We now wish to find the *most probable* of all *coarse-grained distributions* of an assembly of *distinguishable* particles. This will evidently be the distribution whose "star" occupies the largest volume within the energy shell in Γ-phase space, or which will make the thermodynamic probability G_c for that distribution a maximum under the conditions that ν and E_0 have given constant values.

Taking logarithms on both sides of (14.1.4) and assuming all the ζ_j to be large numbers, so that Stirling's approximation formula for the factorials can be used, we have

$$\ln G_c = \nu \ln \nu - \sum_j \zeta_j \ln \frac{\zeta_j}{z_j}. \tag{14.2.6}$$

The condition for the left-hand member of this equation to be an extremum for arbitrary variations $\delta\zeta_j$ and under the conditions

$$\delta\nu = \delta \sum_j \zeta_j = 0, \quad \delta E_0 = \delta \sum_j \zeta_j E_j = 0 \tag{14.2.7}$$

is, according to Lagrange's theorem of the calculus of variations,

$$\delta \ln G_c - \alpha\, \delta v - \beta\, \delta E_0 = 0, \tag{14.2.8}$$

where α and β are constants (the Lagrangian multipliers). Substituting here from (14.2.6) and (14.2.7) we obtain

$$\sum_j \delta\zeta_j \left\{ \ln \frac{\zeta_j}{z_j} + 1 + \alpha + \beta E_j \right\} = 0. \tag{14.2.9}$$

In order that this equation be satisfied for arbitrary variations $\delta\zeta_j$ the expressions in curly brackets must be separately zero for all j. Hence the ζ_j must be of the form

$$\zeta_i = z_j \exp(-1 - \alpha - \beta E_j), \tag{14.2.10}$$

which expression becomes identical with (14.2.3) if one puts $e^{-1-\alpha} = C'$ and $\beta = \dfrac{1}{kT}$.

Thus it appears that for an assembly of a very large number of *distinguishable* particles the most probable coarse-grained distribution is again a Boltzmann distribution, or that the *most probable* distribution is at the same time *stationary*. Since in actual fact v is always finite, this statement is not *strictly* true, but the most probable distribution will still be *practically* stationary, which means that it must be overwhelmingly more probable than all the other possible distributions put together. Thus the above-mentioned apparent contradiction is now resolved.

Notice that the preceding argument cannot be applied to *fine-grained* distributions because the numbers v_i of phase points to be found in the various cells of the μ-phase space of the particles of the assembly are in general *not* large.

Let us now assume that the particles of the assembly are truly identical, that is *non-distinguishable*. We can then employ the same method as before if we replace the expression (14.1.4) for G_c by the expression (14.1.6). Again, taking logarithms on both sides of this equation and using Sterling's approximation, we have now

$$\ln G_c = \sum_j z_j \ln z_j - \sum_{jr} z_j^{(r)} \ln z_j^{(r)}. \tag{14.2.11}$$

This has to be made an extremum under the same conditions as before which with the help of (14.1.5) can be written in the form

$$\delta z_j = \delta \sum_j z_j^{(r)} = 0, \quad \delta v = \delta \sum_j \zeta_j = \delta \sum_{jr} r z_j^{(r)} = 0,$$

$$\delta E_j = \delta \sum_j \zeta_j E_j = \delta \sum_{jr} r E_j z_j^{(r)} = 0. \tag{14.2.12}$$

By means of Lagrange's method we obtain from (14.2.11) and (14.2.12)

$$\sum_{jr} \delta z_j^{(r)} \{\ln z_j^{(r)} + \gamma + \alpha r + \beta r E_j\} = 0 \tag{14.2.13}$$

in which now appear three Lagrangian multipliers α, β, γ. Again, if this equation is to be satisfied for arbitrary values of the variations $\delta z_j^{(r)}$, the expressions in the curly bracket must all vanish separately so that

$$z_j^{(r)} = \exp(-\gamma - \alpha r - \beta r E_j) \tag{14.2.14}$$

represents the most probable distribution of the $z_j^{(r)}$.

In order to obtain the most probable distribution of the occupation numbers ζ_j of the domains Δ_j of the coarse-grained distribution, we make use of the relations (14.1.5) which may be written in the form

$$\zeta_j = \sum_r r z_j^{(r)} = z_j \frac{\sum_r r z_j^{(r)}}{\sum_r z_j^{(r)}}. \tag{14.2.15}$$

Substituting here from (14.2.14) and introducing for brevity the quantities

$$y_j = -\alpha - \beta E_j \tag{14.2.16}$$

and

$$S_j = \sum_r \exp(r y_j), \tag{14.2.17}$$

we get, finally,

$$\zeta_j = z_j \frac{\sum_r r \exp(r y_j)}{\sum_r \exp(r y_j)} = \frac{z_j}{S_j} \frac{dS_j}{dy_j} = z_j \frac{d(\ln S_j)}{dy_j}. \tag{14.2.18}$$

In section 8.1 it was shown that there existed two types of particles for one of which the wave function Ψ of an assembly of such particles was symmetric in the coordinates of the particles and for the other antisymmetric. In the first case all distributions of the phase points of the particles in the μ-phase space have the same ***a priori*** probability and no restrictions are imposed on the numbers r of phase points which may occupy a certain cell i, except for the condition that the energy rE_i must not be larger than the total energy E_0 of the assembly; this condition will be practically always satisfied if the number ν of particles of the assembly is large. In the second case not more than one particle can be in the same quantum state, which means that any of the cells can only be either unoccupied or occupied by one phase point, so that r can only assume the values 0 or 1.

In the first case, where the wave function of the assembly is *symmetric* the sum in (14.2.17) extends practically to all values of r from 0 to infinity and is thus equal to

$$S_j = \frac{1}{1 - e^{y_j}}. \tag{14.2.19}$$

Consequently from (14.2.16) and (14.2.18)

$$\zeta_j = \frac{z_j}{\exp(\alpha + \beta E_j) - 1}. \tag{14.2.20}$$

If one approaches classical conditions the number of states within a finite domain of the μ-phase space becomes very large and hence ζ_j/z_j becomes very small, which means that α must be very large, and consequently the second term in the denominator of (14.2.20) becomes negligibly small compared with the first, and the expression (14.2.20) goes over into the classical expression (14.2.10) for the Boltzmann distribution. We can then, as before, identify the constant β with $1/kT$ and, setting $\alpha/\beta = -\mu$, write (14.2.20) in the form

$$\zeta_j = \frac{z_j}{\exp[(E_j - \mu)/kT] - 1}, \tag{14.2.21}$$

where the quantity $\mu(T)$ is determined by the condition (14.1.3).

The distribution law (14.2.21) is known as the "*Bose–Einstein distribution*", and particles, for which it holds, are frequently referred to as "*bosons*", because a special version of this law applying to photons (see section 15.3) was first formulated by Bose in 1924 and its extension to gases of material particles was suggested by Einstein the same year. From what has been stated in section 8.1 it appears that bosons are particles with integer spin.

In the case where the wave function of the assembly is *antisymmetric* in the coordinates of the particles the sum (14.2.17) is restricted to the values $r = 0$ and $r = 1$, and therefore is equal to

$$S_j = 1 + e^{y_j}. \tag{14.2.22}$$

From (14.2.16) and (14.2.18) we now have

$$\zeta_j = \frac{z_j}{\exp(\alpha + \beta E_j) + 1}, \tag{14.2.23}$$

which again goes over into the Boltzmann distribution law under classical conditions. Using the same argument as before we can write it in the form

$$\zeta_j = \frac{z_i}{\exp[(E_j - \mu)/kT] + 1}, \tag{14.2.24}$$

where the temperature-dependent quantity μ is determined by the condition (14.1.3).

The distribution law (14.2.24) is called the "*Fermi–Dirac distribution law*", because it was proposed independently by Fermi and Dirac in 1926. Particles which obey this law are usually called "*fermions*"; according to section 8.1 they have *half-integer* spins. A comparison of the formulae (14.2.21) and (14.2.24) shows that they only differ in the sign of the second term in the denominator and that they turn into the Boltzmann distribution law if that term is absent altogether.

For very low temperatures the exponential function in (14.2.24) is zero or infinite according to whether $E_j < \mu_0$ or $E_j > \mu_0$, where μ_0 is the value of μ for $T = 0$.

$$\left.\begin{aligned} \zeta_j/z_j &= 1 \quad \text{for} \quad E_j < \mu_0, \\ \zeta_j/z_j &= 0 \quad \text{for} \quad E_j > \mu_0. \end{aligned}\right\} \tag{14.2.25}$$

μ_0 is called the "*Fermi-level*" of energy. The relations (14.2.25) indicate that near the absolute zero of temperature all cells in the μ-phase space belonging to energies smaller than μ_0 are occupied by just one phase point and the remainder of the cells is empty.

14.3. The grand canonical ensemble and the mean distribution of an assembly

In section 13.3 it was shown that a stationary ensemble of identical systems, which are allowed to exchange energy with a "heat reservoir" of constant temperature T, must obey the canonical distribution law (13.3.21) or (13.3.24). This must, of course, also hold for an *ensemble* of *assemblies* consisting of a *fixed* number ν of particles. It will now be assumed that the members of such an ensemble can not only exchange *energy* but also *particles* with a surrounding consisting of the same type of particles, so that ν is now *no longer fixed*.

In order to find the stationary distribution law for such an ensemble we can proceed in a way that is completely analogous to the one used in section 13.3 [formulae (13.3.14) to (13.3.20)], except that now the sub-systems are allowed to exchange not only energy but also particles; consequently one has to add to the condition (13.3.16) the further condition that the sum of the numbers $\nu^{(l)}$ of the particles in the sub-systems must be equal to the total number ν of particles in the whole system. Thus the conditions are now

$$\sum_l E^{(l)} = E, \quad \sum_l \nu^{(l)} = \nu. \tag{14.3.1}$$

It is easy to see that, as a result of this, the relation (13.3.19) has to be replaced by

$$\ln \varrho^{(l)}\,(E^{(l)}, \nu^{(l)}) = \alpha_l + \beta_l\, \nu^{(l)} - \frac{E^{(l)}}{\theta}, \tag{14.3.2}$$

where α_l, β_l, and θ are constants.

It follows that the stationary distribution law of the ensemble of assemblies has now, instead of (13.3.24), the form

$$\varrho_{k,\nu} = C \exp\left(\frac{\mu\nu - E_k}{\theta}\right). \tag{14.3.3}$$

Here k is the number of the cell in the Γ-phase space of the assemblies, E_k is the energy belonging to that cell, and C is a constant which is determined by the condition

$$\sum_{k,\nu} \varrho_{k,\nu} = 1. \tag{14.3.4}$$

The quantities θ and μ are characteristic constants of the "reservoir" with which the members of the ensemble are in statistical equilibrium. θ, as before, is related to the temperature of the reservoir by the relation (13.4.11), that is $\theta = kT$, and μ is related to the density of the particles in the reservoir.

This consideration can, without difficulty, be extended to ensembles of assemblies which are mixtures of several types of particles with numbers $\nu^{(r)}$, the reservoir also consisting of the same mixture. One then has instead of (14.3.3)

$$\varrho_{k,\nu^{(r)}} = C \exp\left(\frac{\sum_r \mu^{(r)}\nu^{(r)} - E_k}{\theta}\right) \tag{14.3.5}$$

with the normalizing condition

$$\sum_{k,\nu^{(r)}} \varrho_{k,\nu^{(r)}} = 1. \tag{14.3.6}$$

Here, again, $\theta = kT$, and the quantities $\mu^{(r)}$ are related to the concentrations of the particle mixture in the reservoir; it will be shown presently that they are identical with the so-called "*chemical potentials*" of classical thermodynamics.

Returning now to the case of only one type of particles we introduce the quantity

$$Z_g = \sum_{k,\nu} \exp\left(\frac{\mu\nu - E_k}{\theta}\right), \tag{14.3.7}$$

called the "*grand-canonical partition function*". From (14.3.4) it appears that the constant in (14.3.3) is equal to $1/Z_g$.

The ensemble average $\langle\nu\rangle$ of ν is, by definition, from (14.3.3) and (14.3.7)

$$\langle\nu\rangle_g = \sum_{k,\nu} \nu\varrho_{k,\nu} = \frac{1}{Z_g}\sum_{k,\nu} \nu \exp\left(\frac{\mu\nu - E_k}{\theta}\right) = \theta\,\frac{d\,(\ln Z_g)}{d\mu}. \tag{14.3.8}$$

In analogy to (13.4.14) we can now connect the free energy F' of an "open" assembly with the grand partition function of the corresponding grand ensemble by the relation

$$F' = -kT \ln Z_g. \tag{14.3.9}$$

The relations (13.4.17) remain in force for F' to which, in virtue of (14.3.8), one has to add the further relation

$$\langle\nu\rangle = -\frac{\partial F'}{\partial\mu}. \tag{14.3.10}$$

In fact (14.3.10) may be regarded as a special case of the set of relations $\lambda_r = -\partial F/\partial \xi_r$, if the average number of particles of the assembly is taken to be one of the "extensive" parameters of the system; μ is then the corresponding "intensive" parameter.

For assemblies composed of more than one type of particles one has, instead of (14.3.10), a set of relations for the average numbers $\langle \nu^{(r)} \rangle$ of these particles, namely

$$\langle \nu^{(r)} \rangle = -\frac{\partial F'}{\partial \mu^{(r)}}. \tag{14.3.11}$$

These relations are frequently used in classical thermodynamics in problems concerning the equilibrium of mixtures and chemical compounds and, as mentioned before, the parameters $\mu^{(r)}$ are called "chemical potentials". The equality of the chemical potentials in different "phases" are the conditions for thermodynamic equilibrium of these phases.

We shall now make use of the above derived formulae for the grand canonical ensemble for the purpose of obtaining the ensemble averages of the occupation numbers of the cells in the μ-phase space of an assembly of identical particles or, in other words, the *mean fine distribution* of that assembly.

Let ν_i be the number of phase points of the particles in the ith cell of the μ-phase space with energy E_i. The quantities ν, E_k, and θ appearing in the expression (14.3.7) of the grand canonical partition function are then given by

$$\nu = \sum_i \nu_i, \quad E_k = \sum_i \nu_i E_i, \quad \theta = kT. \tag{14.3.12}$$

Thus

$$Z_g = \sum_{\nu_i} \exp\left[\left(\mu \sum_i \nu_i - \sum_i \nu_i E_i\right)/kT\right], \tag{14.3.13}$$

where the sum is to be extended over all permitted combinations of the occupation numbers ν_i.

Setting for brevity

$$x_i = \exp\left(\frac{\mu - E_i}{kT}\right) \tag{14.3.14}$$

we can write Z_g in the form

$$Z_g = \sum_{\nu_i} \prod_i x_i^{\nu_i}. \tag{14.3.15}$$

It is readily seen that the sequence of summation and multiplication in (14.3.15) can be interchanged. For if, for example, i is restricted to two values 1 and 2 only, and ν_i to two values a and b, one has

$$Z_g = x_1^a x_2^a + x_1^a x_2^b + x_1^b x_2^a + x_1^b x_2^b = (x_1^a + x_1^b)(x_2^a + x_2^b).$$

Thus, generally,

$$Z_g = \prod_i \sum_{\nu_i} x_i^{\nu_i}. \tag{14.3.16}$$

The free energy of the assembly is now obtained from (14.3.9)

$$F' = -kT\sum_i \ln\left(\sum_{\nu_i} x_i^{\nu_i}\right) \tag{14.3.17}$$

and from (14.3.10) and the first of the relations (14.3.12) it follows that

$$\langle\nu\rangle = \sum_i \langle\nu_i\rangle = kT\sum_i \frac{\partial}{\partial\mu}\ln\left(\sum_{\nu_i} x_i^{\nu_i}\right)$$

or

$$\sum_i\left\{\langle\nu_i\rangle - kT\frac{\partial}{\partial\mu}\ln\left(\sum_{\nu_i} x_i^{\nu_i}\right)\right\} = 0. \tag{14.3.18}$$

This equation can be identically satisfied only if each term in the sum vanishes separately. Thus the mean occupation numbers $\bar{\nu}_i = \langle\nu_i\rangle$ of the cells in the μ-phase space of the particles of the assembly, which represent the *mean distribution* of the assembly, are given by the relations

$$\bar{\nu}_i = kT\frac{\partial}{\partial\mu}\ln\left(\sum_{\nu_i} x_i^{\nu_i}\right), \tag{14.3.19}$$

where x_i is defined by (14.3.14).

Let us first assume that the particles of the assembly are "bosons". In this case there is no restriction imposed on the occupation numbers ν_i of the cells and one has

$$\sum_{\nu_i} x_i^{\nu_i} = \sum_{n=0}^{\infty} x_i^n = \frac{1}{1-x_i} = \left[1 - \exp\left(\frac{\mu - E_i}{kT}\right)\right]^{-1}. \tag{14.3.20}$$

Substitution from this equation into (14.3.19) yields

$$\bar{\nu}_i = \left[\exp\left(\frac{E_i - \mu}{kT}\right) - 1\right]^{-1}. \tag{14.3.21}$$

Next we assume that the particles are "fermions". The occupation numbers are then restricted to 0 and 1, and instead of (14.3.20) one now has

$$\sum_{\nu_i} x_i^{\nu_i} = 1 + x_i = 1 + \exp\left(\frac{\mu - E_i}{kT}\right). \tag{14.3.22}$$

Thus the mean occupation numbers are, from (14.3.19),

$$\bar{\nu}_i = \left[\exp\left(\frac{E_i - \mu}{kT}\right) + 1\right]^{-1}. \tag{14.3.23}$$

Comparing the formulae (14.3.21) and (14.2.21), and the formulae (14.3.23) and (14.2.23) respectively, one sees that the most probable coarse-grained Bose–Einstein and Fermi–Dirac distributions are obtained by multiplying the mean fine-grained occupation numbers of the cells in the domains Δ_j by the numbers z_j of the cells in these domains. In other words, the *most probable distributions* of *undistinguishable particles* in assemblies of large numbers of such particles are identical with the *mean distributions* of these particles. This is clearly only possible if the most probable distribution is overwhelmingly more probable than all the other distributions contributing to the mean distribution. The same fact had been previously established for the Boltzmann distribution of *distinguishable* particles in section 14.2.

The fine-grained distribution formulae (14.3.21) and (14.3.23) are, of course, quite generally valid, even if the numbers ν of particles in the assemblies are *not* large, which is apparent from the derivation of these expressions.

We finally notice that the constant μ appearing in these distribution functions is equal to the chemical potential of the surroundings with which the assembly is in statistical equilibrium. In the case of the Fermi–Dirac distribution the value of this chemical potential at zero temperature is equal to the "Fermi-level" of energy.

Problems and exercises

P. 14.1. Give a proof of the formula (14.1.2) for the thermodynamic probability of a fine-grained distribution of an assembly of ν particles in μ-phase space.

P. 14.2. Give a proof of the formula (14.1.4) for the thermodynamic probability of a coarse-grained distribution of distinguishable particles over the domains with numbers z_j of cells in μ-phase space.

P. 14.3. According to (14.1.9) and (2.2.3) the mean energy $\bar{E}$ of a stationary assembly of N linear harmonic oscillators in equilibrium with a reservoir of temperature T is equal to

$$\bar{E} = Nh\nu/(e^{h\nu/kT} - 1).$$

Use the distribution law (14.2.1) in connection with the relations (5.1.33) and (5.1.38) for calculating the mean energies of assemblies of two- and three-dimensional oscillators respectively. Comment on the results obtained.

P. 14.4. Consider an ensemble of assemblies each of which consists of n perfectly elastic and non-interacting particles in a field-free container of volume V. Determine the classical energy distribution function $f(E)$ (13.3.23). In the Γ-phase space of this ensemble, calculate the energy E_m for which $f(E)$ has its maximum value and the mean value $\bar{E}$ of E. Show that $E \simeq E_m$ for large n and that it satisfies the equipartition law. Show further that the probability for E to deviate appreciably from E_m is very small for large n.

P. 14.5. Two assemblies of ν respectively ν' distinguishable "particles" (of not necessarily the same type) are combined to form a single closed system of given energy E_0 so that they may exchange energy but not particles. Following a procedure analogous to that used in section 14.2, derive expressions for the most probable coarse-grained distribution of the numbers ξ_i and ξ'_j of particles in the μ-phase space domains with z_i and z'_j cells respectively, and show that the temperatures of the two part-systems must be the same.

P. 14.6. Show in detail how the stationary distribution law (14.3.3) of a grand canonical ensemble is derived.

P. 14.7. Use a method, analogous to that employed for deriving the expressions (13.4.18) for the entropy S of a canonical ensemble, to obtain a formula for the entropy of a grand canonical ensemble in terms of the grand-canonical distribution function (14.3.3). Comment on the result.

P. 14.8. Make use of the relations (14.3.14), (14.3.17), (14.3.22), and (14.3.23) to derive an expression for the entropy S of an assembly of fermions in terms of the mean occupation numbers $\bar{\nu}_i$.

P. 14.9. Show that for $T \to 0$ all the phase points of the particles of an assembly of bosons are to be found in the cell of lowest energy E_0 in the μ-phase space of the particles, and that for small T: $\mu \simeq E_0 - kT/\nu$.

CHAPTER 15

Some Applications of Statistical Mechanics

IN ORDER to illustrate the general principles of statistical mechanics of ensembles and assemblies, derived in the two preceding chapters, we shall apply them in the present chapter to a few problems of particular importance. In sections 15.1 and 15.2 we shall consider the two simplest cases of assemblies of material particles, the "ideal gas", in which the particles are completely independent of one another, and the "ideal solid", in which the particles are arranged in a perfect lattice. In section 15.3 we shall apply statistical mechanics to the electromagnetic radiation in thermodynamic equilibrium with matter, the so-called "black-body radiation". Finally, in section 15.4 the general statistical principles will be applied to the problem of the irregular spontaneous fluctuations of macroscopic parameters to be expected on account of the microscopic structure of matter.

15.1. The ideal gas

An assembly of material particles, which do not interact with one another by forces, and whose total volume is negligibly small compared with the volume V of the container to which the gas is confined, is called an "*ideal gas*". The particles may be different types of atoms and molecules. In the following we shall consider only gases consisting of *one* type of molecules and assume these to behave like "rigid bodies", that is we shall disregard the internal movement of the constituent atoms of the molecules.

A molecule is then a mechanical system of six degrees of freedom, namely three degrees of freedom of translational movement and three degrees of freedom of rotational movement. The former will be represented by the rectangular coordinates x, y, z of the centre of the molecule and the conjugate linear momenta p_x, p_y, p_z. If we assume that the molecule has an axis of symmetry we can represent the rotational degrees of freedom by an azimuth angle φ and the associate angular momentum P_1 about this axis, and two other angles determining the direction of this axis with respect to the frame of reference and the corresponding angular momentum components. The molecule is thus effectively a superposition of a one-dimensional rotator with angular momentum P_1 and a two-dimensional rotator with angular momentum P_2 about an axis normal to the symmetry axis.

If there are no external forces acting on the gas the energy of a molecule is simply its kinetic energy

$$K = \frac{p^2}{2m} + \frac{P_1^2}{2I_1} + \frac{P_2^2}{2I_2}, \tag{15.1.1}$$

where m is the mass of the molecule and I_1 and I_2 are the moments of inertia about the symmetry axis and an axis normal to it respectively. We shall also consider cases where the gas is subjected to a homogeneous field of force. We shall then let the direction of that force coincide with the z-axis of the co-ordinate system and call θ the angle between $\mathbf{P}$ and the z-axis, so that $P_2 = P\cos\theta$. The potential energy of a molecule is then a function $v(\mathbf{r}, \theta)$ and its classical Hamiltonian function is

$$H = \frac{p^2}{2m} + \frac{P_1^2}{2I_1} + \frac{P_2^2}{2I_2} + v(\mathbf{r}, \theta). \tag{15.1.2}$$

We now assume that the gas is in statistical equilibrium with the walls of the container which are kept at a constant temperature T. The distribution of the assembly of gas molecules is then stationary and, under "classical" conditions, given by Boltzmann's law (14.2.4). Thus the probability density function w can, on account of (15.1.1) and (15.1.2), be written in the form

$$w = f_1(\mathbf{p})\, f_2(P_1, P_2)\, f_3(\mathbf{r}, \theta), \tag{15.1.3}$$

where

$$f_1(\mathbf{p}) = \text{const} \exp\left(-\frac{p^2}{2mkT}\right), \tag{15.1.4}$$

$$f_2(P_1, P_2) = \text{const} \exp\left[-\left(\frac{P_1^2}{I_1} + \frac{P_2^2}{I_2}\right)\frac{1}{2kT}\right], \tag{15.1.5}$$

$$f_3(\mathbf{r}, \theta) = \text{const} \exp\left(-\frac{v(\mathbf{r}, \theta)}{kT}\right). \tag{15.1.6}$$

From the expression (4.1.28) for the energy levels of the "particle in the box", namely

$$E_n = \frac{n^2 h^2}{8ml^2},$$

we see that for sufficiently large quantum numbers n, that is for sufficiently high energies of the particles, the relative differences between neighbouring levels become very small, namely

$$(E_n - E_{n-1})/E_n \simeq 2/n,$$

so that the energy spectrum becomes practically continuous. A numerical estimate for ordinary gases shows that this condition is indeed always satisfied except for very low temperatures for which, as we shall see presently, the Boltzmann distribution law does not apply anyhow.

Thus we may assume the momentum p in the expression (15.1.4) for the *translational* distribution function to be continuous, so that the probability density for the components p_x, p_y, p_z of $\mathbf{p}$ takes the form

$$w(p_x, p_y, p_z) = \frac{\exp\left(-\dfrac{p_x^2 + p_y^2 + p_z^2}{2mkT}\right)}{\iiint \exp\left(-\dfrac{p_x^2 + p_y^2 + p_z^2}{2mkT}\right) dp_x\, dp_y\, dp_z}$$

$$= (2\pi mkT)^{-3/2} \exp\left(-\frac{p_x^2 + p_y^2 + p_z^2}{2mkT}\right). \tag{15.1.7}$$

This is equivalent to Maxwell's velocity distribution law.

Integrating the right-hand side of (15.1.7) over a spherical shell of thickness dp and volume $4\pi p^2\, dp$ in momentum space one obtains for the probability density for the *magnitude* p of $\mathbf{p}$

$$f(p) = \sqrt{\frac{2}{\pi}}\, (mkT)^{-3/2}\, p^2 \exp\left(-\frac{p^2}{2mkT}\right). \tag{15.1.8}$$

Thus the part U_t of the internal energy of the gas due to the translational movement of the molecules is, according to (14.1.11),

$$U_t = \nu \int_0^\infty \frac{p^2}{2m} f(p)\, dp = \frac{3}{2} \nu kT, \tag{15.1.9}$$

in agreement with the equipartition theorem (13.4.12) for a system of 3ν degrees of freedom. The corresponding contribution to the specific heat at constant volume per mole of the gas is

$$c_t = \frac{d}{dT}\left(\frac{3}{2} kT\mathcal{N}\right) = \frac{3}{2} R. \tag{15.1.10}$$

Turning now to the distribution function (15.1.5) for the angular momentum of the molecules we have to take into account that the *angular momenta* P_1 and P_2 are quantized by the quantization rules (7.3.6) and (7.3.21) for the one- and the two-dimensional rigid rotator respectively. Thus the distribution function $f_2(P)$ is discontinuous and has the form

$$f_2(u, l) = C(2l + 1) \exp\left[-\frac{\hbar^2}{2kT}\left(\frac{u^2}{I_1} + \frac{l(l+1)}{I_2}\right)\right], \tag{15.1.11}$$

the factor $g_l = 2l + 1$ being due to the fact that the quantum states l of the two-dimensional rotator are $(2l + 1)$-fold degenerate.

The part U_r of the internal energy of the gas due to the rotational movement of the molecules is, according to section 14.1 from (15.1.11),

$$U_r = \nu \frac{\sum_{u,l} \left(\frac{\hbar^2 u^2}{2I_1} + \frac{\hbar^2 l(l+1)}{2I_2} \right) (2l+1) \exp\left[-\frac{\hbar^2}{2kT} \left(\frac{u^2}{I_1} + \frac{l(l+1)}{I_2} \right) \right]}{\sum_{u,l} (2l+1) \exp\left[-\frac{\hbar^2}{2kT} \left(\frac{u^2}{I_1} + \frac{l(l+1)}{I_2} \right) \right]}. \tag{15.1.12}$$

Setting for brevity

$$\beta_1 = \frac{\hbar^2}{2I_1 kT}, \quad \beta_2 = \frac{\hbar^2}{2I_2 kT}, \tag{15.1.13}$$

we can re-write this expression in the form

$$U_r = \nu kT \left\{ \beta_1 \frac{\sum_{u=1}^{\infty} u^2 \exp(-\beta_1 u^2)}{\sum_{u=1}^{\infty} \exp(-\beta_1 u^2)} + \beta_2 \frac{\sum_{l=1}^{\infty} l(l+1)(2l+1) \exp[-\beta_2 l(l+1)]}{\sum_{l=1}^{\infty} (2l+1) \exp[-\beta_2 l(l+1)]} \right\}. \tag{15.1.14}$$

For high temperatures, that is small values of β_1 and β_2, the sums in (15.1.14) may be replaced by integrals over a continuous variable ξ:

$$U_r \simeq \nu kT \left\{ \beta_1 \frac{\int_0^\infty \xi^2 e^{-\beta_1 \xi^2} d\xi}{\int_0^\infty e^{-\beta_1 \xi^2} d\xi} + \beta_2 \frac{\int_0^\infty \xi^3 e^{-\beta_2 \xi^2} d\xi}{\int_0^\infty \xi e^{-\beta_2 \xi^2} d\xi} \right\} = \nu kT \left(\frac{1}{2} + 1 \right) = \frac{3}{2} \nu kT, \tag{15.1.15}$$

a result to be expected from the classical equipartition theorem. The corresponding contribution to the specific heat per mole of the gas is then

$$c_r = \tfrac{3}{2} R, \tag{15.1.16}$$

equal to the contribution c_t from the translational movement (15.1.10).

For large values of β_1 and β_2 (15.1.14) reduces to

$$U_r \simeq \frac{\nu \hbar^2}{2} \left\{ \frac{1}{I_1} \exp\left(-\frac{\hbar^2}{2I_1 kT} \right) + \frac{6}{I_2} \exp\left(-\frac{\hbar^2}{I_2 kT} \right) \right\}, \tag{15.1.17}$$

and the rotational specific heat per mole becomes

$$c_r \simeq \frac{\mathcal{N}\hbar^4}{2kT^2}\left\{\frac{1}{2I_1^2}\exp\left(-\frac{\hbar^2}{2I_1kT}\right)+\frac{6}{I_2^2}\exp\left(-\frac{\hbar^2}{I_2kT}\right)\right\} \tag{15.1.18}$$

which is seen to decrease exponentially to zero on approaching $T = 0$.

Numerical estimations of the expressions (15.1.13) for β_1 and β_2 show that for *monatomic* molecules at normal temperatures both these quantities are in fact large and the rotational specific heat is thus negligibly small at these temperatures. For *diatomic* molecules β_1 is also *large* at normal temperatures, but β_2 is *small* so that at these temperatures the rotational molar heat is equal to R. At low temperatures it decreases below this value and becomes zero at $T = 0$.

We finally discuss the distribution function f_3 for the coordinates of the molecules in a field of force. We first consider the distribution in space of the molecules in a gas that is subjected to a homogeneous gravitational field in the negative z-direction. In this case $v(z) = mgz$ is a continuous function of z alone and (15.1.6) takes the familiar form of the distribution function for the mass density

$$\varrho(z) = \varrho_0 \exp\left(-\frac{mgz}{kT}\right) = \varrho_0 \exp\left(-\frac{\mathcal{N}mgz}{RT}\right), \tag{15.1.19}$$

where ϱ_0 is the density at the bottom of the gas column at $z = 0$.

From the way this formula was derived it appears that it must also apply to "ideal" solutions of particles of equal size in a liquid, in particular to the distribution in height of the number density of the particles of a "colloidal" solution, when mg has to be replaced by the "apparent weight" of the particles. By making measurements of this distribution Perrin made it possible for the first time (1908) to determine directly Avogadro's number $\mathcal{N}$.

As a second example we wish to calculate the magnetic susceptibility of a paramagnetic gas, that is a gas whose molecules carry permanent magnetic dipoles of moment μ, resulting from the orbital movement of its atomic electrons (6.2.23) and the spins of these electrons. Assume that a homogeneous magnetic field of magnitude H in the direction of the z-axis acts on the gas. From (7.3.36) we know that there exists a simple relationship between the angular momentum of an orbiting electron and the magnetic moment due to that movement, and from (7.3.18) it follows that the z-component of this angular momentum is quantized. Thus the z-component M_z of the magnetic moment of a molecule is also quantized. More generally it may be assumed that this is always the case, irrespective of the origin of the magnetic moment. Let thus M_z^s be the value of M_z in a quantum state with quantum number s.

The potential energy v_s of a molecule in the magnetic field is, according to (7.3.37),

$$v_s = -M_z^s \mathsf{H} \tag{15.1.20}$$

For sufficiently *weak* fields we can expand the exponential function in (15.1.6) as follows

$$\exp\left(-\frac{v}{kT}\right) = 1 + \frac{M_z^s \mathsf{H}}{kT} + \cdots \tag{15.1.21}$$

and neglect all terms higher than of the first order in **H**. The magnetic polarization **M** of the gas has the direction of the z-axis, and its magnitude is, according to (14.1.9), given by

$$\mathsf{M} \simeq \nu \frac{\sum_s M_z^s (1 + M_z^s \mathsf{H}/kT)}{\sum_s (1 + M_z^s \mathsf{H}/kT)}, \tag{15.1.22}$$

where ν is now the number of molecules per unit volume of the gas and the summation is to be extended over all allowed quantum states s. Now $\sum_s M_z^s$ vanishes by reasons of symmetry and, in first approximation, the denominator is simply equal to the total number of quantum states. Thus further

$$\mathsf{M} \simeq \frac{\nu \mathsf{H}}{kT} \overline{(M_z^s)^2}, \tag{15.1.23}$$

and the magnetic susceptibility is

$$\chi = \frac{\nu}{kT} \overline{(M_z^s)^2}. \tag{15.1.24}$$

If θ is the angle between the direction of **M** and the z-axis then under classical conditions, when θ can assume all values between 0 and π continuously, one has

$$\overline{(M_z^s)^2} = \mu^2 \frac{\int_0^\infty \cos^2\theta \sin\theta \, d\theta}{\int_0^\infty \sin\theta \, d\theta} = \frac{\mu^2}{3}, \tag{15.1.25}$$

and from (15.1.24) one gets

$$\chi = \frac{\mu^2 \nu}{3kT} \quad \text{(classical)}. \tag{15.1.26}$$

This relationship was first derived by Langevin (1905). The proportionality of χ to $1/T$ is known as "*Curie's law*".

If, at the other extreme, μ is entirely due to a resultant spin 1/2, s can only assume the two values $+\frac{1}{2}$ and $-\frac{1}{2}$ (see section 7.3), and thus $\overline{(M_z^s)^2} = \mu^2$. It follows that now

$$\chi = \frac{\mu^2 \nu}{kT} \quad \text{(spin 1/2)}. \tag{15.1.27}$$

Our next task is to derive an expression for the free energy F of an ideal gas, taking only the translational movement of the molecules into account. For this purpose we assume the system to be a member of a canonical ensemble (see section 13.3) whose density distribution function $\varrho(\mathbf{r})$ in Γ-space is given by (13.3.21) where $H(\mathbf{r})$ is now the Hamiltonian function of the whole assembly of ν molecules. In the absence of external forces this is simply the total kinetic energy of the molecules, that is

$$H = \sum_{i=1}^{\nu} \frac{p_i^2}{2m}, \tag{15.1.28}$$

where p_i is the magnitude of the momentum of the ith molecule.

The partition function Z of the ensemble is now given by the expression (13.4.16) with $N = 3\nu$. However, as the molecules are supposed to be identical, all permutations of particles represent the same distribution (as discussed in section 14.1), and thus the right-hand member of the expression (13.4.16) has to be divided by the number $\nu!$ of permutations. The integrand in (13.4.16) is, according to (15.1.28), equal to

$$\exp\left(-\sum_{i=1}^{\nu} \frac{p_i^2}{2mkT}\right) = \prod_{i=1}^{\nu} \exp\left(-\frac{p_i^2}{2mkT}\right)$$

and the integral over the momentum variables is, with the help of (15.1.7),

$$\underbrace{\int \cdots \int}_{\nu} \prod_{i=1}^{\nu} \exp\left(-\frac{p_i^2}{2mkT}\right) dp_x^{(1)} \cdots dp_z^{(\nu)}$$

$$= \left[\iiint \exp\left(-\frac{p_x^2 + p_y^2 + p_z^2}{2mkT}\right) dp_x dp_y dp_z\right]^{\nu} = (2\pi mkT)^{3\nu/2}.$$

The integral over the position variables is simply

$$\underbrace{\int \cdots \int}_{\nu} dx^{(1)} \cdots dz^{(\nu)} = \left[\int\int\int dx\, dy\, dz\right]^{\nu} = V^{\nu}.$$

Thus

$$Z = \frac{V^{\nu}}{\nu!} \left(\frac{2\pi mkT}{h^2}\right)^{3\nu/2}. \tag{15.1.29}$$

Using Sterling's approximation formula for $\nu!$ we obtain, finally, from (13.4.14), the following expression for the free energy per mole of the gas:

$$F = -kT \ln Z = -RT \ln\left(\frac{V}{\mathcal{N}} T^{3/2}\right) + F_0, \tag{15.1.30}$$

F_0 being independent of volume and temperature.

The contribution to the *molar entropy* of the gas from the translational movement of the molecules can now be derived from the first of the equations (13.4.17)

$$S = -\frac{\partial F}{\partial T} = R \ln \left(\frac{T^{3/2}}{\varrho}\right) + \text{const}, \qquad (15.1.31)$$

where ϱ is the molar density. The *equation of state* of the ideal gas, that is the relationship between pressure p, volume V, and temperature T, is further obtained from the third of the equations (13.4.17)

$$p = -\frac{\partial F}{\partial V} = \frac{RT}{V}. \qquad (15.1.32)$$

So far we had supposed that the energies of the gas molecules were distributed according to the Boltzmann distribution law. We shall now consider a gas consisting of particles which are "Fermions", that is whose resulting spin s is half-integer (see section 14.2), and which therefore obey the Fermi–Dirac distribution law (14.2.24)

$$\zeta_j = \frac{z_j}{\exp\left[(E_j - \mu)/kT\right] + 1}, \qquad (15.1.33)$$

where ζ_j is the number of phase points in the domain Δ_j with average energy E_j in the μ-phase space of the particles, and z_j the number of cells in that domain.

It was already remarked in section 14.2 that under classical conditions, that is for large values of $e^{-\mu/kT}$, the second term in the denominator of (15.1.33) becomes negligibly small compared with the first, and hence the distribution goes over into the Boltzmann distribution. But for small values of $e^{-\mu/kT}$ the two distributions differ very considerably, and consequently the thermodynamic properties of the gas deviate strongly from those of the classical ideal gas. One calls such a gas "*degenerate*" and the quantity μ/kT may be taken as a measure for the "degree of degeneracy": the larger μ/kT the stronger is the degeneracy.

In the following we restrict ourselves to the translational movement of the particles. Let us first calculate the number z_j of cells in the six-dimensional μ-phase space of a particle with cells of volume h^3. If the domain Δ_j covers the region between p_j and $p_j + \Delta p$ in momentum space and the volume V in ordinary space the volume of the domain is

$$4\pi p_j^2 \, \Delta p \cdot V = 2\pi \, (2m)^{3/2} \sqrt{E_j} \, \Delta E \cdot V,$$

and hence

$$z_j = \frac{2\pi \, (2m)^{3/2}}{h^3} V \sqrt{E_j} \Delta E. \qquad (15.1.34)$$

In section 7.3 we showed that a state with spin quantum number s is $G = (2s + 1)$-fold degenerate, because the spin axis can assume $2s + 1$ directions with respect to a fixed direction in space. Hence each of the cells can accommodate not one but G phase points, so that (15.1.33) is to be multiplied by G. Thus we have

$$\zeta_j = \frac{2\pi G\,(2m)^{3/2}}{h^3} V \frac{\sqrt{E_j}\,\Delta E}{\exp\,[(E_j - \mu)/kT] + 1}. \tag{15.1.35}$$

For sufficiently small values of ΔE we may regard $f(E)\,\Delta E = \zeta_j/\nu$ as a quasi-continuous function of E. From (15.1.35) one obtains for the probability density function

$$f(E) = \frac{2\pi G\,(2m)^{3/2}}{h^3 \nu} V \frac{\sqrt{E}}{\exp\,[(E - \mu)/kT] + 1}, \tag{15.1.36}$$

which, of course, must satisfy the condition

$$\int_0^\infty f(E)\,dE = 1. \tag{15.1.37}$$

The relation (15.1.36) is the quantum equivalent of the classical Maxwell distribution function (15.1.8) for degenerate gases. The quantity μ which, as pointed out in section 14.3, has the character of a "chemical potential", is a function of volume and temperature determined by the condition (15.1.37). For $T = 0$, that is for maximum degeneracy, when $\mu/kT \to \infty$, this condition takes, in view of (14.2.25), the form

$$1 = \frac{2\pi G\,(2m)^{3/2} V}{h^3 \nu} \int_0^{\mu_0} \sqrt{E}\,dE,$$

yielding

$$\mu_0 = \frac{h^2}{2m}\left(\frac{3\varrho}{4\pi G}\right)^{2/3}, \tag{15.1.38}$$

where ϱ is the number density of the particles in the gas. It can be shown that for low but finite temperatures μ is approximately equal to

$$\mu = \mu_0\left[1 - \frac{\pi^2}{12}\left(\frac{kT}{\mu_0}\right)^2\right]. \tag{15.1.39}$$

These relations demonstrate that the degree of degeneracy increases with decreasing temperature and increasing particle density and will be particularly significant for gases consisting of particles of very small mass. He^3 is an atom with spin $\frac{1}{2}$ and is therefore a Fermion. Numerical computations show that at normal pressure and temperature the gas is not degenerate but that at temperatures of about 4°K near the liquefaction point $e^{-\mu/kT} \simeq 6$, so that degeneracy already sets in, though only to a slight extent. Electrons are also Fermions with spin $\frac{1}{2}$ and the "free" conduction electrons in a metal may be regarded as an electron gas. Here, however, μ/kT is at normal temperatures of the order 200, so that this gas is very nearly completely degenerate.

A schematic diagram to illustrate the changeover from the classical Maxwellian to the strongly degenerate distribution function is shown in Fig. 29.

We shall now derive an expression for the translational part of the internal energy of a strongly degenerate gas in analogy to the classical procedure used in (15.1.9),

$$U_t = \nu \int_0^\infty E\, f(E)\, dE. \tag{15.1.40}$$

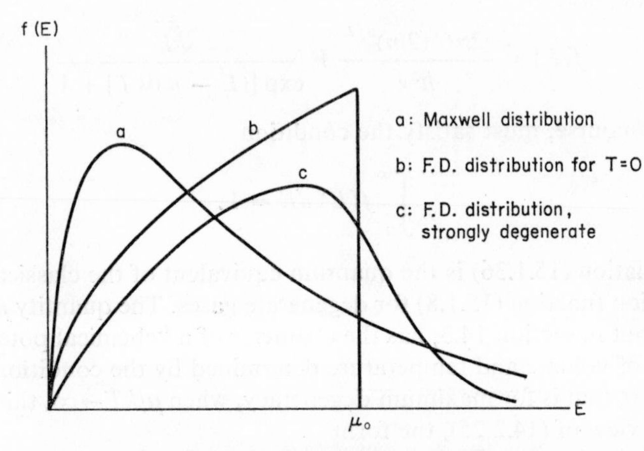

FIG. 29. The energy distribution function for an ideal classical and a degenerate gas of fermions.

In the limiting case $T = 0$ one obtains in this way from (15.1.36) and (15.1.38)

$$U_0 = \frac{2\pi G\,(2m)^{3/2}}{h^3}\, V \int_0^{\mu_0} E^{3/2}\, dE = \frac{3}{5}\,\mu_0 \nu. \tag{15.1.41}$$

It is seen from this formula that the internal energy of the gas is *finite* at zero temperature, in sharp contrast to the behaviour of a classical ideal gas. U_0 is called the "*zero-point energy*" of the gas. It can be shown that for low but finite temperatures the same procedure leads to the approximate formula

$$U = U_0 \left[1 + \frac{5\pi^2}{12} \left(\frac{kT}{\mu_0} \right)^2 \right]. \tag{15.1.42}$$

The translational contribution to the specific heat per mole of a degenerate gas is therefore

$$c_r = \frac{dU}{dT} = \frac{\pi^2 k^2 \mathcal{N}}{2\mu_0}\, T = \frac{3}{2}\, R \left(\frac{\pi^2}{3} \frac{kT}{\mu_0} \right). \tag{15.1.43}$$

This shows that the specific heat decreases linearly with decreasing temperature to reach zero at $T = 0$ in agreement with the demand of the third law of thermodynamics. It is further seen that the expression for the molar heat is obtained from the classical expression (15.1.10) by multiplication with the factor $\frac{\pi^2}{3}\frac{kT}{\mu_0}$ which for an electron gas at normal temperatures in a metal is of the order 1/100. This explains the fact that under these circumstances the contribution of the free electrons to the specific heat of a metal is only of the order of 1 per cent.

15.2. The ideal solid

In the preceding section we considered assemblies of particles whose mutual interaction forces were supposed to be negligibly small. We now go to the other extreme by assuming the interaction forces between the particles to be so strong that, in the absence of external disturbances, they are held in stable equilibrium positions about which they can perform oscillations if the system is *slightly* disturbed. We shall call such an assembly an "*ideal solid*" if the equilibrium positions of the particles form a perfect, regular "*space lattice*", constituting a *perfect crystal.* In the following we shall be only concerned with the translational, vibrational movement of the particles which we assume to be of one type only. If their total number is N the system has $3N$ degrees of freedom.

The thermal movement in a solid consists of irregular oscillations of the particles about their ideal lattice sites. For the purpose of the application of statistical mechanics to this thermal movement it is, however, more convenient to regard the movement as a superposition of *elastic waves* which propagate through the body in various directions with different frequencies, amplitudes, and phases, provided that the displacements are so small that the waves can be assumed to be *harmonic.*

In analogy to wave motion in a continuous medium, which was briefly treated in section 1.2, a state of stationary wave motion in a finite discontinuous lattice under given boundary conditions is characterized by "*normal modes of vibration*" consisting of stationary waves of various frequencies ν_i and wave vectors $\mathbf{k}_i$. However, whereas in a continuum the number of normal modes is unlimited this is not so in a discrete lattice, and it can be shown quite generally that, whatever the boundary conditions, the total number of normal modes must be equal to the number $3N$ of degrees of freedom of the system.

We shall demonstrate the correctness of this statement on a simple special example. Assume that the particles are situated at the corners of a three-dimensional network of cubes of side a, a so-called "simple cubic lattice". Let a parallelepipedon with sides l_1, l_2, l_3 be cut out of the crystal parallel to the crystallographic axes. One then has evidently $l_1 l_2 l_3 = a^3 N$. If the boundary condition consists in keeping the atoms on the surface of the body fixed, then the components k_x, k_y, k_z of the wave vectors of the stationary waves, constituting the normal modes of vibration, must satisfy the conditions

$$k_x = \frac{n_1}{2l_1}, \quad k_y = \frac{n_2}{2l_2}, \quad k_z = \frac{n_3}{2l_3}, \tag{15.2.1}$$

where n_1, n_2, n_3 are integers. As the shortest possible wavelength cannot be smaller than $2a$, these numbers run from 1 to $\frac{l_1}{a}, \frac{l_2}{a}, \frac{l_3}{a}$ respectively so that the total number of possible values of the wave vector $\mathbf{k}$ is equal to $\frac{l_1 l_2 l_3}{a^3} = N$. To each of these waves belong three modes of vibration, namely one longitudinal and two transverse at right angles to one another. Thus the number of normal modes of vibration of the system is indeed $3N$, in accordance with the previously mentioned general theorem.

The frequencies ν of the normal modes of vibration form a "line spectrum", but the separation between neighbouring lines is so small that it can very closely be approximated by a continuous spectrum with a "*spectral density function*" $g(\nu)$, $g(\nu)\Delta\nu$ being the number of modes with frequencies between ν and $\nu + \Delta\nu$. Because the total number of normal modes is $3N$ the function $g(\nu)$ must satisfy the normalizing condition

$$\int_0^\infty g(\nu)\, d\nu = 3N. \tag{15.2.2}$$

Each normal mode of vibration is equivalent to a linear harmonic oscillator of frequency ν, and its energy E can therefore be quantized according to the rule (5.1.14)

$$E_\nu = (n + \tfrac{1}{2})\, h\nu \quad (n = 1, 2, 3, \ldots). \tag{15.2.3}$$

We first wish to calculate the internal energy U of an ideal solid. This is the sum of the static potential energy Φ of the lattice due to the interatomic and external forces, and the mean energy of the vibrations of the particles. The average of the energy E_ν is, from (14.2.1) and (15.2.3), equal to

$$\bar{E}_\nu = \frac{\sum_{n=1}^{\infty} E_\nu \exp(-E_\nu/kT)}{\sum_{n=1}^{\infty} \exp(-E_\nu/kT)} = \frac{h\nu}{\exp(h\nu/kT) - 1} + \frac{h\nu}{2}. \tag{15.2.4}$$

Here the first term is equal to the original Planck expression (2.2.3), and the second is the zero point energy of the harmonic oscillator in its lowest quantum level $n = 1$. Thus the total vibrational energy of the lattice is

$$U_v = \int_0^\infty \bar{E}_\nu g(\nu)\, d\nu = \int_0^\infty \frac{h\nu}{\exp(h\nu/kT) - 1} g(\nu)\, d\nu + \frac{h}{2}\int_0^\infty \nu g(\nu)\, d\nu. \tag{15.2.5}$$

The second term in (15.2.5) can, on account of (15.2.2), be written in the form

$$U_0 = \frac{3Nh}{2} \frac{\int_0^\infty \nu g(\nu)\, d\nu}{\int_0^\infty g(\nu)\, d\nu} = \frac{3Nh}{2} \bar{\nu}, \tag{15.2.6}$$

where $\bar{\nu}$ is the average vibration frequency over the whole spectrum of the normal modes. U_0 represents the total *zero-point energy* of the lattice. The existence of the zero point vibrations in crystal lattices is proved by the results of experiments on the scattering of X-rays by crystals at low temperatures.

The total internal energy of the ideal solid is now given by the expression

$$U = \Phi(V) + U_0 + \int_0^\infty \frac{h\nu}{\exp(h\nu/kT) - 1} g(\nu)\, d\nu. \tag{15.2.7}$$

For low temperatures the third term tends towards zero and for high temperatures it tends to the classical value (2.1.5)

$$kT \int_0^\infty g(\nu) d\nu = 3NkT.$$

The molar specific heat at constant volume of the solid

$$c_v = \left(\frac{\partial U}{\partial T}\right)_v \tag{15.2.8}$$

is thus seen to tend to the classical Dulong–Petit value $3R$ [see (2.1.6)], at high temperatures and to zero on approaching $T = 0$, as demanded by the third law of thermodynamics. Generally one has from (15.2.7) and (15.2.8)

$$c_v = k \int_0^\infty \frac{(h\nu/kT)^2 \exp(h\nu/kT)}{[\exp(h\nu/kT) - 1]^2} g(\nu)\, d\nu. \tag{15.2.9}$$

In order to determine the dependence of the specific heat of a solid on temperature it is necessary first to evaluate the frequency spectrum of its normal modes of vibration. However, for an approximate treatment one can use simplified "models", defined by particularly simple frequency spectra. We shall consider two such models briefly here, namely the "*Einstein model*" and the "*Debye model*".

In the Einstein model (1907) the spectrum is reduced to a single line with frequency ν_E, called the "*Einstein frequency*". In this case $g(\nu)$ is a "δ-function", that is it is zero everywhere except at $\nu = \nu_E$, and in virtue of (15.2.2) the expression (15.2.9) becomes

$$c_v = 3Nk\varphi_E\left(\frac{\theta_E}{T}\right). \tag{15.2.10}$$

Here the function $\varphi_E(x)$, the so-called "*Einstein function*", is defined by

$$\varphi_E(x) = \frac{x^2 e^x}{(e^x - 1)^2}, \tag{15.2.11}$$

and θ_E is a constant of the dimension of temperature, called "*Einstein temperature*",

$$\theta_E = \frac{h\nu_E}{k}. \tag{15.2.12}$$

The Einstein treatment of the specific heat of solids has already been discussed in section 2.2, and formulae (2.2.8) and (2.2.9) derived there are indeed identical with those obtained here. A graph of the Einstein function is shown in Fig. 30, which demonstrates how φ_E decreases from unity for $x = 0$ and reaches zero exponentially on approaching $x = \infty$.

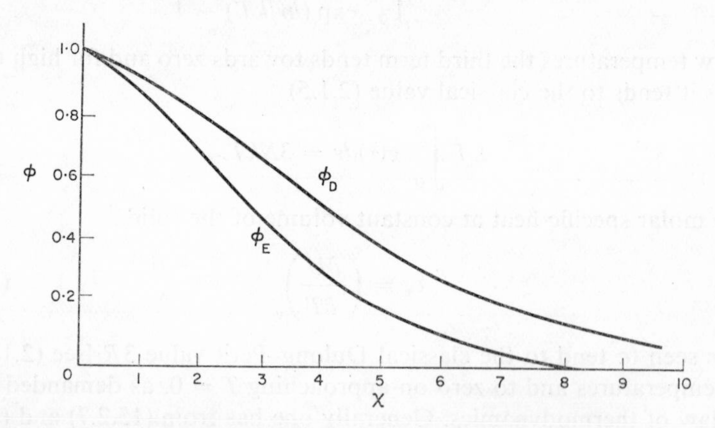

FIG. 30. The Einstein and the Debye functions for the dependence of the specific heat of ideal solids on temperature.

In the Debye model (1912) it is assumed that the end points of the wave vectors of the normal modes of vibration are uniformly distributed over the interior of a sphere of radius k_m in an abstract "k-space", and that the wave velocity u of these waves is independent of frequency. The ratio of the number of normal modes between k and $k + \Delta k$ to the total number of normal modes is therefore equal to the ratio of the volumes of a shell of radius k and thickness Δk to the total volume of the sphere, namely

$$4\pi k^2 \, \Delta k \Big/ \frac{4\pi}{3} k_m^3 = 3\nu^2 \, \Delta\nu / \nu_m^3,$$

where ν_m is the highest frequency that can occur, which is called the "*Debye frequency*". On the other hand, according to the definition of the spectral density function and as a consequence of (15.2.2), this ratio is equal to $g\,\Delta\nu/3N$. Thus the "*Debye spectrum*" is described by the relations

$$\left.\begin{aligned} g(\nu) &= \frac{9N}{\nu_m^3}\nu^2 \quad \text{for } \nu \leqslant \nu_m, \\ g(\nu) &= 0 \qquad\quad \text{for } \nu > \nu_m. \end{aligned}\right\} \tag{15.2.13}$$

For the specific heat we now obtain from (15.2.9)

$$c_v = 3Nk\varphi_D\left(\frac{\theta_D}{T}\right), \tag{15.2.14}$$

where the function φ_D, called "*Debye function*", is defined by

$$\varphi_D(x) = \frac{3}{x^3}\int_0^x \frac{\xi^4 e^\xi}{(e^\xi - 1)^2}\,d\xi \tag{15.2.15}$$

and where θ_D, called the "*Debye temperature*", is a characteristic constant of the substance concerned which is related to the Debye frequency ν_m by the relation

$$\theta_D = \frac{h\nu_m}{k}. \tag{15.2.16}$$

It is easily seen that in the limiting case of high temperatures, that is small values of x, the function $\varphi_D(x)$ tends to unity, and hence c_v again tends to the classical Dulong–Petit value. For low temperatures, on the other hand, the integral in (15.2.15) may be extended to infinity and thus becomes a constant. Consequently $\varphi_D(x)$ becomes proportional to x^{-3} and $c_v(T) \sim T^3$. This relationship is known as "*Debye's law*". The general shape of the Debye function is shown in Fig. 30.

The most convenient way to account for the other thermodynamic properties of an ideal solid is to evaluate the free energy F as a function of volume and temperature from the canonical distribution function Z, as outlined in section 13.4, from which the entropy and the equation of state are derived by means of the relations (13.4.17).

As the normal modes of vibration are equivalent to a system of *independent* harmonic oscillators the free energy F of the whole system is equal to the sum of the free energies F_j of the separate modes with frequencies ν_j. From (13.4.15) we obtain for the partition sum Z_j of the jth mode with the help of (15.2.3)

$$\begin{aligned} Z_j &= \sum_{n=0}^{\infty} \exp(-E_n/kT) = \sum_{n=0}^{\infty} \exp\left[-\left(n + \frac{1}{2}\right)(h\nu_j/kT)\right] \\ &= \frac{\exp(-h\nu_j/2kT)}{1 - \exp(-h\nu_j/kT)}, \end{aligned} \tag{15.2.17}$$

since all the quantum states n are non-degenerate ($g_n = 1$).

From (13.4.14) we have now

$$F_j = -kT \ln Z_j = \frac{h\nu_j}{2} + kT \ln\left[1 - \exp\left(-\frac{h\nu_j}{kT}\right)\right], \quad (15.2.18)$$

and the contribution to the free energy of the body from the thermal vibrations of its constituent atoms is the sum of the F_j over all $3N$ normal modes. In order to get the total free energy one has to add to this the potential lattice energy Φ. Thus, finally,

$$F(V, T) = \Phi(V) + \sum_{j=1}^{3N} \left\{\frac{h\nu_j}{2} + kT \ln\left[1 - \exp\left(-\frac{h\nu_j}{kT}\right)\right]\right\}. \quad (15.2.19)$$

The entropy S of the body follows from (15.2.19) by means of the first of the relations (13.4.17) which results in

$$S = -\left(\frac{\partial F}{\partial T}\right)_V = -k \sum_{j=1}^{3N} \left\{\ln\left[1 - \exp\left(-\frac{h\nu_j}{kT}\right)\right] - \frac{h\nu_j/kT}{\exp(h\nu_j/kT) - 1}\right\}. \quad (15.2.20)$$

Combining the expressions (15.2.19) and (15.2.20) we obtain again a formula for the internal energy U

$$U = F + ST = \Phi + \sum_{j=1}^{3N} \left\{\frac{h\nu_j}{2} + \frac{h\nu_j}{\exp(h\nu_j/kT) - 1}\right\}. \quad (15.2.21)$$

As the expressions in curly brackets are, according to (15.2.4), the mean energies of the various modes of vibration the formula simply reiterates the fact that U is the sum of the potential lattice energy Φ and the total vibrational energy of all the $3N$ modes of vibrations of the lattice.

For very high temperatures the expression (15.2.20) reduces to

$$S \simeq -k \sum_{j=1}^{3N} \left\{\ln \frac{h\nu_j}{kT} - 1\right\} \simeq 3Nk \ln T + \text{const},$$

so that for a mole of the substance

$$dS \simeq \frac{3R\, dT}{T} = \frac{c_v\, dT}{T} = \frac{dU}{T},$$

which is the classical definition for dS at constant volume. This is a necessary condition for the expression (15.2.20) to be correct. For very low temperatures, on the other hand, one sees that $S \to 0$ for $T \to 0$, as demanded by the third law of thermodynamics.

In order to derive the equation of state of an ideal solid from its free energy we make use of the third of the relations (13.4.17) and obtain from (15.2.19)

$$p = -\left(\frac{\partial F}{\partial V}\right)_T = -\frac{d\Phi}{dV} - h\sum_{j=1}^{3N}\left\{\frac{1}{2} + \frac{1}{\exp\left(\frac{h\nu}{kT}\right) - 1}\right\}\frac{d\nu_j}{dV}. \tag{15.2.22}$$

The first term on the right-hand side of this equation is independent of temperature and simply represents the *elastic* pressure due to a uniform compression of the body. The second term is dependent on temperature but would be zero if the frequencies of the thermal vibrations of the atoms were independent of volume. Thus it appears that the phenomenon of *thermal expansion* of a solid is due to the dependence of these frequencies on volume. This explanation of the thermal expansion of solids was first given by Grüneisen (1912).

If one adopts the simple "Einstein model" of the solid all the ν_j are equal to ν_E and (15.2.22) reduces, with the help of (15.2.4), to

$$p + \frac{d\Phi}{dV} = -3N\bar{E}_\nu \frac{d(\ln \nu_E)}{dV}. \tag{15.2.23}$$

Following Grüneisen one can define a quantity

$$\gamma = -\frac{d(\ln \nu_E)}{d(\ln V)} \tag{15.2.24}$$

which, under plausible assumptions about the interatomic forces in the lattice, is a constant, called the "*Grüneisen constant*". Under these circumstances (15.2.23) assumes the simple form

$$\left(p + \frac{d\Phi}{dV}\right)V = \gamma U_t(T), \tag{15.2.25}$$

where U_t is the thermal part of the internal energy, including the zero-point energy. This equation is known as the "*Grüneisen–Mie*" equation of state.

The rigorous solution of the problem of deriving the thermodynamic properties of an ideal solid of given lattice structure with given interatomic force law is very complicated. It was first attempted by Born on the basis of his "dynamic lattice theory of crystals" (1914).

15.3. The black-body radiation

So far we have in this chapter applied the general principles of statistical mechanics to *material bodies* only; in this section we shall apply them to a *non-material* system, namely the electromagnetic radiation field in the interior of an enclosure. We shall assume that the material of which the enclosure is made is capable of absorbing radiation of all frequencies and that it is artificially kept at a constant temperature. Under these conditions radiation will continuously be absorbed and re-emitted from the interior surface of the enclosure; we shall assume that eventually an equilibrium between enclosure material and radiation inside the enclosure will establish itself when the electromagnetic energy density in the whole interior is uniform and the radiation acquires a characteristic continuous spectrum. Such a radiation is usually called "*black-body radiation*" because, if a small hole is made in the enclosure, the radiation issuing from this hole will be the same as that emitted by a "perfectly black" body, that is a body which absorbs all impinging radiation completely.

The problem of the theoretical derivation of the spectral intensity distribution of the black-body radiation has already been given a preliminary treatment in sections 2.1 and 2.2, in connection with the difficulties encountered by trying to solve it by the methods of classical physics and the overcoming of these difficulties by the advent of quantum theory. We shall now attack this problem on a broader front by means of the general methods of statistical mechanics and quantum theory along three different lines.

The idea of the first method is due to Jeans (1905) and consists in resolving the radiation in the interior of the enclosure into normal modes of vibration in the form of stationary electromagnetic waves, in close analogy to the procedure adopted in the preceding section for the ideal solid.

For the sake of simplicity we shall, as in section 15.2, assume that the enclosure has the shape of a parallelepipedon with sides l_1, l_2, l_3, and that the boundary condition consists in keeping the amplitudes of the waves zero at the walls of the enclosure. In that case the allowed values of the components of the wave vector of the standing waves are given by the equations (15.2.1), that is

$$k_x = \frac{n_1}{2l_1}, \quad k_y = \frac{n_2}{2l_2}, \quad k_z = \frac{n_3}{2l_3}, \tag{15.3.1}$$

where n_1, n_2, n_3 are integers. It follows that the equation

$$4k^2 = \frac{4\nu^2}{c^2} = \frac{n_1^2}{l_1^2} + \frac{n_2^2}{l_2^2} + \frac{n_3^2}{l_3^2} \tag{15.3.2}$$

must be satisfied.

Now imagine an abstract space in which the sets of triplets (n_1, n_2, n_3) are represented by the points of a cubic lattice. If x_1, x_2, x_3 are rectangular coordinates in that space, then

$$\frac{x_1^2}{l_1^2} + \frac{x_2^2}{l_2^2} + \frac{x_3^2}{l_3^2} = \frac{4\nu^2}{c^2}$$

is the equation of an ellipsoid with semi-axes $\frac{2l_1\nu}{c}, \frac{2l_2\nu}{c}, \frac{2l_3\nu}{c}$ whose volume is equal to $\frac{32\pi}{3}\frac{\nu^3}{c^3}V$ where $V = l_1l_2l_3$ is the volume of the enclosure. The number of lattice points with *positive* integer values of n_1, n_2, n_3 satisfying (15.3.2) is therefore $\frac{4\pi}{3}\frac{\nu^3}{c^3}V$. Since electromagnetic waves are transverse there are two independent states of polarization belonging to each triplet of values n_1, n_2, n_3, and hence the total number of normal modes of vibration with frequencies smaller or equal to ν in the enclosure is $\frac{8\pi}{3}\frac{\nu^3}{c^3}V$. It follows that the spectral density function for the black-body radiation is given by

$$g(\nu) = 8\pi\nu^2 V/c^3. \tag{15.3.3}$$

In order to determine the distribution of energy density in the spectrum of the black-body radiation it remains to assign to each of its monochromatic components a certain average energy. Jeans argued that this should be the equipartition value for the harmonic oscillator kT. This would lead to a spectral energy density proportional to $\nu^2 T$, that is Rayleigh's law (2.1.4), which, as was previously pointed out, does not agree with experiment and, moreover, results in the total electromagnetic energy density in the enclosure being infinitely large, because the spectrum has no high frequency "cut-off" like the Debye spectrum (15.2.13).

Debye therefore suggested in 1910 to "quantize" the radiation in the enclosure by assigning to each mode the mean energy value (2.2.3), following from Planck's quantization rule for the harmonic oscillator. In this way one obtains for the spectral energy density $u(\nu)$ per unit volume of the enclosure from (15.3.3)

$$u(\nu) = \frac{h\nu}{\exp(h\nu/kT) - 1}\frac{g(\nu)}{V} = \frac{8\pi h\nu^3}{c^3}\frac{1}{\exp(h\nu/kT) - 1}, \tag{15.3.4}$$

which is "*Planck's radiation law*".

For high temperatures T (15.3.4) takes the form

$$u(\nu) \simeq \frac{8\pi\nu^2}{c^3}kT, \tag{15.3.4a}$$

which can also be obtained direct by multiplying the expression (15.3.3) for $g(\nu)$ by the equipartition value kT of the energy of a harmonic oscillator. This is the Rayleigh radiation law which was previously discussed in section 2.1 and has the form (2.1.4). For low temperatures, on the other hand, (15.3.4) becomes

$$u(\nu) \simeq \frac{8\pi h\nu^3}{c^3}e^{-h\nu/kT}. \tag{15.3.4b}$$

This is known as "*Wien's radiation law*".

The total energy density u follows from (15.3.4) by integration over all ν from zero to infinity:

$$u = \int_0^\infty u(\nu)\, d\nu = \frac{8\pi(kT)^4}{(hc)^3} \int_0^\infty \frac{\xi^3}{e^{\xi-1}}\, d\xi = \frac{8\pi^5}{15} \frac{(kT)^4}{(hc)^3}, \tag{15.3.5}$$

which is the "*Stefan–Boltzmann*" law.

Although one obtains the right result by means of this procedure, it is open to serious objection. For the correct quantum mechanical treatment of the quantization problem of the linear harmonic oscillator leads in fact to the expression (15.2.4) for the mean energy which entails the zero-point term $h\nu/2$. Evidently integration over ν results in an infinitely large contribution to the energy density in the enclosure which is physically unsatisfactory. However, since only *changes* in the electromagnetic field can give rise to observable effects, one can argue that the zero-point energy may be disregarded and that one can "normalize" the energies of the monochromatic components of the radiation by choosing $h\nu/2$ as the zero levels.

One is, of course, not really justified in applying quantization rules, derived from the fundamental laws of quantum *mechanics*, to non-material systems like electromagnetic *fields*, and one should instead derive them from *quantum electrodynamics*, a branch of quantum theory which lies outside the scope of this textbook. However, even quantum field theory is beset with the same kind of difficulties, and the methods of "re-normalization" currently used for resolving these difficulties are rather artificial and not likely to be the final answer.

We now turn to the second method for deriving the black-body radiation law which is due to Bose (1924) and which subsequently gave rise to the development of the Bose–Einstein statistics. The idea is to regard the radiation in the interior of the enclosure as a "*photon gas*" and, as photons are particles with integer spin, to apply to them the distribution law (14.2.21).

However, one has to take into account that photons, in contrast to "material fundamental particles" are not permanent and can be annihilated by absorption by the walls of the enclosure and created by emission from the walls. Thus the first of the equations (14.2.7) which was used in the derivation of the Bose–Einstein distribution law, does not hold here. Consequently the constant α does not appear in the analogue for the expression (14.2.10) for the occupation number ζ_j of the domain Δ_j in the μ-phase space of the photons and in all subsequent formulae, in particular (14.2.20). That means that the constant μ in (14.2.21) has to be suppressed and the coarse-grained distribution law for photons takes the form

$$\zeta_j = \frac{z_j}{\exp(E_j/kT) - 1}. \tag{15.3.6}$$

The volume in momentum space of a domain corresponding to a narrow energy interval between E and $E + \Delta E$ is $4\pi p^2 \Delta p$, and since, according to (2.2.5),

$$E = h\nu, \quad \mathbf{p} = h\mathbf{k} \quad \left(k = \frac{\nu}{c}\right), \tag{15.3.7}$$

the volume in μ-phase space belonging to photons with frequencies between ν and $\nu + \Delta\nu$ is

$$\frac{4\pi h^3}{c^3} \nu^2 \Delta\nu V.$$

However, for each given energy and momentum the spin of the photon can have two opposite directions with respect to the propagation direction (corresponding to the two possible states of polarization) so that the number of cells of volume h^3 in the domain is equal to

$$z_j = \frac{8\pi\nu^2 V}{c^3} \Delta\nu. \tag{15.3.8}$$

From (15.3.6), (15.3.7), and (15.3.8) we now have

$$\zeta(\nu) = \frac{8\pi\nu^2 V}{c^3} \frac{\Delta\nu}{\exp(h\nu/kT) - 1}, \tag{15.3.9}$$

which finally leads to the following expression for the spectral energy density distribution of the radiation:

$$u(\nu) = \frac{\zeta(\nu)\, h\nu}{V \Delta\nu} = \frac{8\pi h\nu^3}{c^3} \frac{1}{\exp(h\nu/kT) - 1}, \tag{15.3.10}$$

which is seen to be identical with the expression (15.3.4) obtained by the first method. The advantage of the photon gas theory is that, in contrast to the theory using the quantization of the normal modes of vibration of the enclosure, it does not lead to an (infinite) zero-point energy and therefore seems to be more adequate to deal with problems of this kind.

The photon theory is based on the fundamental quantum mechanical aspect of the equivalence of waves and particles, photons being equivalent to plane electromagnetic waves, and particles with finite rest mass being equivalent to plane De Broglie waves. The success of the photon gas theory suggests to apply similar principles to the statistical mechanics of solids, which in the preceding section was dealt with by the method of the normal modes of vibration of the thermal movement of the atoms in a solid. Thus one may imagine the mechanical (or acoustic) waves, which constitute these modes, to be equivalent to hypothetical particles without rest mass, usually called "*phonons*", obeying relations similar to (15.3.7), except that **k** is now the wave vector of a plane *acoustic* wave.

If it is further assumed that these particles are "bosons", then the distribution law (15.3.6) holds for a "phonon gas" as well, and the vibrational energy of a solid body is

$$U_v = \sum_j h\nu_j\zeta_j = \sum_j \frac{h\nu_j z_j}{\exp(h\nu_j/kT) - 1}. \qquad (15.3.11)$$

Here z_j, as before, is the number of cells in the domain corresponding to phonons with frequencies between ν_j and $\nu_j + \Delta\nu$ in the μ-phase space of the phonons. We can thus put $z_j = g(\nu_j)\,\Delta\nu$, and going to the limit of infinitesimally small intervals $\Delta\nu$ we obtain from (15.3.11)

$$U_v = \int_0^\infty \frac{h\nu}{\exp(h\nu/kT) - 1}\, g(\nu)\, d\nu. \qquad (15.3.12)$$

This expression is identical with the first term of the formula (15.2.5), previously obtained by the method of normal modes of vibration if $g(\nu)$ is identified with the spectral density function; for this reason $g(\nu)$ is frequently referred to as the "*phonon spectrum*".

This example shows that the phonon theory can successfully be applied to the statistical mechanics of solid bodies, although the fact that the expression (15.3.12) does not contain the experimentally established zero-point contribution points to certain shortcomings of the phonon gas concept. The idea of phonons behaving like particles capable of interacting with one another and with other particles, like electrons and photons, by "collision", has proved extremely useful in the theoretical treatment of scattering phenomena of these particles in solids. However, one has to keep in mind that the notion of phonons is only an artifice; for whereas the photon theory assumes that the energy of a light wave is actually concentrated in the associated photon, the mechanical waves in a solid body consist in the vibrations of the atoms, and the energy of such a wave must therefore be located in the vibrating atoms and cannot be concentrated in another "particle".

We now return to the subject of the black-body radiation. Following Einstein we shall now show that Planck's radiation law can be derived by a third method from certain assumptions concerning the statistical properties of the interaction processes between matter and radiation. Einstein's basic propositions, pronounced before the advent of quantum mechanics in 1917, are as follows: (1) there exists a fixed "transition probability rate" A_{ij} for the *spontaneous* transition of an atom from a quantum state i with energy E_i to a state j with energy $E_j < E_i$, resulting in the emission of a photon of frequency $\nu = (E_i - E_j)/h$; (2) there further exists a transition probability rate $B_{ij}u(\nu)$ for the *induced* transition between the same levels under the influence of a radiation field with spectral density $u(\nu)$, again resulting in the emission of a photon of frequency ν; (3) an induced transition probability rate $B_{ji}u(\nu)$ in the opposite direction, resulting in the absorption of a photon of frequency

ν by the atom. Whereas assumptions (1) and (3) are the quantum analogues of the classical mechanism of emission and absorption of radiation by electric oscillators, assumption (2) is essentially quantum theoretical in nature. Its correctness has only comparatively recently been proved by experiments which also revealed that the induced radiation is *coherent* with the primary one. This fact is the basis for the function of the devices known as "*masers*" and "*lasers*" which have since become of the greatest practical importance for telecommunication and have opened completely new fields of research.

Let us now consider a canonical ensemble of material systems in equilibrium with a radiation field with spectral density distribution function $u(\nu)$. The relative numbers of systems in a quantum state i is then given by the relation (13.3.24)

$$\varrho_i = C \exp(-E_i/kT). \tag{15.3.13}$$

If Einstein's propositions are extended to hold for arbitrary material systems the principle of microscopic reversibility (13.2.4) demands that for equilibrium to be maintained the conditions

$$\varrho_i (A_{ij} + B_{ij} u(\nu)) = \varrho_j B_{ji} u(\nu) \tag{15.3.14}$$

and

$$B_{ij} = B_{ji} \tag{15.3.15}$$

must be fulfilled. From (15.3.13), (15.3.14), and (15.3.15) then follows

$$\frac{A_{ij}}{B_{ij} u(\nu)} + 1 = \exp\left(\frac{E_i - E_j}{kT}\right) = \exp\left(\frac{h\nu}{kT}\right)$$

and thus

$$u(\nu) = \frac{A_{ij}}{B_{ij}} \frac{1}{\exp(h\nu/kT) - 1}. \tag{15.3.16}$$

This is as far as one can go on Einstein's assumptions alone. If one wishes to derive the complete black-body radiation formula one has to evaluate the factor A_{ij}/B_{ji} which can be done only by using quantum mechanics proper. In the special case of atoms capable of emitting a "dipole radiation" the quantity A_{ij} is given by the expression (6.2.11a), and the quantity B_{ij} is the factor of $u(\nu)$ in the expression (6.2.11b), so that in the present notation

$$A_{ij} = \frac{16\pi^4 e^2}{3c^3 h} |x_{ij}|^2 \nu^3, \quad B_{ij} = \frac{2\pi^3 e^2}{3h^2} |x_{ij}|^2, \tag{15.3.17}$$

and thus

$$\frac{A_{ij}}{B_{ij}} = \frac{8\pi h \nu^3}{c^3}. \tag{15.3.18}$$

The latter expression is seen to be independent of the special emission mechanism because it contains, apart from the frequency ν, only universal constants.

Hence it must be generally valid, for the radiation formula (15.3.16) must hold independently of the material of the enclosure walls. Indeed, if the expression (15.3.18) is substituted into (15.3.16) one obtains again Planck's law (15.3.4).

15.4. The statistical mechanics of fluctuations

It was shown in section 13.1 that the *ensemble* averages $\langle a \rangle$ of macroscopic observable quantities A and their variances σ_a^2 can be calculated on the basis of the general principles of the statistical mechanics of ensembles. It was further pointed out that the expressions obtained in this way also represented the *time* averages of these parameters and their spontaneous fluctuations in time about the averages, measured by means of repeated observations of the stochastic time series of individual systems. In this section general formulae for the magnitude of these fluctuations will be derived and applied to a few special cases.

It will be shown that in general the fluctuations of macroscopic parameters are very small; nevertheless, it has been possible not only to detect them by means of very sensitive measuring devices but also to verify the formulae quantitatively. This is not only relevant from the theoretical point of view, for it proves decisively the correctness of the statistical interpretation of the classical laws of thermodynamics. But it is also of great practical significance because the existence of the fluctuations sets a natural limit to the accuracy of the measurement of macroscopic observables, and in particular makes it impossible to detect very weak regular "signals" in telecommunication above the background of irregular "noise".

According to (13.1.1) the probability for finding the quantity A to have a value between a and $a + \delta a$ at a time t is

$$f(a, t)\,\delta a = \int_{\omega} \cdots \int \varrho\,(\mathbf{r}, t)\,dV, \tag{15.4.1}$$

where ϱ is the distribution function of the ensemble in question and the integral is to be extended over the region ω of the Γ-phase space bounded by the hypersurfaces $A = a$ and $A = a + \delta a$.

We now assume the ensemble to be stationary and canonical. In that case, and under "classical" conditions, we have from (13.3.21)

$$\varrho(\mathbf{r}) = C \exp\,(-H(\mathbf{r})/kT). \tag{15.4.2}$$

Provided the interval δa is sufficiently narrow we obtain, on substitution of (15.4.2) into (15.4.1),

$$f(a)\,\delta a = \exp\,(-\langle E(\mathbf{r})\rangle/kT)\,\omega = \exp\,(-U(a)/kT)\,\omega \tag{15.4.3}$$

where $U(a)$ is the internal energy of the system as a function of the parameter A.

For a fixed value of a, and hence of $U(a)$ Boltzmann's theorem (13.4.22),

$$S = k \ln G \tag{15.4.4}$$

holds, which connects the entropy S of the system with the number G of ways in which the state of the system, defined by A being between a and $a + \delta a$, can be realized. G is thus equal to the number of cells in the region ω of the Γ-phase space of our system and consequently equal to

$$G = h^n \omega. \tag{15.4.5}$$

Hence from (15.4.4) and (15.4.5)

$$S = \frac{U - F}{T} = k \ln \omega + \text{const}$$

or

$$\omega = \text{const} \exp \left(\frac{U - F}{kT} \right). \tag{15.4.6}$$

If the expression (15.4.6) is substituted into (15.4.3) one obtains the following formula for the probability density function for the parameter A:

$$f(a) = C \exp(-F(a)/kT), \tag{15.4.7}$$

where the constant C is determined by the condition

$$\int_{-\infty}^{+\infty} f(a)\, da = 1. \tag{15.4.8}$$

Since $F(a)$ is a function which reflects only *macroscopic* properties of the system concerned, one sees that the fluctuation properties of macroscopic parameters only depend on the macroscopic characteristics of the observed systems and are completely independent of the special microscopic mechanism responsible for the fluctuations as long as classical conditions prevail. The only indication that the fluctuations are caused by the thermal movement of the constituent particles of the system is the appearance of the quantity kT in the expression (15.4.7). This remarkable fact makes it possible to derive formulae for the magnitude of the spontaneous fluctuations of a system solely on the basis of a knowledge of its macroscopic properties.

Let a_0 be the *equilibrium* value of a and let F_0 be the corresponding value of the free energy. We can then write the relation (15.4.7) in the form

$$f(\alpha) = f(0) \exp \left(\frac{F_0 - F}{kT} \right) = f(0) \exp(-W(\alpha)/kT), \tag{15.4.9}$$

where $W(\alpha)$ is the work to be done from outside on the system in order to remove it from its equilibrium state to a state $\alpha = a - a_0$ (Einstein, 1905 b). This shows that the equilibrium is stable if $W(\alpha) > 0$ for all α, for then the equilibrium state has maximum probability. We also see that unless $W(\alpha)$ is of the order of magnitude kT, that is unless the deviation α from equilibrium is very small, $f(a)/f(a_0)$ will also be small, so that, indeed, as mentioned before, very sensitive measuring instruments must be used for detecting the fluctuations unless the systems themselves are of microscopic dimensions.

Provided that $W(\alpha) = F - F_0$ is positive definite, we can expand this quantity in a Taylor series as follows:

$$F - F_0 = \frac{1}{2}\left(\frac{d^2F}{da^2}\right)_0 \alpha^2 + \cdots \tag{15.4.10}$$

and neglect the higher terms in α for sufficiently small values of α. Thus, under these conditions, we can transform (15.4.9) into

$$f(\alpha) = f(0) \exp\left(-\frac{b}{2kT}\alpha^2\right), \tag{15.4.11}$$

where

$$b = \left(\frac{d^2F}{da^2}\right)_0. \tag{15.4.12}$$

The function $f(\alpha)$ is thus seen to be a *Gaussian error function*, and hence the variance of the fluctuation magnitude of A or the mean square of α is simply given by

$$\sigma_a^2 = \overline{\alpha^2} = \frac{kT}{b}. \tag{15.4.13}$$

We shall now illustrate the general results expressed in the relations (15.4.7) to (15.4.13) by a few simple examples.

(a) First we consider a system consisting of a small colloidal particle in a fluid with "apparent weight" w. According to the laws of macroscopic mechanics and thermodynamics, the particle ought to sink down with constant speed and, once it has reached the bottom of the vessel, remain there in static equilibrium. In actual fact, owing to the Brownian movement of the particle,

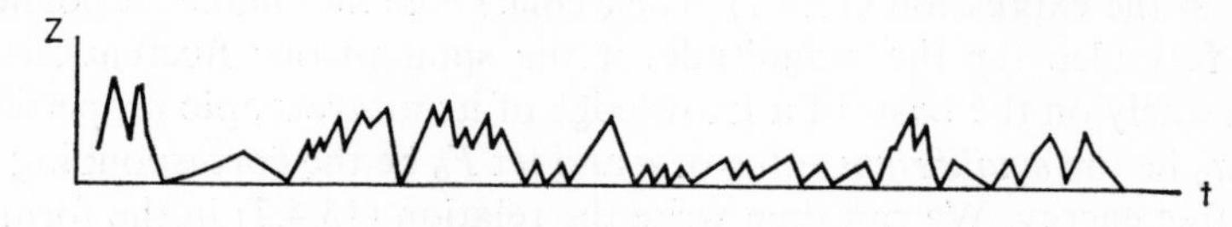

FIG. 31. Record of the statistical time series of the height of a particle above the bottom of a vessel owing to its Brownian movement.

it will, after having reached the bottom, rise again and move downwards and upwards in an irregular manner as indicated by Fig. 31 (Fürth, 1917) in which the height z of the particle above the bottom is plotted as a function of time.

If one puts $\alpha = z$ in (15.4.9), one has, in this case, evidently $W(z) = wz$ and thus, with the help of (15.4.8), one obtains for the time distribution function (13.1.9) for z,

$$\theta(z) = \frac{w}{kT} \exp\left(-\frac{wz}{kT}\right) \quad (z \geqslant 0). \tag{15.4.14}$$

From records like the one shown in Fig. 31 this formula can indeed be verified. It is seen that the *time* distribution law (15.4.14) has the same form as the *space* distribution law (15.1.19) for a colloidal solution under the action of gravity, as one would expect on the basis of the equivalence between stationary assemblies of non-interacting particles and stationary stochastic time series of individual particles.

For the average height $\bar{z}$ of the particle above the bottom one has from (15.4.14) by definition

$$\bar{z} = \int_0^\infty z\theta(z)\, dz = \frac{kT}{w}. \tag{15.4.15}$$

(b) As a second example we consider a system of one macroscopic degree of freedom which is held in stable equilibrium in a position $x = 0$ by an "elastic force", like a pendulum or a torsion balance. Observations of such systems with very sensitive measuring devices reveal that they in fact perform an irregular movement about the equilibrium position. This is illustrated by Fig. 32 (Kappler, 1931) in which the displacement angle of a tiny mirror suspended by a fine quartz fibre is plotted against time.

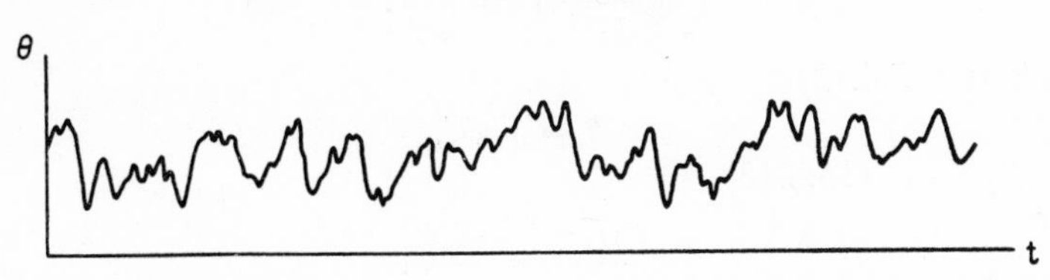

FIG. 32. Record of the statistical time series of the angular position of a torsional system, owing to its rotational Brownian movement.

In this case one has $\alpha = x$ and $W(\alpha) = bx^2/2$, where b is the "restoring force" (or the "restoring couple") so that the time distribution law has the form (15.4.11)

$$\theta(x) = \sqrt{\frac{b}{2\pi kT}} \exp\left(-\frac{b}{2kT}x^2\right), \tag{15.4.16}$$

and the variance of x is from (15.4.13)

$$\sigma^2 = (\overline{x^2}) = kT/b. \tag{15.4.17}$$

This shows that in order to be able to detect the fluctuations the restoring force must be made very small indeed.

(c) Our general formulae also hold if the parameters concerned are of a non-mechanical character. For example, it is to be expected that in an electric circuit, consisting of a capacitor of capacitance C and a conductor of high resistance, a fluctuating electric voltage v will develop spontaneously across the condenser plates. In this case one has $w(v) = \frac{C}{2} v^2$, and from (15.4.13) one obtains at once for the magnitude of the voltage fluctuations

$$\sigma_v^2 = \overline{v^2} = kT/C. \tag{15.4.18}$$

(d) As a last example we consider a material body of volume V in thermodynamic equilibrium with its surroundings. We expect that V, which is an "extensive" macroscopic parameter, will be subject to spontaneous fluctuations. Let $\alpha = \delta V$ be the deviation of V from its equilibrium value. If the compressibility of the substance in the neighbourhood of equilibrium is $\varkappa$ then one has, by definition,

$$\frac{1}{\varkappa} = -V\frac{dp}{dV},$$

and according to the thermodynamic relation

$$p = -\left(\frac{\partial F}{\partial V}\right)_T,$$

$$\left(\frac{\partial^2 F}{\partial V^2}\right)_0 = \frac{1}{\varkappa V}.$$

Hence from (15.4.12)

$$b = 1/\varkappa V,$$

and finally from (15.4.13)

$$\frac{\overline{(\delta V)^2}}{V^2} = \frac{\varkappa kT}{V}. \tag{15.4.19}$$

The mean square of the relative spontaneous volume fluctuations is thus seen to be proportional to the compressibility and inversely proportional to the volume itself.

The phenomenon can also be described as a fluctuation of mass density ϱ of the body since $\delta\varrho/\varrho = -\delta V/V$, so that from (15.4.19)

$$\frac{\overline{(\delta\varrho)^2}}{\varrho^2} = \frac{\varkappa kT}{V}. \tag{15.4.20}$$

This relation shows that for a body of macroscopic dimensions the *bulk* density fluctuation will be negligibly small. But one can always imagine a macroscopically homogeneous body to be subdivided into small domains which are still large compared with the molecular dimensions, but small enough to be subject to appreciable density fluctuations. Thus, at any moment, the substance will appear to exhibit *spatial* fluctuations of density, giving it the appearance of irregular inhomogeneity or "graininess".

In a transparent substance this gives rise to a scattering of light, an effect which is experimentally well established. Equation (15.4.20) shows that this effect is proportional to the compressibility of the substance. It is therefore extremely weak in transparent crystals but quite appreciable in gases and very conspicuous in the neighbourhood of the critical point of a gas where the compressibility is large. Measurements of the intensity of scattered light have made it possible to determine Boltzmann's constant k independently in agreement with other determinations of this constant, and thus to verify the theory of the fluctuations of macroscopic parameters.

Problems and exercises

P. 15.1. Derive expressions for the most probable value v_m, the mean value $\bar{v}$ and the root mean square value $\sqrt{\overline{v^2}}$ of the speed of the molecules of mass m in a gas of temperature T according to Maxwell's distribution law. Also calculate the "spread" $\delta = \dfrac{\sqrt{\overline{v^2} - (\bar{v})^2}}{\bar{v}}$ of the distribution of the values of v about their mean.

P. 15.2. An ideal gas with ν molecules per unit volume, carrying permanent electric dipoles of magnitude Π, is subjected to a homogeneous electric field of field strength $\mathbf{E}$. Assume that the "effective" field $\mathbf{E}^*$, acting on such a dipole, composed of $\mathbf{E}$ and the "inner field" produced by the other dipoles, is given by the expression $\mathbf{E}^* = \mathbf{E} + \gamma\mathbf{P}$. Evaluate the dielectric susceptibility of the gas under classical conditions, following the same procedure as used in section 15.1 for the analogous magnetic problem.

P. 15.3. Prove that the sub-volume in coordinate space, available to $\mathcal{N}$ molecules of finite size ω in the Γ-phase space of a gas in a container of volume V, is equal to $V(V-\omega)\times \times(V-2\omega)\cdots(V-(\mathcal{N}-1)\,\omega)$. Assuming the total volume $\mathcal{N}\omega$ of the molecules to be small compared with V show, following the procedure given in section 15.1, that the equation of state of such a gas has the form $p = RT/(V-b)$, where $b = \dfrac{\mathcal{N}\omega}{2}$.

P. 15.4. Assume that the interaction potential between two equal particles at a distance r is given by $\varphi = -\dfrac{a}{r^m} + \dfrac{b}{r^n}$, where a, b, m, n $(n > m)$ are constants. Discuss the shape of the function $\varphi(r)$ and derive a formula for the equilibrium distance r_0 of the particles. Use this formula to express $\varphi(r)$ in terms of a, r_0, m, n. Give an expression for the static potential energy $\Phi(V)$ of a cubic lattice consisting of $\mathcal{N}$ such particles (assuming that only interaction between nearest neighbours need be taken into account) in terms of $V = \mathcal{N}r^3$, $V_0 = \mathcal{N}r_0^3$ and $A = \mathcal{N}^{(m/2)+1}$. Hence calculate the molar energy H_s of sublimation of the substance and its isothermal compressibility, defined by $\varkappa = -\dfrac{1}{V_0}\left(\dfrac{\partial V}{\partial p}\right)_0$, and state the relationship between H_s and $\varkappa$.

P. 15.5. A particle of mass μ is held in equilibrium by two other fixed particles at distances r to the right and to the left under the interaction potential used in P. 15.4. Show that, if the equilibrium is slightly disturbed, the particle will perform a harmonic oscillation about the equilibrium position. Calculate the frequency ν of these oscillations in terms of the constants a, r_0, m, n. Show that if $n \gg m$, the quantity (15.2.24) is a constant and its value is equal to $\gamma = (n+2)/6$.

P. 15.6. Assuming the equation of state (15.2.25) with a constant γ to apply to a certain solid substance, derive an expression for the coefficient of thermal expansion $\alpha = \frac{1}{V}\left(\frac{\partial V}{\partial T}\right)_p$ at constant pressure in terms of the specific heat c_v at constant volume and the isothermal compressibility, making use of the thermodynamic relation

$$\left(\frac{\partial p}{\partial T}\right)_V = -\left(\frac{\partial V}{\partial T}\right)_p \bigg/ \left(\frac{\partial V}{\partial p}\right)_T.$$

P. 15.7. Transform equation (15.3.4) so as to obtain the distribution of energy density $U_\lambda(\lambda)$ of the black-body radiation in terms of the wavelength λ, $u_\lambda(\lambda)\, d\lambda$ representing the energy per unit volume of the enclosure between the wavelengths λ and $\lambda + d\lambda$. Use this relation for deriving an expression for the wavelength λ_m at which U_λ is a maximum, and show that it obeys Wien's displacement law. Compare the results with those of P. 2.1.

P. 15.8. Verify the thermodynamic relationship

$$F = -T\int_0^T \left(\frac{U}{T^2}\right) dT$$

from which the free energy can be obtained as a function of V and T if $U(V, T)$ is known. Use this relationship to derive expressions for the free energy and the entropy of a volume V of black-body radiation and the radiation pressure on the walls of the enclosure.

P. 15.9. Show how the formula for the radiation pressure, derived in P. 15.8, can also be obtained from classical electromagnetic theory or from the theory of photons, taking into account that the pressure on the container walls is the result of the transfer of momentum by the impinging radiation on the wall.

P. 15.10. Show that, if the process of induced transition from a higher to a lower energy level under the influence of a radiation field is not taken into account, Wien's radiation law (15.3.4b) follows from the consideration in section 15.3 instead of Planck's law (15.3.4).

P. 15.11. Use formula (15.4.20) to show that the fluctuations in the numbers ν of particles in a small volume ω of an *ideal* gas, containing on the average $\bar{\nu}$ particles, satisfy the relationship $\frac{\overline{(\nu - \bar{\nu})^2}}{\bar{\nu}^2} = \frac{1}{\bar{\nu}}$. Prove that this relation must hold if there exists a fixed probability p for any of the gas molecules to be found within ω at any particular time. Show further that in a *real* gas the relationship takes the form $\frac{\overline{(\nu - \bar{\nu})^2}}{\bar{\nu}^2} = \frac{\varkappa}{\varkappa_0}\frac{1}{\bar{\nu}}$, where $\varkappa$ is the compressibility of the real gas and $\varkappa_0$ that of an ideal gas.

P. 15.12. In analogy to example (c) in section 15.4, derive a formula for the mean square of the electric charge Q which will accumulate spontaneously on the plates of the capacitor in the circuit described there. Show that the mean electric energy due to this phenomenon is equal to $kT/2$.

P. 15.13. Show that in a circuit with self-inductance L the mean square of the electric current flowing spontaneously in the circuit is equal to $\overline{I^2} = kT/L$ and that the mean magnetic energy due to this current is equal to $kT/2$.

APPENDIX

Solutions to Problems

P. 1.1.

$$K = \frac{m}{2}\frac{ds^2}{dt^2} = \frac{m}{2}(\dot{r}^2 + r^2\dot{\theta}^2 + r^2\sin^2\theta\cdot\dot{\varphi}^2),$$

$$p_r = \frac{\partial K}{\partial r} = m\dot{r}^2, \quad p_\theta = \frac{\partial K}{\partial \theta} = mr^2\dot{\theta}, \quad p_\varphi = \frac{\partial K}{\partial \varphi} = mr^2\sin^2\theta\cdot\dot{\varphi}.$$

P. 1.2.

$$H = \frac{p_\theta^2}{2mr^2} + \frac{p_\varphi^2}{2mr^2\sin^2\theta} + U(\theta, \varphi),$$

$$\dot{p}_\theta = -\frac{\partial H}{\partial \theta} = \frac{p_\varphi^2}{mr^2}\cot\theta\,(1 + \cot^2\theta) - \frac{\partial U}{\partial \theta},$$

$$\dot{p}_\varphi = -\frac{\partial H}{\partial \varphi} = -\frac{\partial U}{\partial \varphi}.$$

For U independent of φ: $\dot{p}_\varphi = 0$, $p_\varphi = \text{const.}$

P. 1.3.

$$K = \frac{1}{2}\sum_i m_i r_i^2 \sin^2\theta_i \dot{\varphi}^2 = \frac{I}{2}\dot{\varphi}^2,$$

where $I = \sum_i m_i r_i^2 \sin^2\theta_i$ is the constant "moment of inertia".

$$H = \frac{I}{2}\dot{\varphi}^2 + U(\varphi), \quad P = \frac{\partial H}{\partial \dot{\varphi}} = I\dot{\varphi}, \quad \dot{P} = -\frac{\partial H}{\partial \varphi} = -\frac{\partial U}{\partial \varphi} = M(\varphi).$$

Hence $I\ddot{\varphi} = M(\varphi)$ is the equation of motion.

P. 1.4. The general solution of the differential equation for $r \geqslant R$, which satisfies the boundary condition, is $\Phi = B/r$. For $r \leqslant R$ the first integral which satisfies the condition of continuity at $r = R$ is

$$\frac{d(r\Phi)}{dr} = 2\pi\varrho\,(R^2 - r^2).$$

A second integration under the condition: Φ finite at $r = 0$ yields

$$\Phi = 2\pi\varrho\,(R^2 r - r^3/3).$$

The condition of continuity at $r = R$ again demands $B = \frac{4\pi}{3} R^3 \varrho$, hence, finally,

$$\Phi = 2\pi\varrho r\,(R^2 - r^2/3) \quad \text{for} \quad r \leqslant R,$$

$$\Phi = \frac{4\pi}{3}\varrho\frac{R^3}{r} \qquad \text{for} \quad r \geqslant R.$$

P. 1.5. The proposed solution yields

$$\frac{\partial^2\Phi}{\partial x^2} = f''(x - ut) + g''(x + ut), \qquad \frac{\partial^2\Phi}{\partial t^2} = u^2 f''(x - ut) + u^2 g''(x + ut).$$

Hence (1.2.2) is satisfied, and as the solution contains two independent arbitrary functions f and g the solution is the most general under the stated condition. The distribution $\Phi = f(x)$ at $t = 0$ progresses with speed u in the positive x-direction and the distribution $\Phi = g(x)$ at $t = 0$ in the opposite direction simultaneously.

P. 1.6. The wave equation is now

$$\frac{\partial^2 (r\Phi)}{\partial t^2} = u^2 \frac{\partial^2 (r\Phi)}{\partial r^2}$$

and its general solution

$$\Phi = \frac{1}{r}[f(r - ut) + g(r + ut)]$$

represents two waves, one going outwards and one going inwards, whose amplitudes are inversely proportional to r.

P. 1.8. The Hamiltonian is

$$H(r, \theta) = \frac{p_r^2}{2m} + \frac{p_\theta^2}{2mr^2} + \frac{B}{r},$$

and the Hamilton–Jacobi equation

$$\frac{1}{2m}\left(\frac{\partial W}{\partial r}\right)^2 + \frac{1}{2mr} + \left(\frac{\partial W}{\partial \theta}\right)^2 + \frac{B}{r} = E.$$

Setting $W = W_r(r) + W_\theta(\theta)$ one obtains from this by means of separation of variables

$$\frac{dW_\theta}{d\theta} = A \quad \text{and} \quad \left(\frac{dW_r}{dr}\right)^2 = 2mE - \frac{2mB}{r} - \frac{A^2}{r^2},$$

thus

$$W = A\theta + \int \left(2mE - \frac{2mB}{r} - \frac{A^2}{r^2}\right)^{1/2} dr + C,$$

where A and C are arbitrary constants.

P. 1.9. As the speed along s_1 and s_2 is the same the principle of least action demands that $s = s_1 + s_2$ be a minimum. If x is the projection of s_1 on the wall and $L - x$ that of s_2 one has

$$S = (a_1^2 + x^2)^{1/2} + (a_2^2 + (L - x)^2)^{1/2}.$$

From $ds/dx = 0$ follows: $\sin\alpha_1 - \sin\alpha_2 = 0$.

P. 1.10. The eiconal equation is from (1.3.23)

$$\left(\frac{d\Phi_x}{dx}\right)^2 + \left(\frac{d\Phi_y}{dy}\right)^2 = \alpha^2 y^2 \quad (\alpha = \text{const}).$$

By means of separation of variables one obtains

$$\Phi_x = Ax + B \quad \text{and} \quad \frac{d\Phi_y}{dy} = \sqrt{\alpha^2 y^2 - A^2}$$

from which follows

$$\Phi_y = \frac{\alpha y}{2}\sqrt{y^2 - A^2/\alpha^2} - \frac{A^2}{2\alpha}\ln\left(y + \sqrt{y^2 - A^2/\alpha^2}\right) + C.$$

P. 1.11. According to Fermat's principle (1.3.24) the integral $\int_{P_1}^{P_2} \frac{ds}{u}$ must be a minimum for a certain x. Since $u = c/\mu_1$ in the first medium and $u = c/\mu_2$ in the second medium, this reduces to the condition that

$$S = \mu_1 s_1 + \mu_2 s_2 = \mu_1 (a_1^2 + x^2)^{1/2} + \mu_2 (a_2^2 + (L - x)^2)^{1/2}$$

be a minimum. From $dS/dt = 0$ then follows

$$\mu_1 \sin \alpha_1 - \mu_2 \sin \alpha_2 = 0.$$

P. 1.12. Fermat's principle demands that the lengths $s = s_1 + s_2$ of all the rays from F_1 to F_2 be equal, where s_1 is the ray path from F to a point P on the mirror and s_2 the ray path from P to F. This is indeed the case if the cross-section of the mirror with the plane containing s_1 and s_2 is an ellipse.

P. 2.1. Setting $h\nu/kT = x$, the expression (2.2.4) can be written $I_x \simeq T^3 \frac{x^3}{e^x - 1}$ so that $I \simeq \int_0^\infty I_x \, dx \simeq T^4$. Further, as the maximum of I_x occurs at a certain x_m, the corresponding ν_m/T must be a constant.

P. 2.2. The equations of motion are

$$x = a \sin(2\pi\nu_1 t + \varphi_1), \quad y = b \sin(2\pi\nu_2 t + \varphi_2),$$

characterizing a Lissajous figure which is contained within a rectangle with sides $2a$ and $2b$. The quantum conditions are $E_1 = n_1 h$ and $E_2 = n_2 h$ for the x- and y-components of the oscillation, and the total energy $E(n_1, n_2) = h(n_1\nu_1 + n_2\nu_2)$. If ν_1 and ν_2 are commensurate one has $\nu_1 = \nu_0 s_1$, $\nu_2 = \nu_0 s_2$, and hence $E = h\nu_0 (n_1 s_1 + n_2 s_2)$. All quantum states for which $n_1 s_1 + n_2 s_2$ is equal to the same integer have the same energy E.

P. 2.3. From $\omega_n = \frac{m e^4 Z^2}{(n\hbar)^3}$ and equation (2.2.12) follows $v_n = \omega_n r_n = \frac{e^2 Z}{n\hbar}$, and hence from (2.2.19) one has

$$\lambda_n = \frac{h}{m v_n} = \frac{2\pi n \hbar^2}{m e^2 Z} = 2\pi r_1 n.$$

Appendix

P.3.1. From (3.1.11) $\mu_1 = \sqrt{E + eV_1}$, $\mu_2 = \sqrt{E + eV_2}$, and hence from Snell's law $\frac{\sin\alpha_1}{\sin\alpha_2} = \sqrt{\frac{E + eV_2}{E + eV_1}}$. Classically, as the tangential component of the velocity of the electrons at the discontinuity surface is unchanged, one has $v_1 \sin\alpha_1 = v_2 \sin\alpha_2$. Further from the conservation of energy $\frac{v_2^2}{v_1^2} = \frac{E + eV_2}{E + eV_1}$. Thus

$$\frac{\sin\alpha_1}{\sin\alpha_2} = \frac{v_2}{v_1} = \sqrt{\frac{E + eV_2}{E + eV_1}}.$$

P.3.2. From P.1.3 the Hamiltonian function is $H(\varphi) = \frac{p^2}{2I} + U(\varphi)$, and hence the Hamiltonian operator $\mathscr{H}(\varphi) = -\frac{\hbar^2}{2I}\frac{d^2}{d\varphi^2} + U(\varphi)$. Thus the wave equation is $\frac{d^2\psi}{d\varphi^2} + \frac{2I}{\hbar^2}(E - U(\varphi))\,\psi = 0$, which differs from (3.1.6) insofar as x is replaced by φ and m by the moment of inertia I.

P.3.3. From $\mathbf{H} = \operatorname{curl} \mathbf{A}$ and $H_x = H_y = 0$, $H_z = H$ follows

$$A_x = -Hy/2, \quad A_y = Hx/2, \quad A_z = 0.$$

Thus

$$(\nabla \cdot \mathbf{A}) = 0, \quad (\mathbf{A} \cdot \nabla\psi) = \frac{H}{2}\left(-y\frac{\partial\psi}{\partial x} + x\frac{\partial\psi}{\partial y}\right) = \mathsf{H}\frac{\partial\psi}{\partial\varphi}, \quad \mathsf{A}^2 = \frac{\mathsf{H}^2}{4}(y^2 + z^2) = \frac{\mathsf{H}^2}{4}r^2,$$

and from (3.1.27)

$$\nabla^2\psi - \frac{ie}{2\hbar c}\mathsf{H}\frac{\partial\psi}{\partial\varphi} - \frac{e^2}{4\hbar^2c^2}\mathsf{H}^2r^2\psi + \frac{2m}{\hbar^2}E\psi = 0.$$

P.3.4.

$$\varphi(t) = C\int_{-\infty}^{+\infty} e^{-\alpha\nu^2}\, e^{2\pi i\nu t}\, d\nu = C\sqrt{\frac{\pi}{\alpha}}\, e^{-\pi^2t^2/\alpha}$$

$$\int_{-\infty}^{+\infty} \varphi(t)\, e^{-2\pi i\nu t}\, dt = C\, e^{-\alpha\nu^2} = f(\nu),$$

$$\overline{(\Delta\nu)^2} = \overline{\nu^2} = \frac{1}{2\alpha}, \quad \overline{(\Delta t)^2} = \overline{t^2} = \frac{\alpha}{2\pi^2}, \quad \Delta\nu\,\Delta t = \frac{1}{2\pi}.$$

P.3.5.

$$f(\nu) = 0 \quad \text{for} \quad \nu < \nu_0 - \frac{\Delta\nu}{2} \quad \text{and} \quad \nu > \nu_0 + \frac{\Delta\nu}{2},$$

$$f(\nu) = C \quad \text{for} \quad \nu_0 - \frac{\Delta\nu}{2} < \nu < \nu_0 + \frac{\Delta\nu}{2},$$

$$\varphi(t) = C\int_{\nu_0 - \Delta\nu/2}^{\nu_0 + \Delta\nu/2} e^{2\pi i\nu t}\, d\nu = \frac{C}{\pi t}\sin(\pi\,\Delta\nu t)\, e^{2\pi i\nu_0 t}.$$

The signal consists of an oscillation with frequency ν_0 and amplitude $A(t) = \frac{C}{\pi t}\sin(\pi\,\Delta\nu t)$. The function $A(t)$ has the shape of Fig. 3 and consists of a central peak around $t = 0$ terminated by $t = \pm 1/\Delta\nu$. Thus the width of this peak is $\Delta t \simeq 1/\Delta\nu$.

P.3.6. The first diffraction maximum occurs at an angle α with the z-direction for which $\sin\alpha = \lambda/\Delta x$. With $\lambda = h/p$ one has

$$p \sin\alpha\, \Delta x = \Delta p_x \cdot \Delta x = h.$$

P.4.1. Instead of the second of the equations (4.1.17) one now has $\psi_2 = B_1 e^{-2\pi i k' x} + B_2 e^{2\pi i k' x}$, where $k'^2 = 2m(E - U_0)/h^2$. Thus the expression for N_t/N_i is obtained from (4.1.21) if $\varkappa$ is replaced by ik' and $U_0 - E$ by $E - U_0$:

$$\frac{N_t}{N_i} = \left[1 + \frac{U_0^2 \sin^2 (2\pi k' a)}{4E(E - U_0)}\right]^{-1}.$$

The second term in the bracket is zero for $k'a = n/2$, i.e. when the thickness of the wall is equal to an integer multiple of the half-wave length of the De Broglie wave of the particle inside the wall. In this case the wall is completely transparent. The first maximum of N_t/N_i occurs when $k'a = \frac{1}{4}$. In this case $N_t/N_i = [1 + U_0^2/4E(E - U_0)]^{-1}$. For very large $E - U_0$ the transmission is always perfect.

P.4.3. The only difference between this problem and problem P.4.1 is that U_0 has now the opposite sign. Thus now

$$\frac{N_t}{N_i} = \left[1 + \frac{U_0^2 \sin^2 (2\pi k' a)}{4E(E + U_0)}\right]^{-1}, \quad \text{where} \quad k'^2 = 2m(E + U_0)/h^2.$$

Complete transmission will only occur if $k'a = n/2$ or if E is very large.

P.4.4. From symmetry reasons it follows that inside the well

$$\psi_1 = A \sin 2\pi k x \quad \text{or} \quad \psi_1 = A \cos 2\pi k x \quad (k = \sqrt{2mE}/h).$$

Within the right-hand wall one has, as in (4.1.17),

$$\psi_2 = B_1 e^{-2\pi\varkappa x} + B_2 e^{2\pi\varkappa x} \quad (\varkappa = \sqrt{2m(U_2 - E)}/h),$$

and to the right of the wall

$$\psi_3 = C \cos [2\pi k' a + \varphi] \quad (k' = \sqrt{2m(E - U_1)}/h).$$

With the usual boundary conditions for ψ one obtains for the odd case

$$e^{-2\pi\varkappa a}\left(\sin \pi k l - \frac{k}{\varkappa} \cos \pi k l\right) + e^{2\pi\varkappa a}\left(\sin \pi k l + \frac{k}{\varkappa} \cos \pi k l\right)$$

$$= \frac{2C}{A} \cos [\pi k' (l + 2a) + \varphi], \tag{a}$$

$$e^{-2\pi\varkappa a}\left(\sin \pi k l - \frac{k}{\varkappa} \cos \pi k l\right) - e^{2\pi\varkappa a}\left(\sin \pi k l + \frac{k}{\varkappa} \cos \pi k l\right)$$

$$= \frac{2C}{A} \frac{k'}{\varkappa} \sin [\pi k' (l + 2a) + \varphi]. \tag{b}$$

As C must necessarily tend to zero for large a the terms with $e^{2\pi\varkappa a}$ must vanish, which leads to the quantum condition $\tan(\pi k l) = -k/\varkappa$. From this relation and the equations (a) and (b) follows

$$\frac{C^2}{A^2} = \frac{1}{4}\left(1 + \frac{\varkappa^2}{k'^2}\right) e^{-4\pi\varkappa a}\left(\sin \pi k l - \frac{k}{\varkappa}\cos \pi k l\right)^2$$

$$= \frac{1 - U_1/U_2}{1 - U_1/E}\exp\left(\frac{-4\pi a}{h}\sqrt{2m\,(U_2 - E)}\right).$$

P. 4.5. $$f(x) = \sqrt{\frac{2}{l}}\sum_{n=2,4,\ldots} \alpha_n \sin\frac{n\pi x}{l} + \sqrt{\frac{2}{l}}\sum_{n=1,3,\ldots} \beta_n \cos\frac{n\pi x}{l}.$$

From (4.2.10)

$$\alpha_n = 0, \quad \beta_n = \frac{8a}{n^2\pi^2}\sqrt{\frac{l}{2}} \quad (n = 1, 3, 5, \ldots).$$

Thus

$$f(x) = \frac{8a}{\pi^2}\left\{\cos\frac{\pi x}{l} + \frac{1}{9}\cos\frac{3\pi x}{l} + \frac{1}{25}\cos\frac{5\pi x}{l} + \cdots\right\}.$$

P. 4.8. If the reciprocal matrix to $\mathfrak{A}$ is $\mathfrak{B} = \begin{pmatrix} b_{11} & b_{12} \\ b_{21} & b_{22} \end{pmatrix}$ then by definition

$$\begin{pmatrix} a_{11} & a_{12} \\ a_{21} & a_{22} \end{pmatrix}\begin{pmatrix} b_{11} & b_{12} \\ b_{21} & b_{22} \end{pmatrix} = \begin{pmatrix} 1 & 0 \\ 0 & 1 \end{pmatrix}.$$

This leads to four linear equations for the elements of $\mathfrak{B}$ whose solution is

$$\mathfrak{B} = \frac{1}{\begin{vmatrix} a_{11} & a_{12} \\ a_{21} & a_{22} \end{vmatrix}}\begin{pmatrix} a_{22} & -a_{12} \\ -a_{21} & a_{11} \end{pmatrix}.$$

P. 4.9. $$x_{jk} = \frac{2}{l}\int_{-l/2}^{+l/2} x \begin{Bmatrix} \sin(j\pi x/l) \\ \cos(j\pi x/l) \end{Bmatrix}\begin{Bmatrix} \sin(k\pi x/l) \\ \cos(k\pi x/l) \end{Bmatrix} dx,$$

where the sin-functions are to be used for even quantum numbers and the cos-functions for odd quantum numbers. Evaluation of the integrals gives the result

$$x_{jk} = 0 \quad \text{if } j \text{ and } k \text{ have the same parity}$$

and

$$x_{jk} = \frac{2l}{\pi^2}\left\{\frac{\sin(j+k)\,\pi/2}{(j+k)^2} \pm \frac{\sin(j-k)\,\pi/2}{(j-k)^2}\right\},$$

where the upper sign holds if j is even and k odd and the lower sign if j is odd and k even. One further has

$$|x_{jk}| = \frac{8l}{\pi^2}\frac{jk}{(j^2 - k^2)^2}.$$

As $x_{jk} = x_{kj}$ and the elements are real the matrix is Hermitian. The diagonal elements are zero and hence the expectation value of x is also zero, which is evident from the symmetry of the problem.

P.5.1. For even k: $A_k = \dfrac{(-2)^{k/2}\, n\,(n-2)\cdots(n-k+2)}{k!}$.

For odd k: $A_k = \dfrac{(-2)^{(k-1)/2}\,(n-1)\,(n-3)\cdots(n-k+2)}{k!}$.

P.5.2. The normalized eigenfunctions are:

$$\psi_1(\xi) = \left(\frac{2}{a\sqrt{\pi}}\right)^{1/2} \xi\, e^{-\xi^2/2}, \quad \psi_2(\xi) = -\left(\frac{1}{2a\sqrt{\pi}}\right)^{1/2} (1 - 2\xi^2)\, e^{-\xi^2/2},$$

$$\psi_3(\xi) = -\left(\frac{3}{a\sqrt{\pi}}\right)^{1/2} \xi \left(1 - \frac{2}{3}\xi^2\right) e^{-\xi^2/2},$$

$$\psi_4(\xi) = \left(\frac{3}{8a\sqrt{\pi}}\right)^{1/2} \left(1 - 4\xi^2 + \frac{4}{3}\xi^4\right) e^{-\xi^2/2}.$$

And the corresponding probability densities

$$|\psi_1|^2 = \frac{2}{a\sqrt{\pi}}\frac{x}{a} e^{-x^2/2a_2}, \quad |\psi_2|^2 = \frac{1}{2a\sqrt{\pi}} (1 - 2x^2/a^2)\, e^{-x^2/2a^2},$$

$$|\psi_3|^2 = \frac{3}{a\sqrt{\pi}}\frac{x}{a}\left(1 - \frac{2}{3}\frac{x^2}{a^2}\right) e^{-x^2/2a^2},$$

$$|\psi_4| = \frac{3}{8a\sqrt{\pi}}\left(1 - 4\frac{x^2}{a^2} + \frac{4}{3}\frac{x^4}{a^4}\right) e^{-x^2/2a^2}.$$

P.5.3. For $n = 0$ one must have $i = 0$ and $j = 0$, hence

$$\psi_0(\xi, \eta) = \text{const}\; e^{-\varrho^2/2},$$

where ϱ is the distance of the point ξ, η from the origin of the coordinate system. There is no nodal line. For $n = 1$ one has either $i = 1$ and $j = 0$, or $i = 0$ and $j = 1$, hence

$$\psi_1(\xi, \eta) = (a\xi + b\eta)\, e^{-\varrho^2/2}.$$

There is one nodal line with equation $a\xi + b\eta = 0$. For $n = 2$ there are three possibilities: $i = 2, j = 1$; $i = 1, j = 1$; $i = 0, j = 2$. Hence

$$\psi_2(\xi, \eta) = [a(1 - 2\xi^2) + b(1 - 2\eta^2) + c\xi\eta]\, e^{-\varrho^2/2}.$$

The equation for the nodal lines is $2a\xi^2 + 2b\eta^2 - c\xi\eta = a + b$ which represents either an ellipse or a hyperbola. For $n = 3$ there are four possibilities: $i = 3, j = 0$; $i = 2, j = 1$, $i = 1, j = 2$; $i = 0, j = 3$, hence

$$\psi_3(\xi, \eta) = [a\xi(1 - \tfrac{2}{3}\xi^2) + b\eta(1 - \tfrac{2}{3}\eta^2) + c\xi(1 - 2\eta^2) + d\eta(1 - 2\xi^2)]\, e^{-\varrho^2/2}.$$

The equation for the nodal lines is

$$\frac{2a}{3}\xi^3 + \frac{2b}{3}\eta^3 + 2d\xi^2\eta + 2c\xi\eta^2 - (a + c)\,\xi - (b + d)\,\eta = 0,$$

which represents a third order curve of radial symmetry, consisting of either two or four branches.

P.5.5. Legendre functions:

$$P_3^0 = \xi(1 - \tfrac{5}{3}\xi^2),\quad P_3^1 = (1-\xi^2)^{1/2}(1-5\xi^2),\quad P_3^2 = \xi(1-\xi^2),\quad P_3^3 = (1-\xi^2)^{3/2}.$$

Non-normalized surface harmonics:

$$\psi_3^0 = \cos\theta\,(1 - \tfrac{5}{3}\cos^2\theta),\quad \psi_3^1 = \sin\theta\,(1 - 5\cos^2\theta)\begin{Bmatrix}\cos\varphi\\ \sin\varphi\end{Bmatrix},$$

$$\psi_3^2 = \cos\theta\sin^2\theta\begin{Bmatrix}\cos 2\varphi\\ \sin 2\varphi\end{Bmatrix},\quad \psi_3^3 = \sin^3\theta\begin{Bmatrix}\cos 3\varphi\\ \sin 3\varphi\end{Bmatrix}.$$

Nodal lines:

$$u = 0:\quad \theta = \frac{\pi}{2}\quad\text{and}\quad \cos\theta = \pm\sqrt{\frac{3}{5}};$$

$$u = 1:\quad \theta = \pm\sqrt{\frac{1}{5}}\quad\text{and}\quad \varphi = \frac{\pi}{2}\quad\text{or}\quad \varphi = 0;$$

$$u = 2:\quad \theta = \frac{\pi}{2},\ \varphi = \pm\frac{\pi}{4}\quad\text{or}\quad \varphi = 0\quad\text{and}\quad \varphi = \frac{\pi}{2};$$

$$u = 3:\quad \varphi = \frac{\pi}{2}\quad\text{and}\quad \varphi = \pm\frac{\pi}{6}\quad\text{or}\quad \varphi = \frac{2\pi}{3}\quad\text{and}\quad \varphi = \pm\frac{\pi}{3}.$$

P.5.6. The normalizing condition is

$$\int_{-\pi/2}^{+\pi/2}\int_0^{2\pi} \psi_{l,n}^2(\theta,\varphi)\sin\theta\,d\theta\,d\varphi = 1,$$

and the coefficients of the surface harmonics are from this:

$$C_{00}^2 = \frac{1}{4\pi},\quad C_{10}^2 = \frac{3}{4\pi},\quad C_{11}^2 = \frac{3}{4\pi},\quad C_{20}^2 = \frac{5}{16\pi},\quad C_{21}^2 = \frac{15}{4\pi},\quad C_{22}^2 = \frac{15}{16\pi},$$

$$C_{30}^2 = \frac{63}{16\pi},\quad C_{31}^2 = \frac{21}{32\pi},\quad C_{32}^2 = \frac{105}{16\pi},\quad C_{32}^2 = \frac{35}{32\pi}.$$

P.5.7.

$$A_i = (-1)^i \frac{2^i k(k-1)(k-2)\cdots(k-i+1)}{i!\,(2l+2)(2l+3)(2l+4)\cdots(2l+i+1)}.$$

P.5.8. No zero's of χ for $n - l = 1$; $\chi_2^0 = 0$ for $r/a = 2$;

$$\chi_3^1 = 0\quad\text{for}\quad r/a = 6;\quad \chi_3^0 = 0\quad\text{for}\quad r/a = (3 \pm \sqrt{3})/2.$$

(a) For $n - l = 1$: $w \sim e^{-2r/na}r^{2n}$, from $dw/dr = 0$; $r_n/a = n^2$. Thus according to (2.2.12) the radii r_n at which the probability density is a maximum are in this case identical with the Bohr radii of the old quantum theory.

(b) For $n = 2$, $l = 0$ from (5.3.23) $w \sim e^{-r/a}\left(1 - \frac{r}{2a}\right)^2 r^2$, whose maxima are at $\frac{r}{a} = 3 \pm \sqrt{5}$.

For $n = 3$, $l = 1$: $w \sim e^{-2r/3a}\left(1 - \frac{r}{6a}\right)^2 r^4$ with maxima at $\frac{r}{a} = 3$ and $\frac{r}{a} = 12$.

P.5.10. The wave equation is

$$\frac{d^2g}{dr^2} + \left\{\frac{2m}{\hbar^2}\left[E - \frac{1}{2}\alpha r^2\right] - \frac{l(l+1)}{r^2}\right\} g = 0,$$

which transforms into

$$\frac{d^2g}{dr^2} + \left\{1 - \lambda^2\varrho^2 - \frac{l(l+1)}{\varrho^2}\right\} g = 0.$$

Setting $g = e^{-(\lambda/2)\varrho^2} f(\varrho)$ one obtains the equation

$$f'' - 2\lambda\varrho f' + \left\{1 - \lambda - \frac{l(l+1)}{\varrho^2}\right\} f = 0$$

for the polynomial $f(\varrho)$ which has the form $\varrho^s \sum_{i=0}^{k} A_i\varrho_i$. From the behaviour of this equation for small ϱ it follows that $s = l + 1$ and the polynomial is thus of order $l + 1 + k$. For large values of ϱ the equation assumes the asymptotic form

$$[-2\lambda(l + k + 1) + (1 - \lambda)] A_k = 0$$

and since $A_k \gtrless 0$ one must have $\lambda = \dfrac{1}{2l + 2k + 3}$. Hence the quantum levels are

$$E = h\nu\,(k + l + \tfrac{3}{2}).$$

Here the first term represents the radial oscillations, the second term the rotation of the systems about the centre, and the third term the zero-point energy.

P.6.1. The energy flux in the direction θ is $S = \dfrac{1}{4\pi c^3}\left(\dfrac{\ddot{\Pi}}{r}\sin\theta\right)^2$, and hence for large r the total intensity of the emitted radiation is

$$I = \int_{-\pi/2}^{+\pi/2} \bar{S}\, 2\pi r^2 \sin\theta\, d\theta,$$

where $\bar{S}$ is the average of S over one period. Thus generally

$$I = \frac{2}{3c^3}\overline{(\ddot{\Pi}^2)},$$

and in particular for a harmonic oscillation of frequency ν, where

$$\overline{(\ddot{\Pi}^2)} = \frac{\Pi_0^2}{2}(2\pi\nu)^4; \quad I = \frac{1}{3c^3}(2\pi\nu)^4\, \Pi_0^2.$$

P.6.2. From $E_k = k^2h^2/8ml^2$ one has for the frequencies of the spectral lines emitted by transitions from a quantum state k to a quantum state $j < k$:

$$\nu_{kj} = \frac{h}{8ml^2}(k^2 - j^2).$$

From (6.2.9)

$$I_{kj} = \frac{e^2}{3c^3}(2\pi\nu_{kj})^4\,|x_{kj}|^2$$

and from P.4.9

$$|x_{kj}|^2 = \left(\frac{8l}{\pi^2}\right)^2 \frac{(kj)^2}{(k^2 - j^2)^4}.$$

Thus

$$I_{kj} = \frac{e^2}{12c^3}\left(\frac{h}{m}\right)^4 \frac{1}{l^6}(kj)^2$$

if k and j have different parity, and $I_{kj} = 0$ if k and j have the same parity. As the movement of the electrons is restricted to the x-direction the electric oscillations of the emitted light must also take place in the x-direction. No light is emitted in that direction.

P. 6.3. The spectrum is monochromatic with one line of frequency ν. The non-vanishing matrix elements are given by

$$x_{n,n+1} = x_{n+1,n} = \sqrt{\frac{\hbar(n+1)}{4\pi\nu m}},$$

hence the intensity of the emitted light is from (6.2.9)

$$I = \frac{e^2}{3c^2}(2\pi\nu)^4 \frac{\hbar}{4\pi\nu m}\sum_{n=0}^{\infty}|c_{n+1}|^2 (n+1) N = \frac{e^2\hbar}{6c^3 m}(2\pi\nu)^3 \sum_{n=1}^{\infty}|c_n|^2 n \cdot N,$$

where N is the total number of oscillators and the $|c_n|^2$ are the fractional numbers of those in the quantum state n.

P. 6.4. The loss of diffusing matter from a volume element dV in a time dt is equal to div $\mathbf{Q}\, dV\, dt$, and this must be equal to $-\frac{\partial c}{\partial t} dV\, dt$, hence

$$\frac{\partial c}{\partial t} = -\operatorname{div}\mathbf{Q} = D\,\nabla^2 c.$$

P. 6.5. As a consequence of the solution quoted in the question it follows that, if the total diffusing matter $c_0\, d\xi$ is concentrated at $x = \xi$ at $t = 0$ the concentration at subsequent times must be given by

$$c(x, t) = \frac{c_0\, d\xi}{2\sqrt{\pi D t}}\, e^{\frac{-(x-\xi)^2}{4Dt}}.$$

Hence, if the concentration distribution at $t = 0$ is $c_0(x)$, then, because the diffusion equation is linear, the unique solution is obtained by integrating over ξ:

$$c(x, t) = \frac{1}{2\sqrt{\pi D t}}\int_{-\infty}^{+\infty} e^{\frac{-(x-\xi)^2}{4Dt}}\, c_0(\xi)\, d\xi.$$

P. 6.6. Putting $D = i\hbar/2m$ in the expression given in the question one obtains the solution

$$\varphi = e^{-\frac{i\hbar}{2m}\lambda^2 t}(A\cos\lambda x + B\sin\lambda x)$$

of the equation (6.3.18). If this is to satisfy the boundary conditions one must have $\lambda = n\pi/l$ and thus

$$\varphi = e^{-2\pi i\nu_n t}\left(A\cos\frac{n\pi}{l} + B\sin\frac{n\pi}{l}\right) \quad \text{with} \quad \nu_n = \frac{hn^2}{8\pi l^2} = E_n/h.$$

This represents a standing wave with nodes at $x = \pm l/2$.

P.7.1. From $d\psi \equiv (d\mathbf{r}\,\nabla\psi)$ follows $\psi = \int (d\mathbf{r}\,\nabla\psi)$ and hence $\mathscr{I} = \int (d\mathbf{r}\,\nabla)$. The reciprocal operator to ∇ is therefore the operator $\int d\mathbf{r}$.

P.7.2. The condition (7.1.18) demands that for any two functions f and g one must have

$$(f \cdot \mathscr{F} g) = (\mathscr{F} f \cdot g)$$

if $\mathscr{F}$ is to be Hermitian. If $\mathscr{F}$ is a function F of the coordinates this can only be satisfied if $F = F^*$, i.e. if F is real. Alternatively the matrix representating $\mathscr{F}$ is

$$F_{ik} = \int \cdots \int \psi_i^* \mathscr{F} \psi_k \, dq_1 \cdots dq_N,$$

and hence

$$F_{ki} = \int \cdots \int \psi_k^* \mathscr{F} \psi_i \, dq_1 \cdots dq_N.$$

If $\mathscr{F}$ is to be Hermitian $F_{ki} = F_{ik}^*$ must be satisfied, and if $\mathscr{F}$ is a function of the coordinates F this is only the case if $F = F^*$.

P.7.3. (a) The conditions for the two operators to be commutative, i.e. for $(\mathscr{A}\mathscr{B} - \mathscr{B}\mathscr{A}) = 0$ are

$$\alpha_1 \frac{d\beta_0}{dx} = \beta_1 \frac{d\alpha_0}{dx} \quad \text{and} \quad \alpha_1 \frac{d\beta_1}{dx} = \beta_1 \frac{d\alpha_1}{dx}.$$

If the two operators are to be anticommutative the relation $(\mathscr{A}\mathscr{B} + \mathscr{B}\mathscr{A}) = 0$ would have to be satisfied, which is only possible if at least one of the operators is identically zero.

(b) If $\alpha_0 = a + a'x$, $\beta_0 = b + b'x$ the operators will commute if the constants satisfy the condition

$$\alpha_1 b' = \beta_1 a'.$$

P.7.4. If f and g are two functions of x which vanish at infinity the condition for $\mathscr{A}$ being Hermitian is from (7.1.18)

$$f \cdot \left(\alpha x g + \beta \frac{dg}{dx}\right) = \left(\alpha x f + \beta \frac{df}{dx}\right) \cdot g,$$

which can be transformed into

$$\alpha \int_{-\infty}^{+\infty} x g f^* \, dx - \beta \int_{-\infty}^{+\infty} g \frac{df^*}{dx} \, dx = \alpha^* \int_{-\infty}^{+\infty} x f^* g \, dx + \beta^* \int_{-\infty}^{+\infty} \frac{df^*}{dx} g dx.$$

This can only be satisfied identically if $\alpha = \alpha^*$ and $\beta = -\beta^*$, i.e. if α is real and β is imaginary.

The vector operator $\mathscr{A} = \mathbf{r} + \dfrac{\hbar \Delta t}{im} \nabla$ is, according to this criterion, Hermitian if it belongs to an "observable" $\mathbf{a} = \mathbf{r} + \dfrac{\Delta t}{m} \mathbf{p} = \mathbf{r} + \mathbf{v}\,\Delta t$. Therefore $\mathbf{a}$ is equal to the displacement $\Delta \mathbf{r}$ of the particle of velocity $\mathbf{v}$ in time t.

P.7.5. The eigenvalue equation is $\alpha x \chi + i\beta \dfrac{d\chi}{dx} = a\chi$, which has the general solution $\chi = C \exp\left(\dfrac{\alpha x - \alpha x^2/2}{i\beta}\right)$. If χ is to have the same value at $x = 0$ and $x = l$ the eigenvalues of $\mathscr{A}$ are $a_n = \dfrac{2\pi n \beta}{l} + \dfrac{\alpha l}{2}$, where n is a positive integer.

P. 7.6. Normalization of ψ demands $CC^* = 1/a\sqrt{\pi}$.

$$\overline{(\Delta x)^2} = \overline{x^2} = \frac{a^2}{2}, \quad \overline{(\Delta p_x)^2} = \overline{p_x^2} = \frac{\hbar^2}{2a^2}.$$

Hence $\sqrt{\overline{(\Delta x)^2}} \cdot \sqrt{\overline{(\Delta p_x)^2}} = \hbar/2$. Comparison with (7.2.24) shows that in this case the uncertainty is the smallest possible.

P. 7.7. From the definition of the momentum operator follows

$$p_{jk} = \frac{2}{l} \int_{-l/2}^{+l/2} \begin{Bmatrix} \sin(j\pi x/l) \\ \cos(j\pi x/l) \end{Bmatrix} \frac{k}{i} \frac{d}{dx} \begin{Bmatrix} \sin(k\pi x/l) \\ \cos(\pi k x/l) \end{Bmatrix} dx,$$

where the sin-functions hold for even quantum numbers and the cos-functions for odd quantum numbers.

If j and k have the same parity: $p_{jk} = 0$, and if they have different parity:

$$p_{jk} = \frac{\hbar}{li}(-1)^{\frac{j+k+1}{2}} \frac{4jk}{j^2 - k^2}.$$

Interchanging j and k changes the sign of p_{jk} and as it is purely imaginary $p_{jk} = p_{kj}^*$, i.e. the matrix is Hermitian.

P. 7.8.

$$(\overline{x^2}) = \frac{2}{l} \int_{-l/2}^{+l/2} x^2 \sin^2\left(\frac{n\pi x}{l}\right) dx \quad \text{for even } n,$$

$$(\overline{x^2}) = \frac{2}{l} \int_{-l/2}^{+l/2} x^2 \cos^2\left(\frac{n\pi x}{l}\right) dx \quad \text{for odd } n.$$

Both integrals give the same result

$$(\overline{x^2}) = \frac{l^2}{2n^2\pi^2}\left(\frac{n^2\pi^2}{6} - 1\right).$$

The momentum for the nth quantum state has the fixed value $p_x = \pm \frac{hn}{2l}$.

Thus $\sqrt{\overline{(\Delta x)^2}\,\overline{(\Delta p_x)^2}} = \frac{\hbar}{2}\left(\frac{n^2\pi^2}{3} - 2\right)^{1/2}$ which has the smallest value for $n = 1$:

$$\frac{\hbar}{2}\left(\frac{\pi^2}{3} - 2\right)^{1/2} \simeq 1 \cdot 3 \frac{\hbar}{2}.$$

P. 7.9.

$$\mathscr{P}_x = -\frac{\hbar}{i}\left(\cos\varphi \cot\theta \frac{\partial}{\partial\varphi} + \sin\varphi \frac{\partial}{\partial\theta}\right),$$

$$\mathscr{P}_y = -\frac{\hbar}{i}\left(\sin\varphi \cot\theta \frac{\partial}{\partial\varphi} - \cos\varphi \frac{\partial}{\partial\theta}\right),$$

$$\mathscr{P}_z = \frac{\hbar}{i} \frac{\partial}{\partial\varphi}.$$

P. 8.2. As all the constituents have half-integer spin an interchange of any one of the constituents between two atoms must change the sign of the wave function; this proves the theorem.

P. 8.3. The binding energy between each of the two electrons and the nucleus is $-e^2/a$ and the energy of interaction between the two electrons is $5e^2/8a$; hence the energy to remove one electron from the atom is $3e^2/8a$, and the energy to remove both electrons is $3e^2/8a + e^2/a = 11e^2/8a$, where a is the Bohr radius of the He-ion.

P. 8.4. From (8.2.10) and (8.2.11): $\psi = Cf$, where

$$f = \left\{\exp\left[-\frac{1}{a}(|r_1 - R_1| + |r_2 - R_2|)\right] + \exp\left[-\frac{1}{a}(|r_2 - R_1| + |r_1 - R_2|)\right]\right\}$$

(the positive sign having been chosen because only the symmetric wave function represents an H_2-molecule). Hence: $\psi\psi^* = C^2 f^2$ with

$$f^2 = \left\{\exp\left[-\frac{2}{a}(|r_1 - R_1| + |r_2 - R_2|)\right] + \exp\left[-\frac{2}{a}(|r_2 - R_1| + |r_1 - R_2|)\right]\right.$$
$$\left. + 2\exp\left[-\frac{1}{a}(|r_1 - R_1| + |r_2 - R_2| + |r_2 - R_1| + |r_1 - R_2|)\right]\right\}.$$

From (8.2.9) and (8.2.12) one has for E' the difference between the total energy of the molecule and the internal energies of the two atoms themselves:

$$E' = e^2 \frac{\int \cdots \int f^2 \left(\frac{1}{R} + \frac{1}{|r_1 - r_2|} - \frac{1}{|r_2 - R_1|} - \frac{1}{|r_1 - R_2|}\right) dV_1\, dV_2}{\int \cdots \int f^2\, dV_1\, dV_2}.$$

Here R_1 and R_2 are fixed, $R = |R_1 - R_2|$, and the integration is extended over all positions r_1 and r_2 of the two electrons.

P. 9.1. The result follows from the fact that the rotation vector $\mathbf{w}$ of the earth's rotation can be resolved into two components, one pointing vertically upwards with magnitude $\omega \sin\varphi$, and one in a horizontal direction with magnitude $\omega\cos\varphi$, and that the Coriolis force due to the latter has a vertical direction and therefore does not influence the movement of the pendulum.

P. 9.2. According to (9.1.18) a change of $\mathbf{w}$ produces an inertial force $\mathbf{F}^* = -\dot{\mathbf{w}} \wedge M\mathbf{r}'$, where $\mathbf{r}'$ is the position vector of the mass centre of the body with respect to the centre of rotation. Further if the magnitude of $\mathbf{w}$ remains unchanged $\dot{\mathbf{w}} = \mathbf{w}\dot{\alpha}\mathbf{i}$, where $\dot{\alpha}$ is the change of direction angle of the axis of rotation per unit time and $\mathbf{i}$ is a unit vector in the direction of the force that produces the change of direction. Thus, finally,

$$\mathbf{F}^* = -M\mathbf{w}\dot{\alpha}\,(\mathbf{i} \wedge \mathbf{r}').$$

P. 9.3. (a) The force $\mathbf{F}_1^*$ acts in a horizontal direction at right angles to $\mathbf{v}'$. It points to the right of an observer looking in the direction of $\mathbf{v}'$ in the northern hemisphere and to the left in the southern hemisphere. The force $\mathbf{F}_2^*$ acts in a vertical direction and points upwards when $\mathbf{v}'$ has an easterly component and downwards if it has a westerly component.

(b) In this case $\mathbf{F}_1^* = 0$ and $\mathbf{F}_2^*$ acts in a horizontal direction. It points to the east if the body is falling and hence produces a deflection of the path to the east.

P.9.4. The orbital velocity **V** of the earth can be resolved into the two components $V_x = V\cos\frac{2\pi t}{T}$, $V_y = V\sin\frac{2\pi t}{T}$ (T = one year). If the x-direction is chosen to be at right angles to the direction joining earth and star in the plane of the ecliptic, then the projection of V_x on a plane normal to that direction is equal to V_x and the projection of V_y on this plane is $V_y \sin\varphi$. Thus the apparent displacements of the star are given by

$$x = \frac{V}{c}\cos\frac{2\pi t}{T}, \quad y = \frac{V}{c}\sin\varphi\sin\frac{2\pi t}{T},$$

from which follows that the path is elliptical with semi-axes

$$a = V/c, \quad b = V\sin\varphi/c.$$

P.9.5. The length of path of the ray in the rotating system is $l' = ct = \sqrt{2}\,l\sin\left(\frac{\pi}{2} \pm \frac{\omega t}{2}\right)$ according to whether the ray travels in the direction of rotation or against it. For small ωt one has from this

$$t = \frac{l}{c \mp l\omega/2}$$

and hence

$$t_1 - t_2 = \frac{4l}{c - l\omega/2} - \frac{4l}{c + l\omega/2} \simeq \frac{4l^2\omega}{c^2} = \frac{4a\omega}{c^2}.$$

Finally

$$\Delta = (t_1 - t_2)\,c/\lambda = 4a\omega/c\lambda.$$

P.9.6.

$$\Phi = \gamma\left[\exp\left(-\frac{2\pi i\nu x}{c - V}\right) + \exp\left(\frac{2\pi i\nu x}{c + V}\right)\right] e^{2\pi i\nu t}.$$

Φ will be zero if the expression in the square bracket is zero, i.e. $-\frac{2\pi i\nu x}{c - V} = \frac{2\pi i\nu x}{c + V} \pm 2\pi i n$, where n is an integer. This equation is satisfied if

$$d = \pm\frac{n(c^2 - V^2)}{2\nu c} = \pm\frac{\lambda n}{2}(1 - V^2/c^2).$$

Thus the distances between neighbouring nodal planes are

$$d = \frac{\lambda}{2}(1 - V^2/c^2).$$

P.9.7. The velocity c' of the light ray in the direction α of the arm is related to c (in the direction β) and V the "ether drift" velocity by

$$c\cos\beta - V = c'\cos\alpha, \quad c\sin\beta = c'\sin\alpha$$

from which follows

$$c' = \frac{l}{t_1} = -V\cos\alpha + \sqrt{c^2 - V^2\sin^2\alpha},$$

where t_1 is the time for the signal to run along the arm. For t_2, the time for the backward journey, one obtains similarly

$$\frac{l}{t_2} = V \cos \alpha + \sqrt{c^2 - V^2 \sin^2 \alpha}.$$

Thus

$$t = \frac{2l}{c} \frac{\sqrt{1 - V^2 \sin^2 \alpha}}{1 - V^2/c^2}.$$

In particular for $\alpha = 0$

$$t = t_0 = \frac{2l}{c} \frac{1}{1 - V^2/c^2}.$$

Now if l for $\alpha = 0$ suffers a Lorentz contraction $\sqrt{1 - V^2/e^2}$ and for any finite α a contraction $\sqrt{\frac{1 - V^2/c^2}{1 - V^2 \sin^2 \alpha/c^2}}$ one sees that $t = t_0$ generally, and the effect of the ether drift is zero for all positions if the length of the arm is reduced to

$$l' = l \sqrt{\frac{1 - V^2/c^2}{1 - V^2 \sin^2 \alpha/c^2}} \simeq l \sqrt{1 - V^2 \cos^2 \alpha/c^2}.$$

P.9.8. From Biot–Savart's law H has the direction of the rotation axis and the contribution $d\mathsf{H}$ to H from the charge dQ on a ring of radius r and width dr is

$$d\mathsf{H} = \frac{2\pi \, dI}{c} \frac{r^2}{(r^2 + l^2)^{3/2}}, \quad \text{where} \quad dI = n dQ = \frac{2rQn}{R^2} dr.$$

Thus

$$\mathsf{H} = \frac{4\pi Qn}{cR} \left[2\left(\sqrt{1 + (R/l)^2} - 1\right) - \frac{(R/l)^2}{\sqrt{1 + (R/l)^2}} \right].$$

For large R/l:

$$\mathsf{H} \simeq 4\pi Qn/cl.$$

P.9.9. According to (9.3.21) the bar acquires a uniform electric polarization P in the z-direction of magnitude $\mathsf{P} = VM/c$ and hence an electric field in the same direction of magnitude $\mathsf{E} = 4\pi VM/c$ is set up in the interior of the bar, which causes a potential difference of magnitude $\varphi = Es = \frac{4\pi VMs}{c}$ between the sliding contacts.

P.9.10. The electric field exerts a force of magnitude eE on a particle in the z-direction and therefore produces a deflection

$$z = \frac{eE}{2m} t^2 = \frac{eEl^2}{2mv^2}.$$

The magnetic field produces a force of magnitude evH/c at right angles to the beam in the xy-plane. If the deflection is small enough its magnitude is given by

$$y = \frac{eHl^2}{2mvc}.$$

Eliminating v from these equations one finds

$$\frac{y^2}{z} = \frac{1}{2} \frac{e}{m} \frac{H^2 l^2}{Ec^2}.$$

Thus the curve traced out on the screen in the yz-plane is a parabola.

P. 9.11. It follows from (9.3.24) that the force on the particle is normal to v and normal to $\mathbf{H}$. Hence the orbit must be a circle in a plane normal to $\mathbf{H}$. From the condition of equilibrium between magnetic force and centrifugal force follows $R = mvc/eH$, and as $n = v/2\pi R$ one has $n = e\mathrm{H}/2\pi mc$, independent of v.

P. 9.12. According to (9.3.34) the electric field $\mathbf{E}$ inside the condenser creates a magnetic field $\mathbf{B} = \frac{1}{c}(\mathbf{V} \wedge \mathbf{E})$, and according to (9.3.24) this field exerts a force

$$\mathbf{F}_m = \frac{Q}{c^2}[\mathbf{V}(\mathbf{VE}) - \mathbf{E}V^2]$$

on the charge Q on one plate of the condenser, and a force in the opposite direction and equal magnitude on the other plate with charge $-Q$. Hence because of $\mathbf{E} \wedge \mathbf{a} = 0$ the couple acting on the condenser is

$$\mathbf{T} = \mathbf{F}_m \wedge \mathbf{a} = \frac{Q}{c^2}(\mathbf{V} \wedge \mathbf{a})(\mathbf{VE}).$$

The direction of T is vertical, i.e. about the suspension of the condenser, and its magnitude is

$$T = Qa\mathrm{E}\frac{V^2}{c^2}\sin\theta\cos\theta.$$

As $\mathrm{E} = \Phi/a$, $Q = \Phi C$ one has finally

$$T = \frac{\Phi^2 C}{2}\frac{V^2}{c^2}\sin 2\theta.$$

P. 10.1. The equation of the sphere in the system S is

$$x^2 + y^2 + z^2 = R^2$$

and in the system S'

$$\frac{(x' + Vt')^2}{1 - V^2/c^2} + y'^2 + z'^2 = R^2.$$

The shape of the body will thus appear to be that of an oblate ellipsoid with semi-axes $a = R\sqrt{1 - V^2/c^2}$, $b = c = R$, which moves with speed V in the direction of the negative x-axis.

P. 10.2.

$$x_1' = \beta x_1 + i\beta\frac{V}{c}x_4, \quad x_2' = x_2,$$

$$x_3' = x_3, \quad x_4' = -i\beta\frac{V}{c}x_1 + \beta x_4.$$

In the determinant of the coefficients of the transformation equations the sum of the products of corresponding elements in each pair of rows or columns is zero, and the sum of the squares of the elements in each row or column is equal to unity. Hence the transformation is orthogonal and represents a rotation in four-dimensional space.

P. 10.3. If the world distance between two events at points A and B, of distance l in space and separated by a time interval Δt, is zero it means that Δt is the time a light signal takes to travel from A to B. If D is real $\Delta t < l/c$ and the events can appear to be simultaneous in a suitable frame of reference; if D is imaginary $\Delta t > l/c$ and the event B is always later than the event A. The equation $D = 0$ represents a four-dimensional cone whose three-dimensional cross-sections normal to the ct-axis are spheres of radius ct. All world points with real D lie outside this cone and those with imaginary D inside the cone.

P. 10.4. Conservation of momentum in the x-direction and normal to it in the $\mathbf{v}$, $\mathbf{v}_1$, $\mathbf{v}_2$ plane demands

$$\frac{v}{\sqrt{1-v^2/c^2}} = \frac{v_1}{\sqrt{1-v_1^2/c^2}}\cos\theta_1 + \frac{v_2}{\sqrt{1-v_2^2/c^2}}\cos\theta_2, \quad (a)$$

$$\frac{v_1}{\sqrt{1-v_1^2/c^2}}\sin\theta_1 = \frac{v_2}{\sqrt{1-v_2^2/c^2}}\sin\theta_2 \quad (b)$$

and conservation of energy further demands

$$\frac{1}{\sqrt{1-v^2/c^2}} + 1 = \frac{1}{\sqrt{1-v_1^2/c^2}} + \frac{1}{\sqrt{1-v_2^2/c^2}}. \quad (c)$$

From these equations follows

$$1 + \frac{1}{\sqrt{1-v^2/c^2}} - \frac{1}{\sqrt{1-v_1^2/c^2}} + \frac{vv_1\cos\theta_1/c^2 - 1}{\sqrt{(1-v^2/c^2)(1-v_1^2/c^2)}} = 0,$$

from which v_1 may be calculated for any given v and θ_1. v_2 then follows from (c) and θ_2 from (b).

P. 10.5. The energy of the bodies before collision is $\dfrac{2m^0c^2}{\sqrt{1-v^2/c^2}}$ and that after collision is $2m^{0\prime}c^2$. Thus from (10.2.6)

$$m^{0\prime} = \frac{K}{c^2} + m^0 = m^0 + \frac{Q}{c^2}.$$

P. 10.6. From (10.2.20) one has for constant $\mathbf{F}$

$$p = Ft = \frac{m^0\dot{x}}{\sqrt{1-\dot{x}^2/c^2}} \quad \text{or} \quad \dot{x} = \frac{Ft}{\sqrt{m^{02} + F^2t^2/c^2}},$$

from which follows

$$x = \frac{c^2m^0}{F}\left(\sqrt{1 + \frac{F^2t^2}{m^{02}c^2}} - 1\right).$$

For small t this becomes approximately $x \simeq \dfrac{F}{2m^0}t^2 = \dfrac{a}{2}t^2$, i.e. the classical formula, for large t: $x \to ct$, i.e. the particle approaches the velocity of light.

P. 10.7. From (10.3.8) follows for $V \to c$: $A' \to -1$, $B' \to 0$, $C' \to 0$. The observer would see the stars all crowded together in the neighbourhood of the point in the sky towards which he is moving.

P. 10.8. From (10.3.7)

$$\nu' = \frac{\nu_1}{\sqrt{1-V^2/c^2}}\left(1 - \frac{VA}{c}\right) = \frac{\nu_2}{\sqrt{1-V^2/c^2}}\left(1 + \frac{VA}{c}\right), \quad \text{where } A = \cos\varphi.$$

Thus

$$\frac{\nu_1}{\nu_2} = \frac{1 + VA/c}{1 - VA/c} \quad \text{and} \quad \frac{\Delta\nu}{\nu'} = \frac{2VA}{c}\frac{\sqrt{1 - V^2/c^2}}{1 - V^2A^2/c^2}.$$

P. 10.9. The total momentum before collision is zero if $\frac{h\nu}{c} = \frac{m^0 v}{\sqrt{1 - v^2/c^2}}$. After the collision the particles must move along the same straight line in opposite directions and it follows from the conservation of momentum and energy that ν and v must retain their values. A frame of reference S' in which the electron is at rest before collision must move with velocity $V = v$. From (10.3.7) one has for the frequency ν_1' measured in S' before collision

$$\frac{\nu_1'}{\nu} = \sqrt{\frac{1 + V/c}{1 - V/c}},$$

and for the frequency ν_2' after collision

$$\frac{\nu_2'}{\nu} = \sqrt{\frac{1 - V/c}{1 + V/c}}.$$

Hence

$$\lambda_2' - \lambda_1' = \frac{2V/c}{\sqrt{1 - V^2/c^2}} \frac{c}{\nu},$$

and because of $\frac{V}{\sqrt{1 - V^2/c^2}} = \frac{h\nu}{m^0 c}$: $\Delta\lambda' = \lambda_2' - \lambda_1' = \frac{2h}{m^0 c}$

P. 10.10. For the components of the electric field in a frame of reference that moves with the electron one has

$$\mathsf{E}_x' = \frac{ex'}{r'^2}, \quad \mathsf{E}_y' = \frac{ey'}{r'^2}, \quad \mathsf{E}_z' = \frac{ez'}{r'^2} \quad (r'^2 = x'^2 + y'^2 + z'^2).$$

Further from the Lorentz transformations for a given t

$$x' = \beta(x - X), \quad y' = y, \quad z' = z,$$

so that with the help of (10.4.22)

$$\mathsf{E}_x = \frac{e(x - X)}{\beta^2 s^3}, \quad \mathsf{E}_y = \frac{ey}{\beta^2 s^3}, \quad \mathsf{E}_z = \frac{ez}{\beta^2 s^3}$$

where $s = r\sqrt{1 - V^2 \sin^2\theta/c^2}$ and θ the angle between $\mathbf{r}$ and $\mathbf{V}$. The vector $\mathbf{E}$ therefore has the direction of $\mathbf{r}$, as for a resting electron, but its magnitude depends on θ, namely

$$\mathsf{E} = \frac{e}{\beta^2 r^2}(1 - V^2 \sin^2\theta/c^2)^{-3/2}.$$

The components of $\mathbf{H}$ follow from (10.4.21)

$$\mathsf{H}_x = 0, \quad \mathsf{H}_y = \frac{ezV}{c\beta^2 s^3}, \quad \mathsf{H}_z = \frac{eyV}{c\beta^2 s^3},$$

and as $\mathbf{H}$ is everywhere normal to $\mathbf{V}$ and to $\mathbf{E}$, i.e. to $\mathbf{r}$, the magnetic lines of force are circles normal to the x-axis with their centres on the x-axis. The magnitude of $\mathbf{H}$ is

$$\mathsf{H} = \frac{eV}{c\beta^2 r^2} \sin\theta\,(1 - V^2 \sin^2\theta/c^2)^{-3/2}.$$

P. 11.2. In order that the four homogeneous equations (11.1.26) for the four variables g_k be compatible the determinant of the coefficients must be zero. Setting

$$\frac{\hbar}{\Lambda} + iJ_4 = a, \quad J_1 + iJ_2 = b, \quad J_3 = c$$

one obtains for the determinant

$$D = (|a|^2 + |b|^2 + |c|^2)^2 = m^{02}c^2 + p^2 - E^2/c^2,$$

which is indeed zero according to (11.1.8). Setting first $g_4 = 0$ one obtains the equations

$$ag_1 - cg_3 = 0, \quad ag_2 - bg_3 = 0, \quad bg_1 - cg_2 = 0,$$

from which follows

$$g_1 = \gamma c, \quad g_2 = \gamma b, \quad g_3 = \gamma a.$$

Setting next $g_3 = 0$ one has similarly

$$ag_1 - b^* g_4 = 0, \quad ag_2 + cg_4 = 0, \quad cg_1 + b^* g_2 = 0$$

leading to

$$g_1 = \gamma b^*, \quad g_2 = -\gamma c, \quad g_4 = \gamma a.$$

P. 11.3.

$$\left(\frac{\hbar}{\Lambda} - iJ_4\right) g_1 + J_3 g_3 + (J_1 - iJ_2)\, g_4 = 0,$$

$$\left(\frac{\hbar}{\Lambda} - iJ_4\right) g_2 + (J_1 + iJ_2)\, g_3 - J_3 g_4 = 0,$$

$$-J_3 g_1 - (J_1 - iJ_2)\, g_2 + \left(\frac{\hbar}{\Lambda} + iJ_4\right) g_3 = 0,$$

$$-(J_1 + iJ_2)\, g_1 + J_3 g_2 + \left(\frac{\hbar}{\Lambda} + iJ_4\right) g_4 = 0.$$

For the particular case of a resting particle $J_1 = J_2 = J_3 = 0$, $\dfrac{\hbar}{\Lambda} + iJ_4 = 0$. Thus the first two equations are satisfied if $g_1 = g_2 = 0$, and the third and fourth equations are satisfied irrespective of the values of g_3 and g_4. Hence the relation (11.1.29) is valid.

P. 11.4.

$$\frac{d^2\varphi}{dr^2} + \frac{2}{r}\frac{d\varphi}{dr} = \frac{1}{r}\frac{d^2(r\varphi)}{dr^2} = 0,$$

hence

$$\frac{d\,(r\varphi)}{dr} = A, \quad \varphi = A + \frac{B}{r}.$$

For $\varphi = 0$ at $r = \infty$ one must have $A = 0$; hence $\varphi = \dfrac{B}{r}$.

P. 11.5. In this case one can put

$$\Phi = \sum_k \varphi_k(\varrho_k) \quad \text{where} \quad \varrho_k = |\mathbf{r} - \mathbf{r}_k|.$$

(11.2.2) then reduces to

$$\sum_k \left[\frac{d^2\varphi_k}{d\varrho_k^2} + \frac{2}{\varrho}\frac{d\varphi_k}{d\varrho_k}\right] = 0,$$

which is satisfied if all the terms of the sum vanish separately. Hence the solution for $\varphi = 0$ at $r = \infty$ is $\varphi = \sum_k \frac{e_k}{\varrho_k}$.

P.11.6. (11.2.7) takes the form $\frac{1}{r}\frac{d^2(r\varphi)}{dr^2} = \frac{\varphi}{\beta^2}$, or putting $\varphi r = f(r)$, $\frac{d^2 f}{dr^2} = \frac{1}{\beta^2} f$ the general solution of which is

$$f = Ae^{r/\beta} + Be^{-r/\beta}.$$

If φ is to be finite at $r = \infty$ one must have $A = 0$ and thus

$$\varphi = \frac{B}{r} e^{-r/\beta}.$$

P.12.1. For an equilateral spherical triangle:

$$\cos a = \cos^2 a + \sin^2 a \cos \alpha$$

from which follows

$$\cos \alpha = \frac{1}{\sec a + 1}.$$

Thus $\quad -1 < \cos \alpha < \frac{1}{2}$ and $180° > \alpha > 60°$, $3\pi > 3\alpha > \pi$.

P.12.2. From $r = R\varphi$, $s = 2\pi R \sin \varphi$, $a = 2R^2 (1 - \cos \varphi) \pi$ follows

$$\frac{s}{r} = \frac{2\pi R}{r} \sin (r/R), \quad \frac{a}{r^2} = \frac{2\pi R^2}{r^2} (1 - \cos r/R).$$

Hence $s/r \leqslant 2\pi$, $a/r^2 \leqslant \pi$. By making geodetic measurements with measuring chains and rods on the surface one can determine s/r and hence calculate R.

P.12.3. From the equation of the hemisphere

$$x^2 + y^2 + z^2 = r^2 \qquad z > 0$$

follows for $y = \text{const}$: $(d\xi)^2 = (dx)^2 + (dz)^2 = (dx)^2 (1 + x^2/z^2)$,

and for $x = \text{const}$: $(d\eta)^2 = (dy)^2 + (dz)^2 = (dy)^2 (1 + y^2/z^2)$.

Thus

$$\begin{aligned}(ds)^2 &= (dx)^2 + (dy)^2 + (dz)^2 \\ &= (d\xi)^2 + (d\eta)^2 + \frac{2xy}{[(r^2 - x^2)(r^2 - y^2)]^{1/2}} d\xi\, d\eta,\end{aligned}$$

so that $\quad g_{11} = g_{22} = 1, \quad g_{12} = \frac{xy}{\sqrt{(r^2 - x^2)(r^2 - y^2)}}$.

P.12.4. From (12.1.22) one has

$$\nabla^2 \varphi_g = G \int\int\int \text{div grad} \left(\frac{\varrho_g (\xi, \eta, \zeta)}{r} \right) d\xi\, d\eta\, d\zeta,$$

where the differential operation div grad refers to the coordinates ξ, η, ζ. By Gauss' theorem for any irrotational vector field $\mathbf{A}$,

$$\iiint \text{div}\, \mathbf{A}\, dV = \iint \mathbf{A}_n\, da,$$

where the l.h.s. extends over a closed volume in space and the r.h.s. over the surface of this volume. For a given point $P(x, y, z)$ we choose a domain of integration excluding a small sphere of radius R with P as centre. Thus because of $\varphi_g = 0$ at infinity

$$\iint \mathbf{A}_n \, da = -G \iint \left[\frac{1}{R} \left(\frac{\partial \varrho_g}{\partial r} \right)_{r=R} - \frac{\varrho_g}{R^2} \right] da.$$

In the limit for $R \to 0$ one obtains from this

$$\nabla^2 \varphi_g = \lim_{R=0} \left\{ -G \left[\frac{1}{R} \frac{\partial \varrho_g}{\partial r} - \frac{\varrho_g}{R^2} \right] 4\pi R^2 = 4\pi G \varrho_g, \right.$$

which is Poisson's equation.

P. 12.5. According to the equivalence principle the field may be replaced by a transformation to a coordinate system S' which moves with acceleration $\mathbf{g}$ in the opposite direction to the field. In S the ray will seem to be subjected to an acceleration $g_n = g \sin \alpha$ in a direction normal to the ray. Hence at arrival at B in time t the ray will have acquired a velocity v_n normal to AB of magnitude $v_n = g_n t = g \sin \alpha \cdot t$ in the plane of g and AB, whereas the velocity in the direction of AB is practically unchanged $v_p \simeq c$. Thus

$$\delta\alpha = \frac{v_n}{v_p} \simeq \frac{g \sin \alpha \cdot l}{c^2},$$

and as the difference in gravitational potential φ between A and B is $\delta\varphi = -gl \cos \alpha$, one has, finally, $\delta\alpha = -\tan \alpha \cdot \delta\varphi / c^2$. The direction of the displacement is opposite to g.

P. 12.6. From (12.2.14) $\delta\nu = \delta\varphi/c^2 = -gl \cos \alpha / c^2$. If $\alpha < \dfrac{\pi}{2}$, $\cos \alpha > 0$, and $\delta\nu < 0$, i.e., the frequency will appear to be increased in comparison to the frequency of a spectral line emitted at B.

P. 13.1. By definition

$$\langle x_i \rangle = \underbrace{\int \cdots \int}_{N} x_i \exp[-\tfrac{1}{2} Q(x_1, \ldots, x_N)] \, dx_1 \cdots dx_N \Big/ \underbrace{\int \cdots \int}_{N} \exp[-\tfrac{1}{2} Q(x_1, \ldots, x_N)] \, dx_1 \cdots dx_N.$$

It follows from the central symmetry of the hypersurfaces $Q(x_1, \ldots, x_N) = \text{const}$ that the integral in the numerator extended over a "shell" must be zero. Hence the integral over all space, i.e. over all shells of this type must also vanish. By an orthogonal transformation of the coordinates (rotation in N-dimensional space) one can always transform Q:

$$\sum_{ij} a_{ij} x_i x_j = \sum_k b_k y_k^2.$$

Thus in the new coordinates

$$\varrho(y_1, \cdots, y_N) = \text{const} \prod_k \exp\left(-\frac{b_k}{2} y_k^2 \right).$$

Then by definition

$$\langle y_k^2 \rangle = \int y_k^2 e^{-b_k y_k^2/2} dy_k \Big/ \int e^{-b_k y_k^2/2} \, dy_k = \frac{1}{b_k}$$

so that $\sum_k b_k \langle y_k^2 \rangle = N$, and hence $\sum_{ij} a_{ij} \langle x_i x_j \rangle = N$. This equation is satisfied if one sets $\langle x_i x_j \rangle = A_{ij}/A$, because $\sum_i a_{ij} A_{ij} = A$ and $\sum_{ij} a_{ij} A_{ij} = NA$.

P. 13.2. The probability of finding x between x and $x + dx$ is

$$p(x)\,dx = dx \int_{-\infty}^{+\infty} \varrho\,(x, y)\,dy = \text{const}\,\exp\left[-\frac{1}{2}\left(\alpha - \frac{\beta^2}{\gamma}\right)x^2\right]dx.$$

If $a = g(x)$ or $x = h(a)$ one has

$$f(a)\,da = p(x)\,dx = \text{const}\,\exp\left[-\frac{1}{2}\left(\alpha - \frac{\beta^2}{\gamma}\right)h^2(a)\right]\frac{dh}{da}\,da.$$

In the special case $a = x: \langle a\rangle = \langle x\rangle = 0$ and $\sigma_a^2 = \langle x^2\rangle = \dfrac{1}{\alpha - \beta^2/\gamma} = \dfrac{\gamma}{\alpha\gamma - \beta^2}$. Since $A = \alpha\gamma - \beta^2$ and $A_{11} = \gamma$ one has indeed $\langle x^2\rangle = A_{11}/A$.

P. 13.3. The distribution function for r is by definition

$$f(r)\,dr = \text{const}\,\varrho(r)\,\omega,$$

where ω is the volume of a hyperspherical shell of radius r and thickness dr in N-dimensional space. Thus

$$f(r) = \text{const}\,e^{-\alpha r^2} r^{N-1},$$

from which follows

$$\langle r\rangle = \int_0^\infty e^{-\alpha r^2} r^N\,dr \Bigg/ \int_0^\infty e^{-\alpha r^2} r^{N-1}\,dr = \frac{1}{\sqrt{\alpha}}\,\frac{\Pi\,[(N-1)/2]}{\Pi\,[(N-2)/2]},$$

where $\Pi(x)$ is the generalized factorial function for non-integer x. Similarly

$$\langle r^2\rangle = \frac{1}{\alpha}\,\frac{\Pi\,(N/2)}{\Pi\,(\frac{1}{2}N - 1)} = \frac{N}{2\alpha}.$$

From Sterling's theorem $n! = n^n\,e^{-n}\sqrt{2\pi n}$ one derives

$$\Pi\left(\frac{N-1}{2}\right)\Bigg/\Pi\left(\frac{N-2}{2}\right) = \left(\frac{N}{2}\right)^{1/2}.$$

Thus $\langle r\rangle \simeq \dfrac{1}{\sqrt{\alpha}}\left(\dfrac{N}{2}\right)^{1/2}$ and $\langle r\rangle^2 \simeq \dfrac{N}{2\alpha} = \langle r^2\rangle$ so that $\sigma^2 = \langle r^2\rangle - \langle r\rangle^2 \to 0$.

P. 13.4. The normalizing condition demands $A^2 = 1/2\alpha_0$. The expansion of $\Phi(\alpha)$ in terms of the eigenfunctions (5.2.10) has the form

$$\Phi(\alpha) = \sum_n c_n \cos \alpha n \quad (n = 0, \pm 1, \pm 2, \ldots),$$

where from Fourier's theorem

$$c_0 = A\,\frac{2\alpha_0}{\pi} = \frac{\sqrt{2\alpha_0}}{\pi}, \quad c_n = A\,\frac{2}{n\pi}\sin n\alpha_0 = \frac{\sqrt{2}}{n\pi\sqrt{\alpha_0}}\sin n\alpha_0.$$

Thus

$$\varrho_0 = c_0^2 = \frac{2\alpha_0}{\pi^2}, \quad \varrho_n = c_n^2 = \frac{2}{\pi^2\alpha_0}\,\frac{\sin^2 n\alpha_0}{n^2}.$$

The ensemble average of $p_\alpha = n\hbar$ is $\langle P\rangle = \hbar \sum_n n\varrho_n$, and this is zero because $\varrho_{-n} = \varrho_n$, as it must be owing to the symmetry of Φ_n. The ensemble average of $E_n = \dfrac{n^2\hbar^2}{2I}$ is

$$\langle E\rangle = \frac{\hbar^2}{2I}\sum_n n^2\varrho_n \sim \sum_n \sin^2 n\alpha_0,$$

and this sum diverges on account of the fact that the relative numbers of rotators in the quantum state n decrease at the same rate as their energies increase.

P. 13.5. From P. 4.5 follows that the coefficients c_n in the expansion of $\Phi(\alpha)$ are inversely proportional to n^2 for all odd n and zero for even n. Thus

$$\varrho_n = \frac{1}{n^4}\Big/\sum_n \frac{1}{n^4} \quad \text{for} \quad n = 1, 3, 5, \ldots, \quad \varrho_n = 0 \quad \text{for} \quad n = 2, 4, 6, \ldots.$$

It follows that

$$\langle P \rangle = \hbar \sum_n \frac{1}{n^3}\Big/\sum_n \frac{1}{n^4} = 0, \quad \langle E \rangle = \frac{\hbar^2}{2I}\sum_n \frac{1}{n^2}\Big/\sum_n \frac{1}{n^4} \quad (n = 1, 3, 5, \ldots).$$

Now $\dfrac{1}{1^2} + \dfrac{1}{3^2} + \dfrac{1}{5^2} + \cdots = \dfrac{\pi^2}{8}$ and $\dfrac{1}{1^4} + \dfrac{1}{3^4} + \dfrac{1}{5^4} + \cdots = \dfrac{\pi^4}{96}$,

thus

$$\sum_n \frac{1}{n^2}\Big/\sum_n \frac{1}{n^4} = \frac{12}{\pi^2}. \quad (n \text{ odd})$$

It follows that

$$\langle E \rangle = \frac{6\hbar^2}{\pi^2 I}, \quad \sigma_P^2 = \langle P^2 \rangle - \langle P \rangle^2 = \frac{12\hbar^2}{\pi^2}, \quad \sigma_E^2 = \infty \text{ because } \langle E^2 \rangle = \infty.$$

P. 13.6. From P. 4.5 and the normalizing condition for Φ one has for the expansion coefficients

$$c_k = \frac{8}{\pi^2}\sqrt{\frac{3}{2}}\,\frac{1}{k^2}$$

and thus, from (13.2.8),

$$\varrho_{kj} = \frac{96}{\pi^4 k^2 j^2} \quad (k, j = 1, 3, 5, \ldots).$$

The energy operator of the harmonic oscillator is

$$\mathscr{E} = \frac{\alpha x^2}{2} - \frac{\hbar^2}{2m}\frac{d^2}{dx^2} \quad \left(\nu = \frac{1}{2\pi}\sqrt{\frac{\alpha}{m}}\right).$$

Thus the energy matrix elements are obtained from $E_{kj} = \int_{-l/2}^{+l/2} \psi_k\, \mathscr{E} \psi_j\, dx$ from which follows

$$E_{kj} = (-1)^{(k+j)/2}\frac{2\alpha l^2}{\pi^2}\frac{k^2 + j^2}{(k^2 - j^2)^2} \quad (k \gtrless j),$$

$$E_{kk} = \frac{\alpha l^2}{4}\left[(-1)^k \frac{1}{\pi^2 k^2} + \frac{1}{6}\right] + \frac{\pi^2 k^2 \hbar^2}{2ml^2}.$$

The ensemble average of the energy of the oscillators $\langle E \rangle$ is finally obtained from

$$\langle E \rangle = \sum_{kj} \varrho_{jk} E_{kj}.$$

P. 13.7. The energy of the system is $E_0 = \dfrac{1}{2m}\sum_{s=1}^{n} p_s^2$. Thus the shape of the energy surface in the n-dimensional momentum space is a hypersphere with radius $r = \sqrt{2mE_0}$ and the thickness of the shell is $\delta r = \sqrt{\dfrac{m}{2E_0}}\,\delta E$. In the coordinate space the phase points are confined to the interior of an n-dimensional "hypercube" of side l.

P. 13.8. The energy is connected with x and p by $E = p^2/2m + mgx$, hence

$$p = \sqrt{2m}\sqrt{E_0 - mgx} \quad \text{and} \quad \delta p = \sqrt{2m}\,\frac{\delta E}{2\sqrt{E_0 - mgx}}.$$

As the density of phase points within the shell is constant one has for the relative number of points between x and $x + \delta x$

$$f(x)\,\delta x = \text{const}\ \delta p\ \delta x,$$

from which follows

$$\langle x \rangle_m = \frac{2}{3}\frac{E_0}{mg} = \frac{2}{3}x_{\max} \quad \text{and} \quad \langle |p| \rangle_m = \frac{\sqrt{2mE_0}}{2} = \frac{1}{2}|p_{\max}|.$$

P. 13.9. From (13.3.21)

$$\varrho = \text{const}\exp\left[-\frac{1}{\theta}\left(\frac{p^2}{2m} + mgx\right)\right].$$

Thus $\langle x \rangle_c = \theta/mg$ and $\langle |p| \rangle_c = \sqrt{2m\theta/\pi}$; $\langle |p| \rangle_c^2/2m = \theta/\pi$. Further $\langle U \rangle_c = \theta$ and $\langle K \rangle_c = \theta/2$, thus $\langle E \rangle_c = 3\theta/2$.

P. 13.10.

$$\text{(a)}\ \delta W = -p\,\delta V, \quad \text{hence} \quad dS = \frac{dU + p\,dV}{T},$$

$$dF = \frac{\partial F}{\partial V}dV + \frac{\partial F}{\partial T}dT = dU - S\,dT - T\,dS = -p\,dV - S\,dT.$$

Thus

$$S = -\left(\frac{\partial F}{\partial T}\right)_V, \quad p = -\left(\frac{\partial F}{\partial V}\right)_T.$$

$$\text{(b)}\ \delta W = \tau\,dl, \quad \text{hence} \quad dS = \frac{dU - \sigma\,dl}{T},$$

$$S = -\left(\frac{\partial F}{\partial T}\right)_l, \quad \tau = \left(\frac{\partial F}{\partial l}\right)_T.$$

P. 13.11. If the temperature is kept constant the expression (13.4.16) for the partition function can be written in the form $Z = \left(\dfrac{v}{v_0}\right)^n$, where v is the volume available to the particles and v_0 is a constant of the dimension of a volume. Thus from (13.4.14)

$$S = \frac{U - F}{T} = \frac{U}{T} + kn\ln\left(\frac{v}{v_0}\right).$$

Initially $v = V/2$ and eventually $v = V$. Hence the increase in entropy is

$$\Delta S = kn\left[\ln\frac{V}{v_0} - \ln\frac{V}{2v_0}\right] = kn\ln 2.$$

P. 13.12. As in P.13.5,

$$\varrho_j = \frac{1}{j^4} \Big/ \sum_j \frac{1}{j^4} = \frac{1}{j^4} \frac{96}{\pi^4} \qquad \text{for odd } j,$$

$$\varrho_j = 0 \qquad \text{for even } j.$$

Thus

$$\varrho_j \ln \varrho_j = \frac{1}{j^4} \frac{96}{\pi^4} \left(-4 \ln j + \ln \frac{96}{\pi^4} \right) \qquad \text{for odd } j,$$

$$\varrho_j \ln \varrho_j = 0 \qquad \text{for even } j,$$

and from (13.4.18)

$$\langle S \rangle = \frac{384}{\pi^4} k \left(\frac{\ln 3}{3^4} + \frac{\ln 5}{5^4} + \cdots \right) - k \ln \frac{96}{\pi^4} \simeq 0{\cdot}075\, k.$$

P. 13.13. The thermodynamic probability G for the particles to be all situated in one compartment when the partition is present is evidently $G = 1$. When the partition is removed each particle can be either in one or the other compartment, hence $G = 2^n$. Thus from (13.4.22) the "configurational" entropy in the first case is zero and in the second case $k \ln (2^n)$ so that $\Delta S = kn \ln 2$.

P. 14.1. If $f(\nu)$ is the number of ways in which ν individual objects can be put in a definite sequence then $f(\nu + 1) = (\nu + 1) f(\nu)$. As evidently $f(1) = 1$ one has $f(\nu) = \nu!$. If the sequence ν is subdivided into groups $\nu_1 + \nu_2 + \cdots$ then the number of ways in which the sequence of objects in the group ν_i may be changed while it remains unchanged in all other groups is $\nu_i!$. Thus the number of ways in which ν objects can be distributed over the ν_i groups irrespective of their individuality in the groups is equa lto $\nu!/\sum_i \nu_i!$.

P. 14.2. Each of the ζ_j phase points in the domain j can be in any of the z_j cells in that domain. Thus there are $z_j^{\zeta_j}$ ways of distributing the individual phase points in the domain j over the individual cells in that domain. Altogether for all phase points the number of these distributions is $\sum_j z_j^{\zeta_j}$ for a given distribution of the ν phase points into groups of ζ_i. According to P.14.1 the number of such distributions is $\nu!/\sum_j \zeta_j!$. Thus the total number of ways to distribute the phase points over the domains is $\nu! \sum_j z_j^{\zeta_j} / \sum_j \zeta_j!$.

P. 14.3. According to (5.1.33) the statistical weight of the state of a two-dimensional oscillator with energy $E_n = (n + 1) h\nu$ is $g_n = n + 1$. Thus by definition

$$\bar{E} = N \frac{\sum_n (n + 1)^2 h\nu \exp [-(n + 1) h\nu/kT]}{\sum_n (n + 1) \exp [-(n + 1) h\nu/kT]}$$

from which follows

$$\bar{E} = 2N h\nu/(e^{h\nu/kT} - 1).$$

For the three-dimensional oscillator one has from (5.1.38)

$$E_n = (n + \tfrac{3}{2}) h\nu \quad \text{and} \quad g_n = (n + 1)(n + 2)/2,$$

thus

$$\bar{E} = N \frac{\sum_n (n + \frac{3}{2})(n + 1)(n + 2) \exp [-(n + \frac{3}{2}) h\nu/kT]}{\sum_n (n + 1)(n + 2) \exp [-(n + \frac{3}{2}) h\nu/kT]},$$

from which follows
$$\bar{E} = 3Nh\nu/(e^{h\nu/kT} - 1).$$

The assembly of N two-dimensional oscillators is seen to be statistically equivalent to an assembly of $2N$ linear oscillators, and an assembly of N three-dimensional oscillators equivalent to an assembly of $3N$ linear oscillators.

P. 14.4. According to (13.3.23)
$$f(E)\,\delta E = C\omega(E)\exp(-E/kT),$$
where ω is the product of V^n and the volume of a hypersperical shell of radius p and thickness $\delta p = \sqrt{\dfrac{m}{2E}}\,\delta E$ in the momentum space (see P. 13.7). Hence
$$f(E)\,dE = \text{const}\, e^{-E/kT}E^{(3n/2-1)}\,dE.$$

It follows that the maximum value of E is $E_m = (3n - 2)\,kT/2$ and its mean value $\bar{E} = 3NkT/2$ (i.e. the equipartition value $kT/2$ times the number $3n$ of degrees of freedom). For large n: $\bar{E} \simeq E_m$. In the neighbourhood of the mean value one finds
$$f = f_m\left(1 - \frac{3n}{4}\delta^2\right), \quad \text{where} \quad \delta = \frac{E - E_m}{E_m}.$$

Thus for large n: f is considerably smaller than f_m even for very small δ.

P. 14.5. The thermodynamic probability G_c of the combined systems is in analogy to (14.1.4)
$$G_c = \frac{\nu!}{\prod_i \zeta_i!}\,\frac{\nu'!}{\prod_j \zeta_j'!}\prod_i z_i^{\zeta_i}\prod_j z_j'^{\zeta_j'},$$
which is to be an extremum under the conditions
$$\sum_i \zeta_i = \nu = \text{const}, \quad \sum_j \zeta_j' = \nu' = \text{const}, \quad \sum_i \zeta_i E_i + \sum_j \zeta_j' E_j' = E_0 = \text{const}.$$

By means of the Lagrangian method one obtains for the most probable coarse-grained distribution
$$\zeta_i = Cz_i\,e^{-\beta E_i}, \quad \zeta_j' = C'z_j'\,e^{-\beta E_j'}.$$

As the factors of the exponents in both expressions are equal the temperatures of the part-systems must be the same.

P. 14.7. From (14.3.3) and (14.3.7) one has
$$\varrho_{j,\nu} = \frac{1}{Z_g}\exp\left(\frac{\mu\nu - E_j}{kT}\right)$$
and hence by means of (14.3.9)
$$\sum_{j\nu} \varrho_{j\nu}\ln\varrho_{j\nu} = \frac{1}{kT}\sum_{j\nu}(\mu\nu - E_j)\,\varrho_{j\nu} - \ln Z_g = \frac{\mu\langle\nu\rangle}{kT} - \frac{U}{kT} + \frac{F'}{kT}$$
from which follows
$$S = -k\sum_{j\nu}\varrho_{j\nu}\ln\varrho_{j\nu} + \frac{\mu\langle\nu\rangle}{T}.$$

This differs from (13.4.18) by the additional term $\mu\langle\nu\rangle/T$ which shows that μ is the work to be done on the system to increase the number ν of particles by one.

P. 14.8. From (14.3.17) and (14.3.22)

$$F' = -kT\sum_i \ln(1 + x_i)$$

and from this

$$S = -\frac{\partial F'}{\partial T} = k\sum_i \ln(1 + x_i) + kT\sum_i \frac{\partial x_i/\partial T}{1 + x_i}.$$

From (14.3.14) and (14.3.23) follows

$$1 + x_i = \frac{1}{1 - \bar{\nu}_i}, \quad \frac{\partial x_i}{\partial T} = -\frac{1}{T} x_i \ln x_i = -\frac{1}{T}\frac{\bar{\nu}_1}{1 - \bar{\nu}_i}[\ln \bar{\nu}_1 - \ln(1 - \bar{\nu}_i)].$$

Thus

$$S = -k\sum_i \{\bar{\nu}_i \ln \bar{\nu}_i + (1 - \bar{\nu}_i)\ln(1 - \bar{\nu}_i)\}.$$

P. 14.9. As necessarily $\nu_i \geqslant 0$ one must in (14.3.21) have $E_i - \mu \geqslant 0$ for all i. Hence as $T \to 0$ all $\nu_i \to 0$ except the one for which $E_i = E_0 = \mu_0$, and E_0 must be the lowest energy level. The corresponding cell then tends to contain all the phase points. For small T therefore $\nu = \dfrac{1}{(E_0 - \mu)/kT - 1}$ and as ν is large $\dfrac{E_0 - \mu}{kT}$ must be small so that $E_0 - \mu(T) \simeq kT/\nu$.

P. 15.1. From

$$f(v) = \text{const}\ v^2 \exp\left(-\frac{mv^2}{2kT}\right)$$

follows

$$v_m = \sqrt{\frac{2kT}{m}}, \quad \bar{v} = \sqrt{\frac{8kT}{m\pi}}, \quad \sqrt{(\overline{v^2})} = \sqrt{\frac{3kT}{m}}$$

and

$$\delta^2 = \frac{(\overline{v^2})}{(\bar{v})^2} - 1 = \frac{3\pi}{8} - 1 \simeq 0{\cdot}18.$$

P. 15.2. Using the analogous relations to (15.1.23) and (15.1.25) one obtains for the electric polarization $\mathsf{P} = \dfrac{\nu\Pi^2}{3kT}\mathsf{E}^*$. Eliminating E^* from this relation and $\mathsf{E}^* = \mathsf{E} + \gamma\mathsf{P}$ one obtains further

$$\mathsf{P} = \frac{\nu\Pi^2/3kT}{1 - \gamma\nu\Pi^2/3kT}\mathsf{E},$$

and hence

$$\varkappa = \frac{\mathsf{P}}{\mathsf{E}} = \frac{\nu\Pi^2/3kT}{1 - \gamma\nu\Pi^2/3kT}.$$

P. 15.3. The volume available to the first molecule put into the container is V, that for the second molecule is $(V - \omega)$, etc. Hence the volume in the position Γ-phase space is, because of the independence of the molecules,

$$V(V - \omega)(V - 2\omega)\cdots(V - (\mathcal{N} - 1)\omega) \simeq V^{\mathcal{N}}\left(1 - \frac{\omega\mathcal{N}^2}{2V}\right)$$

$$\simeq \left(V - \frac{\mathcal{N}\omega}{2}\right)^{\mathcal{N}} = (V - b)^{\mathcal{N}}.$$

One therefore has to replace V by $(V-b)$ in the expressions (15.1.29) and (15.1.30) and obtains $p = RT/(V-b)$.

P. 15.4. The minimum of $\varphi(r)$ is at $r_0 = \left(\frac{bn}{am}\right)^{1/(n-m)}$; hence

$$b = \frac{am}{n} r_0^{n-m} \text{ and } \varphi(r) = \frac{a}{r_0^m}\left\{-\left(\frac{r_0}{r}\right)^m + \frac{m}{n}\left(\frac{r_0}{r}\right)^n\right\};$$

for $r = r_0$: $\varphi(r_0) = -\frac{a}{r_0^m}\frac{n-m}{n} < 0$; the equilibrium is thus stable.

From $\Phi = \frac{\mathcal{N}\varphi}{2}$ follows $\Phi(V) = \frac{A}{2V_0^{m/3}}\left\{-\left(\frac{V_0}{V}\right)^{m/3} + \frac{m}{n}\left(\frac{V_0}{V}\right)^{n/3}\right\}$,

thus $\Phi_0 = \Phi(V_0) = -\frac{A}{2V_0^{m/3}}\frac{n-m}{n}$ and $H_s = -\Phi_0 = \frac{A}{2V_0^{m/3}}\frac{n-m}{n}$.

As

$$p = -\frac{d\Phi}{dV},\ \frac{dp}{dV} = -\frac{d^2\Phi}{dV^2} \text{ and } \frac{1}{\varkappa} = V_0\left(\frac{d^2\Phi}{dV^2}\right)_0$$

one obtains

$$\left(\frac{d^2\Phi}{dV^2}\right)_0 = \frac{A(n-m)m}{18V_0^{(m/3+2)}} \quad \text{and} \quad \varkappa = \frac{18V_0^{(m/3+1)}}{Am(n-m)},\quad H_s\varkappa = \frac{9V_0}{nm}.$$

P. 15.5. The potential energy of the movable particle, when it is displaced from the equilibrium position through a small displacement x, is

$$\varphi(x) = \frac{a}{r^m}\left\{-\left(\frac{r_0}{r+x}\right)^m + \frac{m}{n}\left(\frac{r_0}{r+x}\right)^n - \left(\frac{r_0}{r-x}\right)^m + \frac{m}{n}\left(\frac{r_0}{r-x}\right)^n\right\}.$$

For small x one can put $\varphi(x) = \varphi(0) + \alpha x^2/2$. Thus the particle performs a harmonic oscillation under the action of the restoring force

$$\alpha = \left(\frac{d^2\varphi}{dx^2}\right)_0 = \frac{2a}{r_0^m}\left\{-r_0^m\frac{m(m+1)}{r^{m+2}} + r_0^n\frac{n(n+1)}{r^{n+2}}\right\}$$

and the frequency ν follows from

$$\nu^2 = \frac{1}{4\pi^2}\frac{\alpha}{\mu}.$$

For $n \gg m$: $\alpha \simeq \text{const}\ r^{-(n+2)}$ and thus $\nu \simeq \text{const}\ V^{-(n+2)/6}$, from which follows

$$\gamma = -\frac{d\ln\nu}{d\ln V} \simeq \frac{n+2}{6}.$$

P. 15.6. From (15.2.25) one obtains $\left(\frac{\partial p}{\partial T}\right)_V \cdot V = \gamma c_v$,

thus
$$\alpha = -\frac{\gamma c_v}{V^2}\left(\frac{\partial V}{\partial p}\right)_T = \frac{\gamma c_v \varkappa}{V}.$$

P. 15.7. From

$$u_\lambda \, d\lambda = \frac{u_\nu c}{\lambda^2} \, d\lambda$$

one has

$$u_\lambda = \frac{8\pi hc}{\lambda^5} \frac{1}{e^{hc/\lambda kT} - 1}.$$

λ_m is obtained from $(du_\lambda/d\lambda)_m = 0$, which leads to $\lambda_m T = \dfrac{hc}{4{\cdot}965k}$.

P. 15.8. From $S = -\dfrac{\partial F}{\partial T}$ follows $S = \displaystyle\int_0^T \left(\frac{U}{T^2}\right) dT + \frac{U}{T} = -\frac{F}{T} + \frac{U}{T}$, which satisfies the relationship $F = U - ST$. Substituting from (15.3.5) $U = CT^4V$ one has $F = -U/3$ and from this

$$S = \frac{4}{3} CT^3 V, \quad p = -\frac{\partial F}{\partial V} = \frac{CT^4}{3} = \frac{u}{3}.$$

P. 15.9. If u is the electromagnetic energy density in a parallel beam of radiation then the energy flow density is equal to uc and, according to classical electrodynamics, the momentum flow density is equal to u. If it is assumed that one-sixth of the radiation in an enclosure moves in the positive x-direction this transfers an amount $u/6$ of momentum to the unit area of a wall normal to x, and an equal amount is transferred by the radiation emitted by the wall in the $-x$-direction. Hence the pressure is $p = u/3$.

If a photon beam of frequency ν contains n photons per unit volume, the energy density is $nh\nu$. Each of the photons moving in the x-direction has momentum $h\nu/c$, and hence the amount of momentum transferred to a unit area of wall normal to x is $\dfrac{nh\nu c}{c} = u$. As this is independent of ν it must hold for any beam of photons with an arbitrary frequency spectrum.

P. 15.10. If $B_{ij} = 0$ one has instead of (15.3.14)

$$\varrho_i A_{ij} = \varrho_j B_{ji} u(\nu),$$

and thus instead of (15.3.16) $u(\nu) = \dfrac{A_{ij}}{B_{ji}} \exp\left(-\dfrac{h\nu}{kT}\right)$

from which follows

$$u(\nu) = \frac{8\pi h}{c^3} \nu^3 e^{-h\nu/kT}.$$

P. 15.11. As ϱ is proportional to ν one has evidently from (15.4.20) $\dfrac{\overline{(\nu - \bar{\nu})^2}}{(\bar{\nu})^2} = \dfrac{\varkappa kT}{\omega}$.

The compressibility of an ideal gas is $\varkappa_0 = \dfrac{1}{p}$. Thus for an ideal gas

$$\frac{\overline{(\nu - \bar{\nu})^2}}{(\bar{\nu})^2} = \frac{kT}{p\omega} = \frac{1}{\bar{\nu}}.$$

The probability of finding ν molecules in ω is

$$P(\nu) = \frac{\nu!}{N!\,(N - \nu)!} q^\nu (1 - q)^{N-\nu}$$

if N is the total number of molecules in the gas. From this

$$\bar{\nu} = qN, \quad \sigma_\nu^2 = \overline{(\nu - \bar{\nu})^2} = Nq\,(1 - q).$$

Hence $\dfrac{\sigma_\nu^2}{(\bar{\nu})^2} = \dfrac{1 - q}{\bar{\nu}}$ from which follows for small q:

$$\frac{\sigma_\nu^2}{(\bar{\nu})^2} = \frac{1}{\bar{\nu}}.$$

For a real gas one has

$$\frac{\overline{(\nu - \bar{\nu})^2}}{(\bar{\nu})^2} = \varkappa \frac{kT}{\omega} = \frac{\varkappa}{\varkappa_0} \frac{1}{\bar{\nu}}.$$

P. 15.12. The work $W(Q)$ for putting a charge Q on the condenser is $W(Q) = Q^2/2C$, hence from (15.4.13) $\sigma_Q^2 = \overline{Q^2} = kTC$, and $\bar{E} = \overline{Q^2}/2C = kT/2$.

P. 15.13. Here W is the work required to produce the magnetic field belonging to the current which is equal to $W = LI^2/2$. Thus in this case $\sigma_I^2 = \overline{I^2} = kT/L$ and $E = L\overline{I^2}/2 = kT/2$.

List of References

to original publications mentioned in the text

Adam, M.G. (1948) *M.N. Roy. Astr. Soc. London* **108,** 446.
Airy, G.B. (1871) *Proc. Roy. Soc.* **20,** 35.
Airy, G.B. (1873) *Proc. Roy. Soc.* **21,** 121.
Bohr, N. (1913) *Phil. Mag.* **26,** 1.
Born, M. (1915) *Dynamik der Kristallgitter*, Teubner.
Born, M. (1923) *Atomtheorie des festen Zustandes*, Teubner.
Born, M. (1926) *Z. Phys.* **37,** 863.
Born, M. and Jordan, P. (1925) *Z. Phys.* **34,** 858.
Born, M., Heisenberg, W. and Jordan, P. (1926) *Z. Phys.* **35,** 557.
Bose, S.N. (1924) *Z. Phys.* **26,** 178.
De Broglie, L. (1924) thèse, Paris.
De Broglie, L. (1925) *Ann. de phys.* (10), **3,** 22.
Cedarholm, J.P. and Townes, C.H. (1959) *Nature* **184,** 1350.
Debye, P. (1910) *Ann. d. Phys.* **33,** 1427.
Debye, P. (1912) *Ann. d. Phys.* **39,** 789.
Dirac, P.A.M. (1925) *Proc. Roy. Soc.* A, **109,** 642.
Dirac, P.A.M. (1926) *Proc. Roy. Soc.* A, **112,** 661.
Dirac, P.A.M. (1928) *Proc. Roy. Soc.* A, **117,** 610; **118,** 341.
Dyson, F.W., Eddington, A.S. and Davidson, C. (1920) *Phil. Trans.* **220,** 291.
Ehrenfest, P. (1927) *Z. Phys.* **45,** 455.
Eichenwald, A. (1903) *Ann. d. Phys.* **11,** (4), 421.
Einstein, A. (1902) *Ann. d. Phys.* **9,** 417.
Einstein, A. (1903) *Ann. d. Phys.* **11,** 170.
Einstein, A. (1905a) *Ann. d. Phys.* **17,** 132.
Einstein, A. (1905b) *Ann. d. Phys.* **17,** 549.
Einstein, A. (1905c) *Ann. d. Phys.* **17,** 891.
Einstein, A. (1906) *Ann. d. Phys.* **19,** 371.
Einstein, A. (1907) *Ann. d. Phys.* **22,** 180.
Einstein, A. (1909) *Phys. Z.* **10,** 185.
Einstein, A. (1915) *Berl. Ber.* 831.
Einstein, A. (1916) *Ann. d. Phys.* **49,** 69.
Einstein, A. (1917) *Phys. Z.* **18,** 121.
Einstein, A. (1924) *Berl. Ber.* 261.
Einstein, A. *Investigations on the Theory of the Brownian Movement*, Dover Publications.
Einstein, A., Lorentz, H.A., Minkowski, H. and Weyl, H. *The Principle of Relativity*, Dover Publications.
v. Eötvös, R., Pekar, O. and Fekete, B. (1922) *Ann. d. Phys.* **68,** 11.
Fermi, E. (1926) *Z. Phys.* **36,** 902.
Fitzgerald, J. see Lodge, O. (1893) *Phil. Trans.* **184.**
Fizeau, H. (1851) *C.R.* **33,** 349.

List of References

Fock, V. (1926) *Z. Phys.* **38,** 242.
Frank, Ph. and Rothe, H. (1911) *Ann. d. Phys.* **34,** 825.
Fürth, R. (1917) *Ann. d. Phys.* **53,** 177.
Fürth, R. (1965) *Proc. Phys. Soc.* **85,** 131.
Gordon, W. (1926) *Z. Phys.* **40,** 117.
Goudsmit, S. and Uhlenbeck, G.E. (1925) *Naturwissensch.* **13,** 953.
Goudsmit, S. and Uhlenbeck, G.E. (1926) *Nature* **117,** 264.
Grüneisen, E. (1912) *Ann. d. Phys.* **39,** 257.
Heisenberg, W. (1925) *Z. Phys.* **33,** 879.
Heisenberg, W. (1927) *Z. Phys.* **43,** 172.
Heitler, W. and London, F. (1927) *Z. Phys.* **44,** 455.
Jeans, J.H. (1905) *Phil. Mag.* **10,** 91.
Kappler, E. (1931) *Ann. d. Phys.* **11,** 233.
Kaufmann, W. (1902) *Phys. Z.* **4,** 55.
Kennedy, R.J. and Thorndike, E.M. (1932) *Phys. Rev.* **42,** 400.
Klein, O. (1926) *Z. Phys.* **37,** 895.
Langevin, P. (1905) *J. Phys.* **4,** 678.
Lennard-Jones, J.E. (1929) *Trans. Faraday Soc.* **25,** 668.
Littman-Fürth, H. (1954) *Nature* **173,** 80; **174,** 505.
Lorentz, H.A. (1892) *Versl. Akad. Wet. Amsterdam* **1,** 74.
Lorentz, H.A. *Versuch einer Theorie der elektrischen und optischen Erscheinungen in bewegten Körpern*, Leiden 1895.
Lorentz, H.A. (1904) *Proc. Ac. Sc. Amsterdam* **6,** 809.
Lorentz, H.A. *The Theory of Electrons and its Application to the Phenomena of Light and Radiant Heat*, Dover Publications.
Mach, E. (1883) *Die Mechanik in ihrer Entwicklung, historisch-kritisch dargestellt.*
Michelson, A.A. (1881) *Am. J. Sc.* **22,** 120.
Michelson, A.A. and Gale, H.G. (1925) *Nature* **115,** 566.
Michelson, A.A. and Morley, E.W. (1887) *Am. J. Sc.* **34,** 333.
Minkowski, H. (1908) *Nachr. Ak. Wiss. Göttingen* **53.**
Pauli, W. (1925) *Z. Phys.* **31,** 765.
Perrin, J. (1908) *C.R.* **146,** 967; **147,** 475, 530, 594.
Perrin, J. (1912) *Rapp. de Conges Solvey*, Paris.
Planck, M. (1900) *Verh. D. Phys. Ges.* **2,** 202.
Planck, M. (1901) *Ann. d. Phys.* **4,** 553.
Roentgen, W.C. (1888) *Wied. Ann.* **35,** 286.
Sagnac, G. (1913) *C.R.* **157,** 708, 1410.
Schrödinger, E. (1926 a) *Ann. d. Phys.* **79,** 361, 489, 734.
Schrödinger, E. (1926 b) *Ann. d. Phys.* **81,** 129.
Schwarzschild, K. (1916) *Berl. Ber.* 189.
Sommerfeld, A. (1915) *Münchner Ber.* 425, 459.
Sommerfeld, A. (1916) *Ann. d. Phys.* **51,** 1.
Trouton, F.T. and Noble, H.R. (1903) *Proc. Roy. Soc.* **72,** 132.
Yukawa, H. (1935) *Proc. Phys. Math. Soc. Japan* **17,** 48.

Index

Index

Index

Index

Index